**Wireless PCS**

## McGraw-Hill Series on Computer Communications (Selected Titles)

| ISBN | AUTHOR | TITLE |
|---|---|---|
| 0-07-005147-X | Bates | *Voice and Data Communications Handbook* |
| 0-07-005669-2 | Benner | *Fibre Channel: Gigabit Communications and I/O for Computer Networks* |
| 0-07-005560-2 | Black | *TCP/IP and Related Protocols, 2/e* |
| 0-07-005590-4 | Black | *Frame Relay Networks: Specifications & Implementation, 2/e* |
| 0-07-006730-3 | Blakeley/Harris/Lewis | *Messaging/Queuing with MQI* |
| 0-07-011486-2 | Chiong | *SNA Interconnections: Bridging and Routing SNA in Heirarchical, Peer, and High-Speed Networks* |
| 0-07-020359-8 | Feit | *SNMP: A Guide to Network Management* |
| 0-07-021389-5 | Feit | *TCP/IP: Architecture, Protocols & Implementation with IPv6 and IP Security, 2/e* |
| 0-07-024043-4 | Goralski | *Introduction to ATM Networking* |
| 0-07-034249-0 | Kessler | *ISDN: Concepts, Facilities, and Services, 3/e* |
| 0-07-035968-7 | Kumar | *Broadband Communications* |
| 0-07-041051-8 | Matusow | *SNA, APPN, HPR and TCP/IP Integration* |
| 0-07-060362-6 | McDysan/Spohn | *ATM: Theory and Applications* |
| 0-07-044362-9 | Muller | *Network Planning Procurement and Management* |
| 0-07-046380-8 | Nemzow | *The Ethernet Management Guide, 3/e* |
| 0-07-049663-3 | Peterson | *TCP/IP Networking: A Guide to the IBM Environments* |
| 0-07-051506-9 | Ranade/Sackett | *Introduction to SNA Networking, 2/e* |
| 0-07-054991-5 | Russell | *Signaling System #7* |
| 0-07-057199-6 | Saunders | *The McGraw-Hill High Speed LANs Handbook* |
| 0-07-057639-4 | Simonds | *Network Security: Data and Voice Communications* |
| 0-07-060363-4 | Spohn | *Data Network Design, 2/e* |
| 0-07-069416-8 | Summers | *ISDN Implementor's Guide* |
| 0-07-063263-4 | Taylor | *The McGraw-Hill Internetworking Handbook* |
| 0-07-063301-0 | Taylor | *McGraw-Hill Internetworking Command Reference* |
| 0-07-063639-7 | Terplan | *Effective Management of Local Area Networks, 2/e* |
| 0-07-065766-1 | Udupa | *Network Management System Essentials* |

# Wireless PCS

## Personal Communications Services

**Rajan Kuruppillai**

**Mahi Dontamsetti**

**Fil J. Cosentino**

**McGraw-Hill**

New York   San Francisco   Washington, D.C.   Auckland   Bogotá
Caracas   Lisbon   London   Madrid   Mexico City   Milan
Montreal   New Delhi   San Juan   Singapore
Sydney   Tokyo   Toronto

**Library of Congress Cataloging-in-Publication Data**
Kuruppillai, Rajan.
    Wireless PCS / Rajan Kuruppillai, Mahi Dontamsetti,
  Fil Cosentino.
       p.    cm. — (McGraw-Hill series on computer communications)
    Includes index.
    ISBN 0-07-036077-4 (hc)
    1. Personal communication service systems.  I. Dontamsetti, Mahi.
  II. Cosentino, Fil.  III. Title.  IV. Series.
  TK5103.2.K87  1997                    96-34731
  384.5—dc20                         CIP

## McGraw-Hill

*A Division of The McGraw·Hill Companies*

ISBN 0-07-036077-4

*The sponsoring editor for this book was Stephen S. Chapman, the editing supervisor was Christine Furry, and the production supervisor was Suzanne W. B. Rapcavage. It was set in Century Schoolbook by North Market Street Graphics.*

*Printed and bound by R. R. Donnelley & Sons Company.*

McGraw-Hill books are available at special quantity discounts to use as premiums and sales promotions, or for use in corporate training programs. For more information, please write to the Director of Special Sales, McGraw-Hill, 11 West 19th Street, New York, NY 10011. Or contact your local bookstore.

 This book is printed on recycled, acid-free paper containing a minimum of 50% recycled de-inked fiber.

# Contents

# Part 3    PCS Technology

## Part 4    PCS Business

## Chapter 10.  PCS Network Design                                        375

# Foreword

The demand by consumers all over the world for mobile communications is booming and will continue for at least the next decade. By the end of 1995, over 30 million people in the United States were using cellular service. Another 30 million or so were also in touch by pager. By the year 2002, almost half of Americans will regularly use some kind of device to keep in touch with friends, family, and their work. Globally, the numbers are growing equally as fast. Over 100 million people were using a mobile service by the end of 1995—that number is expected to grow to 300 million by the year 2000. People are utilizing wireless technology to connect their fixed home and business phones as well, particularly in developing nations around the world. Between 1997 and 1999, 120 million new access lines will be installed by telephone companies around the world. While 100 million will use traditional copper wires, 20 million will connect to the public telephone network via a wireless line that is cheaper, faster to deploy, and easier to maintain.

What's causing all this exciting growth? It's a combination of technology and competition bringing more value to consumers. Phones are smaller, lighter, have a longer battery life, and are affordable now for the mass market. Operators are providing excellent voice quality, innovative services, and roaming across the country or world. Most important, mobility is becoming less expensive for people to use. Around the world, as well as in the United States, governments are licensing additional spectrum for new operators to compete with traditional cellular operators. Competition brings innovation, new services, and lower prices for consumers. This is the vision of personal communications service being rapidly deployed around the world.

The term *personal communications services,* or PCS, is often misused and, therefore, is widely misunderstood. It is often used to describe the new spectrum being licensed at 2 GHz to compete with existing cellular operators operating at 800 MHz or 900 MHz. Others use it to describe advanced digital wireless services being provided by operators at all frequencies to offer people new and flexible ways to communicate with each other, be more productive, and be reached any time and any place. To others, it describes the new competitive environment which is making wireless voice and data service affordable to all. To those new to the industry, this must be quite confusing. Even those of us who are involved every day are not always working from the same level of information. A reference text on the subject, like this book, seems sorely needed.

Surprisingly, mobile radio technology used in cellular or PCS systems is in its infancy. The first cellular operators only started operation in the mid-eighties. When I joined the industry in 1991, I was surprised how little had been written about wireless communications to help me get up to speed. Today, with the industry growing so fast in so many directions, more and more people have the need to understand the technology and business issues behind wireless. Opportunities abound, but experienced resources are in short supply. This book should help fill the information gap and provide the knowledge needed to be quickly productive within the industry.

The vision of a phone in every pocket, a wireless connection from every computer or PDA to the internet or corporate database, and the ability for everyone on the planet to have a telephone makes for a very exciting future. The elements that will create that future are here today and are being tested and deployed by wireless operators in many countries. With the information found in this book, I hope you'll be better prepared to help create that future.

*Matthew J. Desch*
President
Northern Telecom Wireless Networks

# Preface

The PCS industry is relatively new. The books that are available in the market are based on the cellular industry and focus either on the technical or on the business aspect but rarely do they cover both. The PCS industry, from a subscriber point of view only marginally different from cellular, has its own unique characteristics. The PCS industry, unlike the earlier cellular industry, has no common technology standard. There are multiple technology standards available for a PCS operator.

The industry structure of PCS is also distinctly different from early cellular. The cellular industry started with a duopoly structure in each service area. The ownership structure in the cellular industry was such that no single owner had a regional/national footprint. However, with subsequent consolidation and mergers, regional/national footprints started to be formed. The PCS industry, however, starts with a regional/national footprint with an oligopoly market structure in each service area.

The PCS operator, unlike the cellular operator, starts with a different set of parameters that impacts the business case. These parameters, such as industry structure, payment for spectrum, and relocation of microwave users, impact the business case for the new PCS operator. The wireless market has moved from a scarce spectrum situation to that of an abundant spectrum situation. This change in availability of spectrum will require new strategies for success probably quite different from those used in the early days of cellular.

This book focuses on the technical aspects, as well as the business aspect, of the PCS industry. The book is divided into four parts. Part 1 is an overview of wireless telecommunications. It provides information on the wireless telecommunication activities in the different regions of the world. It also provides the key technical concepts of wireless telecommunications. These concepts will provide the fundamentals of the industry and will help readers unfamiliar with wireless concepts to understand the rest of the book.

Part 2 discusses the complementary and substitute services that impact the PCS industry. These services include cellular, paging, and personal communications satellite services (also called mobile satellite services). These substitute services could enhance as well as negatively impact the PCS business case.

Part 3 focuses on the major technology that may be deployed for PCS. It provides an overview of the PCS industry and its market structure. This is followed by technical details of the four major technologies that could be deployed in PCS networks. The technologies discussed are GSM, DAMPS (also called U.S. TDMA), CDMA, and PACS.

Part 4 develops the business case for new PCS operators. It first provides the major elements of network design. An actual network design for the Dallas–Fort Worth BTA is provided. This is followed by a simulation model of the competitive structure that a new PCS operator for the Dallas–Fort Worth BTA may encounter. The simulation models, based on key inputs such as price and coverage, develop the business case for the new PCS operator. This business case is followed by a discussion of the future wireless network. The regulatory and subscriber trends, as well as some of the enabling technology that may shape the future networks, are presented.

The wireless industry is constantly changing. The industry players and the major players are changing due to mergers, acquisition, and splits. I have tried to capture much of the latest information, but it is very likely that the reader will encounter names and other information that may not be correct due to changes in the industry since writing the manuscript. The key underlying concepts and elements of the technology and business case should remain the same, independent of the names of the players.

## Acknowledgments

I would like to thank the many friends, vendors, and operators who helped me in various ways with the writing of this book. They helped identify the lack of books discussing the technology and business aspects of the PCS industry. In addition they provided me with the materials and information required for writing this book.

I would like to thank Robert Goodman and Tom Berger (Celcore) for their support and encouragement; Matthew Desch (Nortel) for writing the foreword and providing me with the artwork; Scott Baxter (RF Engineering Consultant and Training) for the PCS network design chapter; Darren Dattalo (Mobile Systems International) for performing the network design for the Dallas–Fort Worth BTA; and Edward

Jungerman III (Impulse Telecommunications Corporation) for the generous use of the WIST simulation tool to develop the business case for the PCS operator.

I would like to thank my coauthors, Mahi Dontamsetti and Fil Cosentino, for their contribution and for tolerating the frequent abuses in the form of editing and deadline changes. I am greatly obliged to my wife, Sangeeta Kuruppillai, and coauthors' wives, Manju Dontamsetti and Jeanne Cosentino, for their generous support and understanding.

*Rajan Kuruppillai*

# Introduction

# 1

# Fundamentals of Wireless Telecommunications

## 1.1 Introduction

Humankind, since Alexander Graham Bell invented the telephone in the 1870s, has desired untethered communication in order to exchange information wherever and whenever they want, as freely as the air they breathe. It took more than 100 years to learn to effectively communicate with each other. This desire for untethered communication has always been inherent. The smoke signals, petragraphs, drums, birds, and other means used to communicate in the early days were manifestations of a desire to communicate without the requirement of a fixed wire between the communicators. In the play *Agamemnon,* the Greek dramatist Aeschylus (525–456 B.C.) describes how beacons relayed news of the fall of Troy from Asia Minor—across a distance of 600 km—to Agamemnon's palace in Greece, in approximately 1200 B.C. These fire-beacon signals had limitations, as the historian Polybius describes in his work *The Histories:*

> Now in former times, as fire signals were simple beacons, they were for the most part of little use to those who used them. It was possible for those who had agreed on this to convey information that a fleet had arrived at Oreus, Peparethus, or Chalcis, but when it came to some of the citizens having changed sides or having been guilty of treachery or a massacre having taken place in the town, or anything of the kind, things that often happen, but cannot all be foreseen—and it is chiefly unexpected occurrences which require instant consideration and help—all such matters defied communication by fire signal.

To overcome the problem, Polybius developed a code in which the characters of the Greek alphabet were divided into five groups. The telegraph consisted of two large screens hiding five torches apiece. Torches would be raised to indicate a pair of numbers. Optical telegraph networks based on the torch concept sprang up in Sweden, England, Germany, Spain, Australia, and the United States.

This desire for untethered communication, however, is strongly manifested in the twentieth century. The growth in the transportation network led to increased use of cars and airplanes for business and pleasure. In the eighteenth century the employment patterns were highly centralized. The industrial complexes were usually centrally located. Workers migrated to these locations for jobs. With the growth of labor and the transportation network, workers commuted to and from their homes to employment centers. This led to increased traffic congestion and as much time being spent on the road as at the home and office. It was natural that these workers desired to access services such as voice, fax, and data to effectively use their time on the road. Further, a shift in management focus from manufacturing to finance to marketing has led to increased mobility of employees. Marketing employees tend to be mobile, unlike the manufacturing employees who are usually fixed and concentrated. Marketing, being externally oriented, requires that employees spend less and less time at the head office and more and more time with customers.

*Wireless telecommunications* could be simply defined as the ability to communicate without the use of wires, the links between communicators in the current telephone network. The technology, as its stands today, does not allow the space between the end users to be completely wireless in a cost-effective manner. There are several segments between the end users that are still wired. The current wireless networks use the atmosphere as a means of communicating between the user and the network. Many of the segments between the network nodes are still wired. The future networks such as Iridium, discussed in Chap. 3, have an increased number of network segments that are wireless.

The invention of the telephone by Alexander Graham Bell and the principle of wireless communications introduced by Heinrich Rudolf Hertz are the key technical ingredients for wireless telecommunications. Hertz was the first to show that energy could be transmitted through air and captured at a distant location with a suitable receiver. Marconi, toward the end of the nineteenth century, demonstrated the principles of wireless communications. The important milestones on the road to wireless telecommunications are shown in Table 1.1. These milestones have contributed significantly in humankind's movement toward untethered communication.

The next few sections will provide an overview of the global cellular market as well as the major vendors and operators in the different

**TABLE 1.1**   **Wireless Telecommunications Milestones**

| Date | Event |
|------|-------|
| 1870 | Alexander Graham Bell's invention of telephone |
| 1888 | Hertz discovers energy can be transmitted by air |
| 1895 | Guglielmo Marconi demonstrates wireless communications |
| 1905 | Reginald Fessenden transmits speech and music through air |
| 1928 | Transmission of broadcasts from central location at the police department in Detroit |
| 1933 | Two-way mobile application using amplitude modulation (AM) at the police department in Bayonne, N.J. |
| 1935 | Edwin Armstrong introduces frequency modulation (FM) technology |
| 1940 | Mobile application using FM at the Connecticut State Police at Hartford |
| 1946 | Mobile radio connection with Public Switched Telephone Network (PSTN) at St. Louis |
| 1947 | Cellular concept originated at Bell Laboratories |
| 1964 | Improved Mobile Telephone Service (IMTS) introduced in the United States |
| 1978 | NAMTS introduced in Japan |
| 1981 | NMT introduced in Norway, Sweden, Denmark, and Finland |
| 1983 | Advanced Mobile Phone Service (AMPS) introduced in the United States |
| 1984 | Total Access Communication System (TACS) introduced in the United Kingdom |
| 1985 | CNETZ system introduced in Germany, Radiocom 2000 introduced in France |
| 1987–1995 | New air interface protocols such as TDMA, NAMPS, GSM, CDMA introduced |

regions and countries of the world. This discussion will be followed by a high-level overview of wireless telecommunications technology. The basics of radio frequency (RF) and cellular telephony are described, followed by an overview of the major analog and digital access technologies. The digital technologies are discussed in detail in subsequent chapters. Analog technology, except for AMPS, is not discussed because of its movement toward obsolescence. AMPS will also eventually be replaced by digital technology; however, two factors will ensure its continuance for several more years. The first factor is the lack of a government mandate in the United States—unlike in many other countries—to force existing operators or new operators to use digital technology. The second factor is related to the installed base. AMPS has the largest number of subscribers among both analog and digital technology. The installed base has ensured its high volume and low cost resulting from the learning curve effect. The history and legacy of the AMPS system, in the absence of a clear economic and government mandate, will ensure that the transition from AMPS to other digital technologies will be slow, if not an evolutionary process. AMPS is discussed in detail in Chap. 2.

## 1.2  Global Market

The global market section will describe the wireless telecommunications market in the different regions and countries of the world as of

1993, the technology selected by the operators in these countries, as well as the vendors and suppliers of equipment. The rapid changes in the wireless telephony arena due to technical, regulatory, and commercial forces will ensure that most of this information is dated. This information, albeit dated, is provided here to give a glimpse of the historical activity and momentum in the different countries of the world. The countries with the least wireless telecom activity, e.g., India and China, probably do not have historical weight of analog systems, and therefore can leapfrog other countries by deploying digital technology upon initial implementation of a wireless telecommunications network.

The number of wireless telecommunications subscribers is growing rapidly. In 1986, there were approximately 1.5 million subscribers. At the end of 1994, there were more than 50 million subscribers, a growth rate of more than 50 percent per year. There are 120 countries with at least one wireless telecommunications system operational. Most of the subscribers are in North America, followed by Western Europe and Asia Pacific. The RF access technology used in each of these countries is different. The main technologies could be divided into analog and digital technologies. In *analog technology,* the media (air) is varied continuously with the variations of human speech. In *digital technology,* the human speech is first converted to a series of coded pulses. These coded pulses (i.e., 0s and 1s) allow human speech to be stored, copied, and manipulated by computer.

In addition there are two world standards for PSTN networks: American National Standards Institute (ANSI) for the United States and Canada, and Consultative Committee for International Telephone and Telegraph (CCITT) for virtually the rest of the world. Each system defines *pulse code modulation* (PCM) differently, and the two types, E1 for the CCITT and T1 for ANSI, are incompatible. The companding algorithm for digitizing is also different. ANSI compands using μ-Law, whereas CCITT compands using A-Law. Both in-band and out-of-band signaling also differ. CCITT countries use primarily R2 signaling with each country's Department of Postal Telephone and Telegraph (PTT) having variations for their own country's call completion, networking, and billing requirements. Also, in regard to out-of-band signaling (i.e., TUP and ISUP), each country has its own variations for access to the network and enhanced features. In summary, ANSI and CCITT networks are incompatible, but conversion and interface devices are available so that calls can pass from one domain to the other. This book is focused on the wircless telecommunications. It will not focus on the differences between ANSI and CCITT, but rather on the air interface.

There are several types of analog systems. The main types are as follows:

1. Advanced Mobile Phone System (AMPS)
2. Nordic Mobile Telephone (NMT)
3. Total Access Communications System (TACS)

Within the digital systems, the main types are as follows:

1. Global System for Mobile communications (GSM)
2. Time-Division Multiple Access (TDMA)   .
3. Code-Division Multiple Access (CDMA)

Many countries, and especially the developing countries, have embarked on cellular telecommunications as a substitute for the local landline networks because of a pent-up demand for basic Plain Old Telephony Service (POTS). For these countries, cellular offers a relatively quick, effective, and economical way to provide telecommunications services to their people who otherwise have to wait, in some cases more than five years. As a result, many of these developing countries have deregulated their telecommunications networks allowing foreign companies to build and operate their wireless networks. In the developed countries, the growth of cellular is fueled by the lowering equipment and service prices due to competition. The global market could be divided into the following regions:

1. Africa
2. Asia and Pacific
3. Caribbean and Latin America
4. Europe
5. North America

### 1.2.1   Africa

Africa reported less than 100,000 wireless telecommunications subscribers in 1994. This represents a penetration of less than 0.03 percent. The number of telephones for every 100 people is less than 3 percent. Africa has been slow to liberalize its telecommunications and cellular networks. Most of the cellular systems are operated by the PTTs. However, systems in Ghana, Mauritius, Nigeria, South Africa, Tanzania, and Yemen are operated by private companies or by a partnership between the private companies and the PTT. Different countries have different system types. Currently, there are six countries with AMPS systems, four countries with NMT systems, six countries with TACS systems, and three with GSM systems. Some countries have multiple systems of different types, for example, Ghana has both TACS and AMPS systems. The major equipment providers are Alcatel, AT&T, Ericsson, Motorola, NEC, Nokia, and Siemens. The cellular market information in 1993 for each country with operational cellular networks is shown in Table 1.2.

### 1.2.2  Asia and Pacific

The Asia and Pacific region is the largest telecommunications market in the world. It consists of more than 50 countries and has a population greater than 3 billion, with most of the population in China and India. This region contains less than 17 percent of the world's telephones, though the percentage of world population is more than 60 percent. Thus, the latent demand, due to insufficient landline networks, is enormous. Many of these countries have embarked on liberalization programs to attract foreign capital and operating expertise. Different countries in this region embark on cellular for different reasons. Countries such as Hong Kong, Singapore, Israel, Bahrain, and Saudi Arabia view cellular as an integral part of their telecommunications network and have developed cellular coverage as an overlay to their existing landline network, while countries such as China and India are considering cellular as a means to provide basic telephone service to their people.

The number of subscribers in the Asia and Pacific region was approximately 10 million in 1994. Most of the networks are privately owned or operated in partnership with the PTT. However, networks in Singapore, Taiwan, China, and most of the Middle East countries are owned and controlled by the PTT. There are multiple system types within the

TABLE 1.2  Cellular Networks in Africa (1993)

| Country | Operator | System type | Infrastructure vendor |
|---------|----------|-------------|------------------------|
| Algeria | SONATTE (PTT) | NMT 900 | Nokia |
| Angola | ENATEL (PTT) | AMPS | Motorola |
| Cameroon | PTT | GSM | Siemens |
| Egypt | ARENTO (PTT) | AMPS | Matsushita |
| Gabon | OPT (PTT) | AMPS | Motorola |
| Gambia | GAMTEL (PTT) | TACS | Motorola |
| Ghana | Millicom Ghana | TACS | Motorola |
| Kenya | KPTC (PTT) | TACS | NEC |
| Mauritius | EMTEL | ETACS | NovAtel |
| Morocco | ONPT (PTT) | NMT 450 | Ericsson |
| Nigeria | MTS (PTT) | TACS | Ericsson |
| South Africa | SAPO (PTT) | NMT 450 | Siemens |
| Tunisia | PTT | NMT 450 | Ericsson |
| Zaire | TELECEL | AMPS | Plexys |

Asia and Pacific region; however, most countries, with the exception of Japan and South Korea, have adopted the GSM as the digital standard. This was due to aggressive marketing by the European countries and aggressive implementation of GSM in Australia. Japan is pursuing its personal digital cellular system, a form of TDMA, while South Korea is pursuing ventures with Qualcomm, the CDMA technology. The major suppliers of equipment to this region are Ericsson and Motorola. The country deployment by vendor is shown in Table 1.3.

### 1.2.3 Caribbean and Latin America

The Caribbean market, which consists of several islands that are recognized countries and oil platforms, has less than 2 million subscribers with an average penetration of 0.4 percent. This market is

**TABLE 1.3    Cellular Network Deployment in Asia and Pacific (1993)**

| Country | Operator | System type | Infrastructure vendor |
|---|---|---|---|
| Australia | Telecom MobileNet, Optus, Arena GSM | AMPS, GSM | Alcatel, Ericsson, Nokia, Northern Telecom |
| Bahrain | BATELCO | TACS | NEC |
| Bangladesh | Bangladesh Telecom | AMPS | Motorola |
| Brunei | Jabatan Telekom Brunei (PTT) | AMPS, GSM | AT&T, Motorola, NEC |
| Cambodia | Camtel, Casacom, CamShin | AMPS, NMT 450, NMT 900 | Motorola, Nokia |
| China | PTT | AMPS, ETDMA, GSM, TACS | Alcatel, AT&T, Ericsson, Hughes, Motorola, Northern Telecom, NovAtel |
| Hong Kong | Communication Services LTD., Hutchison, Pacific Link Communication, Smartcom | AMPS, ETACS, GSM, TACS TDMA | Ericsson, Motorola, NEC, Nokia |
| Indonesia | PT Telkom | AMPS, NMT 450 | Ericsson, Motorola |
| Israel | BEZEQ | AMPS | Motorola |
| Japan | DDI, IDO, NTT | JTACS, NTACS, NTT, PDC, TACS | Motorola, NEC |
| Kuwait | MTSC | TACS | Ericsson |
| Laos | EPTL | AMPS | Systar |
| Lebanon | Spacetel | AMPS | NovAtel |
| Macau | CTM | TACS | Ericsson |
| Malaysia | Celcom, Syarikat Telekom Malaysia | ETACS, NMT 450 | Ericsson |
| Myanmar | Myanma P&T | AMPS | Ericsson |
| New Zealand | BellSouth New Zealand, Telecom Cellular, Telecom Mobile Communications | AMPS, GSM, TDMA | Ericsson, Nokia |
| Oman | GTO | NMT 450 | Ericsson |
| Pakistan | Pakcom, Paktel | AMPS | Ericsson |
| Philippines | Express Telecoms Inc., PLDT, Piltel, Smart Information Tech. | AMPS, TACS | AT&T, Ericsson, Motorola, NEC |

**TABLE 1.3    Cellular Network Deployment in Asia and Pacific (1993) (Continued)**

| | | | |
|---|---|---|---|
| Saudi Arabia | Sauditel | NMT 450 | Ericsson, Philips |
| Singapore | Singapore Telecom | AMPS, ETACS | Ericsson, NEC |
| South Korea | KMT | AMPS | AT&T, Motorola |
| Sri Lanka | Call Link, Celltel Lanka, Mobitel | AMPS, TACS | Ericsson, Motorola, NEC |
| Taiwan | DGPT | AMPS | Ericsson |
| Thailand | Advanced Information Services, Communication Authority of Thailand, Telephone Organization of Thailand, Total Access Communications Co. | AMPS, NMT 450, NMT 900 | Ericsson, Motorola, Nokia |
| United Arab Emirates | Etisalat | TACS | Ericsson |
| Vietnam | DGPT, Saigon Mobile Telephone Company | AMPS, GSM | Alcatel, Ericsson |

more focused on providing maritime teleservice for major shipping lanes between North and South America and traffic coming through the Panama Canal. Boat owners and tourists represent a significant percentage of traffic on the cellular networks in the Caribbean countries. The largest operator of cellular service for the tourists and maritime population is Boatphone. Most countries in Latin America have embarked on liberalization and privatization policies. Many private companies, including BellSouth, Millicom, GTE, McCaw, Telefonica, and Southwestern Bell, and Bell Canada are very active in these markets. The dominant cellular subscribers in this region are Brazil, Mexico, Argentina, and Venezuela. The Caribbean and Latin America region is characterized by uniformity in the system type. All have AMPS type systems. None of the Latin American countries have expressed interest in GSM, though some in the Caribbean Islands might deploy GSM networks. The major equipment vendors are Northern Telecom, Plexys, Motorola, and Ericsson. The cellular market information in 1993 for each country with an operational cellular network is shown in Table 1.4.

## 1.2.4    Europe

The European region, due to historical and political reasons, consists of two distinct markets: Western Europe with a well-developed telephone infrastructure and Eastern Europe with its antiquated switching network and infrastructure. In Western Europe, the growth in cellular stems from reduced subscriber costs, while in Eastern Europe the growth in cellular stems from its ability to substitute for land telephony. Many Eastern European countries are using cellular as a temporary substitute for the public network to meet the pent-up

**TABLE 1.4    Cellular Network in the Caribbean and Latin America (1993)**

| Country | Operator | System type | Infrastructure vendor |
|---------|----------|-------------|----------------------|
| Argentina | Compania de Telefonos, Miniphone, Movicom | AMPS | Ericsson, Motorola |
| Aruba | Setar | AMPS | Northern Telecom |
| Bahamas | BATELCO | AMPS | Northern Telecom |
| Barbados | BARTEL | AMPS | Northern Telecom |
| Bermuda | Bermuda Telco | AMPS | Northern Telecom |
| Belize | Belize Telecom | AMPS | Northern Telecom |
| Bolivia | Telefonica Cellular | AMPS | Motorola |
| Brazil | CTMR, Telebrasilia, Telemig and Brasil Central, Telepar and Sercomtel, Telerj, Telesp | AMPS | AT&T, Ericsson, Motorola, NEC, Northern Telecom Plexys |
| Cayman Islands | Cable & Wireless | AMPS | Plexys |
| Chile | Cidcom, CTC, Telecom Chile, VTR Celular | | Motorola, NEC, Northern Telecom, NovAtel |
| Costa Rica | Millicom Costa Rica | AMPS | NovAtel |
| Cuba | Cubacel | AMPS | Ericsson |
| Curacao | PTT | AMPS | Ericsson |
| Dominican Republic | Codetel, Tricom | AMPS | AT&T, Motorola, Northern Telecom |
| El Salvador | Telemovil | AMPS | Ericsson |
| Guatemala | Comcel | AMPS | Motorola |
| Jamaica | PTT | AMPS | NEC, Northern Telecom |
| Mexico | Baja Celular, Cedetel, Iusacell, Movitel, Norcel, Portatel, Telcel | AMPS | Ericsson, Motorola, Northern Telecom |
| Nicaragua | Nicacel | AMPS | Motorola |
| Paraguay | Telefonica Celular | AMPS | Unknown |
| Peru | CPT, Entel, Tele 2000 | AMPS | Northern Telecom, NovAtel |
| St. Maarten | East Caribbean Cellular | AMPS | Plexys |
| Trinidad | TSTT | AMPS | Northern Telecom |
| Uruguay | Movicom | AMPS | Motorola |
| Venezuela | CANTV, Telecel | AMPS | Ericsson, Motorola |

demand for basic telephone service. The introduction of cellular in Eastern Europe occurred about the same time as the realization of market liberalization and deregulation. This, coupled with the need for foreign capital—unlike in Western Europe—has led to cellular networks that are either privately owned or in partnership with the PTT. In several countries of Western Europe, the cellular network is still controlled by the PTT or some administrative body under the control of the PTT.

Many systems in Eastern Europe, as of 1993, are NMT 450, due mainly to the lack of 800- and 900-MHz spectrum, which were dedicated to military applications. Newer operators in Russia, Hungary, Poland, and Ukraine are planning to set up GSM networks. The Western European cellular networks are of different types (NMT 450, NMT 900, TACS, ETACS, CNETZ, Radicom 2000), but ETACS/TACS has the largest market share. All countries in this region have plans or have

**TABLE 1.5   Cellular Network in Europe (1993)**

| Country | Operator | System type | Infrastructure vendor |
|---|---|---|---|
| Austria | PTT | CNETZ | Motorola |
| Belgium | Belgacom | NMT 450 | Nokia |
| Belorussia | Belcel | NMT 450 | Ericsson |
| Cyprus | CYTA | NMT 900 | Ericsson |
| Czech & Slovak Republics | Eurotel | NMT 450 | Nokia |
| Denmark | TDM | GSM, NMT 450, NMT 900 | Ericsson, Nokia |
| Estonia | EMT | NMT 450 | Ericsson |
| Finland | Telecom Finland, Radiolinja | GSM, NMT 450, NMT 900 | Ericsson, Nokia, Siemens |
| France | France Telecom, SFR | GSM, NMT 900, Radiocom | Alcatel, Ericsson, Matra, Nokia, Siemens |
| Germany | D2 Mobil, Mannesman | CNETZ, GSM | Alcatel, Ericsson, Motorola, Siemens |
| Greece | STET, Panafon | GSM | Ericsson |
| Hungary | Westel | NMT 450 | Ericsson, Nokia |
| Iceland | PTT | NMT 450 | Ericsson |
| Ireland | Eircell | TACS | Ericsson |
| Italy | SIP | GSM, TACS 900 | Ericsson, Italtel |
| Latvia | LMT | NMT 450 | Nokia |
| Lithuania | Comliet | NMT 450 | Nokia |
| Luxembourg | PTT | NMT 450 | Ericsson |
| Malta | Telemalta | TACS 900 | Ericsson |
| Netherlands | PTT | NMT 450, NMT 900 | Ericsson, Siemens |
| Norway | Netcom, Tele-Mobil | GSM, NMT 450, NMT 900 | Ericsson, Motorola |
| Poland | Centertel | NMT 450 | Nokia |
| Portugal | TMN, Telemovel | CNETZ, GSM | Ericsson, Motorola, Siemens |
| Romania | TR | NMT 450 | Ericsson |
| Russia | Delta Telecom, MCC | NMT 450 | Ericsson, Nokia |
| Spain | Telefonica | NMT 450, TACS | Motorola |

**TABLE 1.5   Cellular Network in Europe (1993) (Continued)**

| | | | |
|---|---|---|---|
| Sweden | Comvik, Nordictel, Telia | GSM, NMT 450, NMT 900 | Ericsson, Motorola, Nokia |
| Switzerland | PTT | GSM, NMT 900 | Ericsson |
| Turkey | PTT | NMT 450 | Nokia |
| Ukraine | UMC | NMT 450 | Ericsson |
| United Kingdom | Cellnet, Vodafone | GSM, TACS | Ericsson, Motorola, Nokia |

already installed GSM networks for the next generation. The cellular deployment for each country by vendor is shown in Table 1.5.

### 1.2.5   North America

All operators in the United States and Canada use the same analog technology, i.e., AMPS; however, different operators are planning to take different routes for the second-generation digital market. It seems, unlike in Europe, where operators started with different analog technology but all decided to migrate to the GSM technology for their second-generation wireless infrastructure, in North America the trend is in the opposite direction. In the United States, two cellular licenses were awarded for each of the 305 MSA and 428 RSA areas. One license was set aside for the local wire-line telephone company. In Canada, two licenses were awarded for 23 MSAs. One of the licenses was set aside for the local wire-line telephone company but, unlike in the United States, the non-wire-line license was awarded to a single company (CANTEL) for all the 23 markets so they could effectively compete with the giant Bell Canada.

AT&T, Ericsson, Hughes Network Systems, Motorola, and Northern Telecom (Nortel) are the major equipment providers for the MSA markets; Motorola, NovAtel and Northern Telecom are the top providers for the RSA market. Northern Telecom and Ericsson share equally the Canadian equipment market. There have been other players in the North American cellular market. NEC and Harris were temporary providers of cellular equipment, but were displaced or never got a firm foothold. There is still a need for a small system in both North America and the rest of the world. Celcore is a newcomer in this niche. Harris is resurrecting in this market by the acquisition of NovAtel.

## 1.3   Technology

In this section, the basic concepts of wireless telecommunication are discussed. These concepts form the foundation for the information contained in subsequent chapters. The RF basics section which follows deals with the elements of RF transmission. This cellular basics section presents the concepts and terms used in wireless telecommunications. A brief overview of the analog and digital access technologies appears at the end of the section. AMPS and digital technologies are discussed in detail in subsequent chapters. The concepts presented in this section are discussed in more detail in Chap. 10.

### 1.3.1    RF basics

A basic, generic wireless telecommunications system is shown in Fig. 1.1. This system can be broken down into blocks as shown in Fig. 1.2. The human voice fed to the microphone of a handset is transmitted through the atmosphere to the base station. Similarly, at the network end, the voice information is transmitted from the base station and received by the handset. The handset and base station both have the transmitter/receiver (transceiver) function. The overview of the three major blocks—speech input, transmitter and receiver, and radio medium—are described in the next sections.

**1.3.1.1   Speech input.**   Human speech originates when air passes from the speaker's lungs through the voice box (larynx), a passage in the human throat with the opening obstructed by vocal cords. As air passes over the vocal cords, they vibrate, causing puffs of air to escape into the oral cavity. The resulting speech energy is distributed between 20 Hz

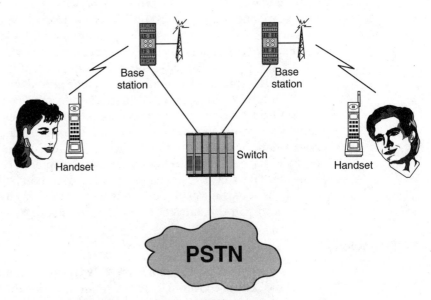

**Figure 1.1**   Generic wireless telecommunications system.

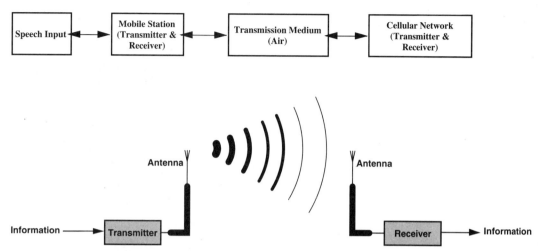

**Figure 1.2** Block diagram of a wireless telecommunications system.

and 20 kHz of frequency, though most of it is concentrated at the lower frequencies as shown in Fig. 1.3.

Further tests have shown that frequencies from 600 to 700 Hz add very little to the intelligibility of speech to the human ear. The dashed curve represents the intelligibility of the speech. Most of the speech can be reproduced accurately by analyzing the 4000-Hz bandwidth.

**1.3.1.2 Transmitter and receiver.** The transmitter and receiver convert information contained in the speech frequency to the frequency that can be transmitted through the desired frequency of the medium. This process is called *mo*dulation, as in *mo*dem. *Modulation* is the process by which some characteristic (information) of one wave is varied in accordance with another wave or signal. The modulation input signal contains the information to be transmitted and is also called the baseband signal. The carrier signal or broadband signal transfers the information. Each signal is distinguished by three parts: amplitude, frequency, and phase (time interval). These three characteristics of the signal lead to three types of modulation as shown in Fig. 1.4.

In amplitude modulation, the amplitude of the carrier is changed to carry the speech signal. This is similar to the amplitude modulation (AM) broadcasts on radio. When digital inputs are modulated by changing the amplitude, the process is called amplitude shift keying (ASK). In frequency modulation (FM) the frequency of the carrier is varied proportionally to the speech signal. This is similar to the FM broadcasts on radio. Voice transmission in AMPS cellular also uses this technique. For digital inputs this type of modulation is called frequency

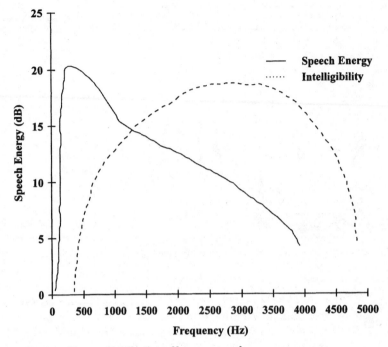

**Figure 1.3**   Energy distribution of human speech.

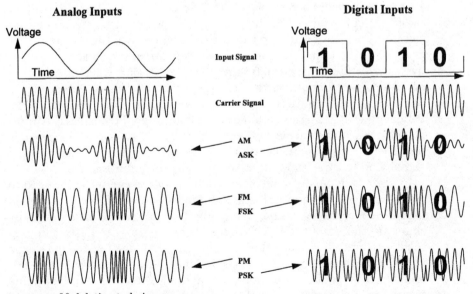

**Figure 1.4**   Modulation techniques.

shift keying (FSK). The control messages in AMPS cellular use the FSK technique. In phase modulation (PM), the phases are altered proportionally to the speech signal. Frequency and phase modulation are close mathematical cousins, and for analog inputs they are virtually indistinguishable. The digital input version is called phase shift keying (PSK). GSM uses this type of modulation.

The modulated signal is then transmitted through the handset antenna to be propagated through the atmosphere so that the base station antenna can pick the transmitted signals. The transmitter variable is the power output. High-power output increases the distance traveled by the signal but it also increases the probability of interference.

The antenna takes the power from the transmitter and radiates it out into space. The receiving antenna intercepts the signal from space and feeds it into a receiver. The antenna is characterized by *gain* and *beam width*. The height and angle of the antenna affects the coverage of the signal. The gain of the antenna is the increase in apparent power due to the focusing action of the antenna. The beam width describes the width of the beam of the antenna's radiation.

The receiver gets a signal from its antenna, which also receives a number of unwanted signals. A filter allows the required signal to pass while tuning out the unwanted signal. The filtered signal is then demodulated and decoded. This is then converted back into an audio signal for the ear to understand. The important characteristics of a receiver are *sensitivity, selectivity,* and *dynamic range.* Sensitivity describes the minimum signal strength needed for acceptable operation. Greater sensitivity of receivers improves the coverage of signals. Selectivity is the ability of the receiver to reject strong out-of-band signals. This is related to the filter design. Dynamic range is the useful range of acceptable signal strengths between the weakest recognizable signal (just above the noise floor) and the strongest readable signal (signal level at which distortion becomes unacceptable).

**1.3.1.3   Radio frequency (RF) medium.**   The radio frequency (RF) waves are part of the electromagnetic spectrum that includes visible light, ultraviolet, and infrared waves as shown in Fig. 1.5.

The relevant range of electromagnetic waves for wireless telecommunications services is shown in Fig. 1.6. The bands at 800 and 1900 MHz are for cellular and PCS respectively. The bands for satellite services are scattered above 2.5 GHz. The 1900 MHz bands for PCS are currently shared by terrestrial microwave users. They are to be relocated to higher bands in the next few years.

The waves are characterized by wavelength and frequency. The distance between the same points on two consecutive waves is the *wavelength* of the radio wave. The frequency is the number of waves emitted

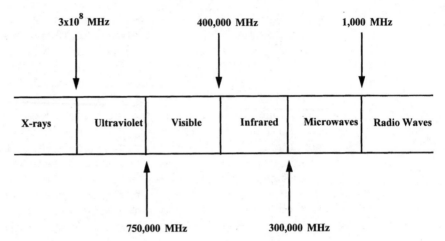

**Figure 1.5**  Radio and other electromagnetic waves.

per second. The wavelength is obtained by dividing the frequency of the wave by the speed of light (300,000 km per second). The wavelengths for cellular and PCS bands are 0.3 m and 0.2 m, respectively. The wavelength impacts several design parameters. It determines the size of the antenna elements and its ability to penetrate or reflect from obstruction in its path.

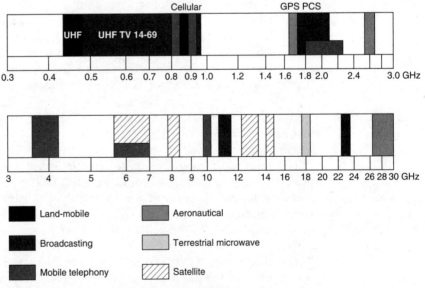

**Figure 1.6**  Wireless telecommunications bands.

In understanding wireless telecommunications systems, one must first determine the frequency range at which the system operates and the characteristics of that frequency. Some of the common parameters to consider are propagation loss, multipath fading, and noise. Frequencies on the lower end of the spectrum tend to have long range with high levels of interference, whereas those on the higher end tend to have shorter ranges with low level of interference and the ability to penetrate structures.

**1.3.1.4  The decibel jargon.**  The *decibel* (dB) is a unit for measuring the relative power. The human ear perceives power logarithmically. If a person estimates a signal to be twice as loud when the transmitter power is increased 10 times from 10 to 100 watts, then the person will estimate that a 10,000-watt power is twice as loud as a 1000-watt power. The decibel is the log of a power ratio. This can be expressed mathematically as shown:

$$dB = 10 \log (P_2/P_1)$$

where $P_1$ and $P_2$ are power levels. The decibel, being a ratio, is a relative unit. If dBs are not used for RF power calculation, then the calculation leads to unwieldy large and small numbers because of multiplication and division. The dBs, however, are log values, hence, instead of multiplying and/or dividing are added and/or subtracted. No change in a signal is a 0 dB. Increasing a signal by 10 times is a 10-dB increase, while increasing a signal by 1000 times is a (10 + 10 + 10 = 30)-dB increase. The addition/subtraction instead of multiplication/ division of power values simplifies computation.

When using decibels to specify an absolute power level, the decibel value must be qualified by a reference level. In radio work, power is often expressed in dBm, i.e., decibels referenced to 1 milliwatt of actual power.

**1.3.1.5  Propagation loss.**  The *propagation loss* is the loss in signal strength when it traverses space or a medium. The propagation loss increases with distance and frequency. The greater the distance traveled, the greater the losses. Similarly, the higher the frequency, the higher the losses. Hence, the propagation losses are higher for the PCS bands at 1900 MHz than for the cellular bands at 800 MHz. Furthermore, at higher frequencies, the outer atmosphere (troposphere/ionosphere) absorbs the waves instead of reflecting them so that the major mode of transmission is the direct wave. At certain frequencies, reflection from buildings and diffraction can extend the range of the transmissions. In free space, there is a 20-dB loss in signal strength when the transmitter/receiver moves by a factor of 10, i.e., from 1 to 10 km. In the mobile radio environment the loss is between 20 and 50 dB.

**1.3.1.6  Multipath fading.**  The mobile is usually surrounded by large buildings and other surroundings, and the wavelength (inverse of frequency) is much less than the sizes of the surrounding structures. This leads to reflection and diffraction. The reflected and diffracted waves will reach mobile at a different time than the direct wave. The result of these multipath waves causes signal fading called *multipath fading*. This fading is also accentuated by the moving vehicle. Fading is also complicated by diffusion. As the wave spreads out, the weaker it becomes. Fading can be affected by fog, humidity, rain, and other elements that intensify diffusion and absorption.

**1.3.1.7   Noise.**  Between the transmission and receipt of a radio signal, the transmitted signal gets distorted due to the addition of unwanted signals called *noise*. Noise can be divided into three categories:

1. Thermal noise

2. Intermodulation noise

3. Impulse noise

*Thermal noise* is proportional to the operating temperature and is due to the movement of electrons in a conductor. It is present in all electronic devices and media. It cannot be eliminated but can be reduced by operating at a lower temperature. In some high-quality receivers, in order to reduce the thermal noise, certain components are operated at liquid-nitrogen temperatures. In many cases, better electroconducting metals are used, e.g., gold and silver conduct electricity with less resistance. Thus, the metal remains cooler and less thermal noise is introduced.

*Intermodulation noise* occurs when different frequencies share the same transmission medium. The effect of the intermodulation noise is to produce spurious signals at frequencies that are equivalent to the sum or difference of the different input frequencies. When two frequencies, $f_1$ and $f_2$, are transmitted through air, they would produce energies at frequencies $f_1 + f_2$ and $f_2 - f_1$, thus interfering with other input signals at $f_1 + f_2$ and $f_2 - f_1$.

*Impulse noise* consists of irregular pulses or noise spikes of short duration and of relatively high amplitude. It is generated from a variety of causes, including external electromagnetic disturbances, such as lightning. This type of noise is not generally a factor for voice communications.

In order for a signal to be detected, the received signal must be higher than the noise level by a respectable margin. The margin depends on the sensitivity of the receiver. The parameter used to determine the performance of the transmission system is the signal-to-noise (S/N) ratio. This is usually reported in decibels. This ratio is measured at the receiver, since it is at this point that an attempt is made to process the signal and eliminate the unwanted noise.

### 1.3.2 Cellular basics

Prior to the cellular concept, the approach to providing mobile services was similar to the approach taken by radio and television stations. The operators set up huge transmitters at the highest point in a town. Then they sent high-power transmissions resulting in a large coverage area. The consequence of this was twofold: one, there was a capacity problem; and two, the mobile stations consumed a large amount of power. Therefore, they were very bulky and expensive. Consider the capacity problem. Let's say there are 25 channels available for voice transmission in the metropolitan area of Dallas. Thus only 25 calls can be made simultaneously on the system.

The solution to this problem is to decrease the power of transmission, thereby reducing the coverage area of a transmitter. Because the range of each cell is small, an area is divided into several smaller areas called *cells*. Each cell has its own antenna, a set of frequencies, and transmitter/receiver radio units. Adjacent cells have different frequencies to prevent cochannel interference, but distant cells could have the same frequencies, i.e., reuse the same frequencies. The cellular concept shown in Fig. 1.7 allows a limited number of frequencies to be reused, thereby increasing the capacity of the system. In addition, due to the low power, the handset became physically smaller and more cost economical.

A system could start with a small number of large cells. As the number of subscribers increased in a particular cell, that cell could be split into several cells, thereby increasing the overall capacity. In theory, this splitting of cells could continue ad infinitum, but there are economic and practical constraints. Each time a cell is split, more base-station equipment and interconnection costs are incurred. In practice, the smallest cells have sizes of about a 1-mile radius in urban areas and 10 miles in rural areas. The process of cell splitting is shown in Fig. 1.8.

In cellular networks, unlike in the old mobile architecture, there are multiple cells covering an area. Hence, calls had to be handed off to other cells as the vehicle moved. This is called *handoff*. As a vehicle moves away from the base station, its signal strength decreases. The

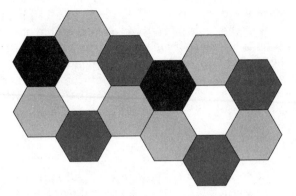

Cells with the same pattern use the same frequencies

**Figure 1.7**  Cellular concept.

base station monitors the signal strength during the duration of the call. When the signal strength falls below a predetermined threshold level, the network will ask all predetermined candidate neighboring cells to report the signal strength of the mobile. If a signal strength in a neighboring cell is stronger by a predetermined amount, then the network will attempt to hand off the call to the candidate neighboring cell. The hand off process is shown in Fig. 1.9.

Two areas of interference that need to considered in the cellular systems are the cochannel interference and adjacent channel interference. Cochannel interference occurs when a mobile in a distant cell is transmitting/receiving on the same frequency. This is shown in Fig. 1.10.

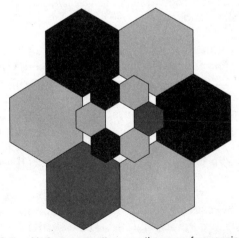

**Figure 1.8**  Cell splitting.

Cells with the same pattern use the same frequencies

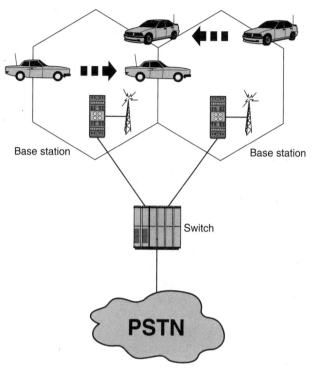

**Figure 1.9** Handoff concept.

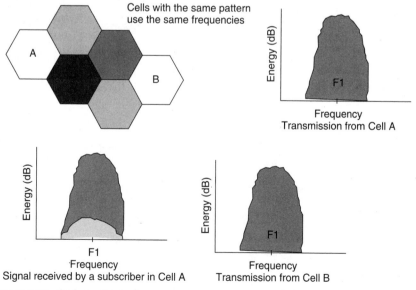

**Figure 1.10** Cochannel interference.

Adjacent channel interference occurs when energies from two adjacent channels overlap as shown in Fig. 1.11. Adjacent interference can be managed by proper frequency planning.

The interference that can be tolerated depends on the modulation type and is measured as a threshold ratio of carrier power to interference called $C/I$. C/I is measured in dB. For AMPS and TACS the acceptable value of C/I is 17 dB while for GSM it is 9.5 dB. This is due to the different technical parameters of the two systems.

### 1.3.3  Cellular access technology

As discussed previously, the cellular access technologies could be analog or digital. Among the analog technologies, AMPS, NMT, and TACS/ETACS are commonly found in cellular networks. The digital technologies are relatively new. Among the several digital technologies, GSM, DAMPS (U.S. TDMA), and CDMA (Qualcomm's CDMA) appear to have garnered the most interest and attention among the operators—both new and existing operators. The basic concepts, such as frequency reuse, are the same in all these access methods. The differences are in the modulation scheme, channel spacing, and other technical parameters.

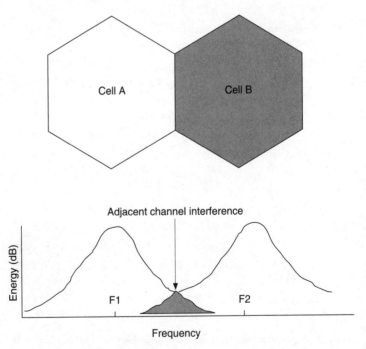

**Figure 1.11**   Adjacent channel interference.

**1.3.3.1 Analog access technology.** The analog access technology in protocols such as NMT, TACS/ETACS, and AMPS is based on Frequency-Division Multiple Access (FDMA). In FDMA, the available radio spectrum is divided into channel bandwidths. Each such channel bandwidth supports one voice channel as shown in Fig. 1.12. Thus, each RF channel carries one conversation.

Though most analog access technology protocols are based on FDMA, their technical specifications differ. Different technical specification was due to different requirements in each country, and in particular, the availability of frequency.

NMT was introduced in Scandinavian countries in the 450-MHz frequency, a band available in these countries. NMT was optimized for the largely rural nature of the Scandinavian countries. The 450-MHz frequency has better propagation characteristics compared with the 800 MHz; therefore, it had large cell sites that were needed to cover large open areas. Thus, NMT was designed and optimized based on coverage requirements. NMT was later modified to operate at 900 MHz for capacity and ease of use with handheld portables.

AMPS and TACS/ETACS were designed and optimized for capacity. They operate at 800- and 900-MHz frequencies. These systems allow smaller cell sites compared to the cell sites at 450 MHz. This is particularly useful for a country like the United Kingdom with concentrated and high-population density areas.

The analog access technology, except for AMPS, is being transitioned to digital, due to either government mandate or market requirements. It is expected AMPS will also eventually be displaced with digital technologies, but not any time in the near future because of the large number of current AMPS subscribers (more than half of the world's cellular subscribers) and the high investment in infrastructure. AMPS is discussed in detail in Chap. 2.

Three frequency bands allowing three users to communicate simultaneously

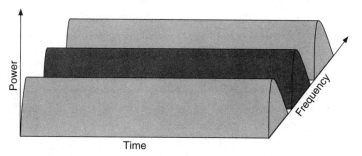

**Figure 1.12** Frequency-division multiple access.

### 1.3.3.2 Digital access technology.

The digital access technologies are based on Time-Division Multiple Access (TDMA) or Code-Division Multiple Access (CDMA). Some of the digital technologies, such as Omnipoint's, are a hybrid combination of both CDMA and TDMA. Many others are modifications or different variations of TDMA or CDMA.

In TDMA, the RF carrier is divided into time slots. The system allocates speech to these time slots. Both DAMPS (commonly referred to as TDMA in the United States) and GSMs are TDMA-based systems. They differ in their technical parameters. Speech information is converted to digital form and compressed by a process called *vocoding* so that the speech samples can fit in the time slots. The transmission occurs during the assigned time slots. The TDMA concept is shown in Fig. 1.13.

The transmitter is active during the assigned time slot; thus, the terminal can either be idle and save power or monitor the environment and take a proactive role in, say, the handoff. In addition, other features such as short message services can be easily introduced. Currently, DAMPS has three time slots in each 30 kHz, while GSM has eight time slots in the 200-kHz bandwidth. DAMPS and GSM are discussed in detail in subsequent chapters.

In CDMA, the signals are spread over a large frequency band called *broadband.* In TDMA and FDMA, the signals are concentrated within the narrow time slots or frequency bandwidth. Each voice signal in CDMA is coded by another signal called a *pseudorandom signal.* Multiple voice signals are sent through the same wide-frequency band. The receiver that has the correct pseudorandom signal will reproduce the original voice signal. The CDMA concept is shown in Fig. 1.14.

In TDMA and FDMA, if all the time slots and frequency are used, then additional traffic cannot be added, i.e., there is a maximum capacity limit. However, additional traffic can be added in CDMA. This additional traffic will increase the noise level not just for the added user but

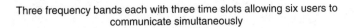
Three frequency bands each with three time slots allowing six users to communicate simultaneously

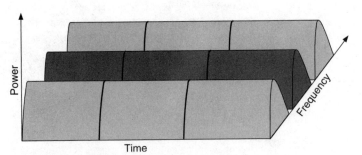

**Figure 1.13** Time-division multiple access.

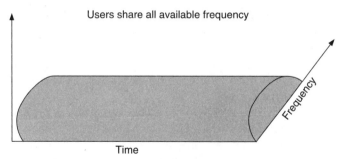

Users share all available frequency

Time

**Figure 1.14**    Code-division multiple access.

for all users of the wide frequency band. Thus, there is no hard-capacity limit. The capacity limit is determined by the user's tolerance to noise. CDMA requires that each signal be received at the same signal strength; otherwise, the strong signal will prevent the reception of other signals. If a mobile is very close to the base station, there is a likelihood of the mobile overwhelming other signals. This is called the *near-far problem*. To overcome the near-far problem, a tight reverse and dynamic power control have to be employed. A detailed discussion of CDMA is presented in subsequent chapters.

### 1.3.4  Basics of frequency planning

The channelized RF air interface technologies (AMPS, U.S. IS-54 TDMA, GSM) rely on frequency reuse to achieve needed network capacity. The reuse must be planned to achieve acceptable levels of reuse, but to contain interference at levels which will not cause user annoyance or dropped calls.

The feasible degree of frequency reuse is generally determined by (1) the degree of inherent vulnerability to interference of the chosen technology, and (2) local propagation characteristics that determine the degree of interference actually delivered.

**1.3.4.1  C/I requirements of channelized technologies.**  Various technologies react to interference differently. AMPS, for example, gives good quality while the desired signal is at least 17 to 18 dB stronger than other interfering signals reaching the receiver on the same channel as the desired signal. If this cochannel C/I ratio worsens, call quality degrades until the range of about 10–14 dB when the voice quality becomes objectionable. Occasional bursts of the interferer's voice may be distinctly recognizable. There is a risk of the call being dropped due to misrecognition of the interferer's signaling messages at the degraded C/I levels.

U.S. IS-54 and IS-136 TDMA have slightly better immunity than AMPS, tolerating about 14 dB C/I. However, at poorer C/I ratios voice quality degrades rapidly due to high-bit error rates, and dropped calls can occur. To provide a cushion against this possibility, most design engineers seek to maintain the same C/I objectives for a U.S. TDMA system as for an AMPS system.

The NAMPS, TACS, JTACS, NMT, and ESMR technologies are all roughly similar to AMPS and U.S. TDMA in their interference characteristics.

GSM has a natural advantage because of its more robust GMSK modulation scheme. GSM systems can tolerate cochannel C/I ratios as low as 9 to 6.5 dB before bit error rates begin to impact voice quality. Frequency hopping, if applied, makes the lower 6.5-dB figure practical.

### 1.3.4.2 Required C/I forces minimum cochannel separation and reuse factor *N*.

These required C/I values, together with the local propagation characteristics, determine the separation which must be maintained between cochannel cells to maintain acceptable call quality.

Demonstrating this philosophy, AMPS cell site A is intended to deliver interference-free service out to the edge of its coverage area. At the edge, its signal will have fallen to a low level; a typical design value might be –95 dBm. In a well-engineered system, the nearest cell (cell B) using the same channel is separated in distance sufficiently that it will deliver no more than –112 dBm at this point, thereby barely preserving the desired 17-dB C/I for calls on cell A. Figure 1.15 illustrates this relationship. R refers to the interference-free coverage radius of cell A. D refers to the separation between cell A and cell B, which barely achieves the desired 17-dB C/I.

Careful study of the local propagation environment is necessary to determine the closest permissible separation between cells to establish the situation described above. This distance is usually expressed in relative terms of distance and radius (D/R) rather than an absolute number of miles or kilometers. In typical terrain, it is often possible to achieve it with the cells separated by a D/R of approximately 4.6. This ratio of D/R is achieved if the cells are arranged in symmetrical clusters of seven cells. This is called an $N = 7$ frequency plan. All of the available frequencies are pooled and divided among the seven cells of the cluster; therefore, each cell receives one-seventh of the total number of available channels.

As the market is covered by repeated, interlocking clusters of seven cells, each cell's cochannel neighbors barely maintain the desired C/I of 17 dB at its edge, and it maintains its neighbor's own desired 17-dB C/I at the neighbor's edges.

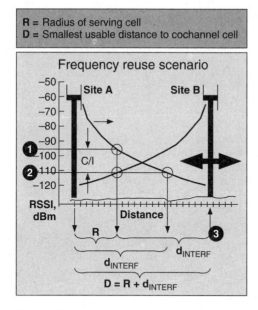

**Figure 1.15**    Frequency reuse.

A GSM system with its less stringent need for only 6.5 to 9 dB C/I might be able to survive with a lower value of $N$ (often four) in typical terrain. This would allow one-fourth of the available channels to be assigned in each cell in the four-cell cluster, not one-seventh as in an $N = 7$ system. All other conditions being equal, an $N = 4$ cell will have 1.75 times more channels than an $N = 7$ cell.

CDMA systems, unlike the channelized RF technologies, do not need frequency planning. The subscriber traffic is carried on a single RF channel that is reused in every sector and cell. Each subscriber's traffic has to be differentiated from other subscribers' traffic by codes. These codes need to be assigned among the different cells. Thus, in CDMA systems, frequency planning is replaced by code planning.

### 1.3.5    Traffic analysis

Traffic engineering is the process of administering and dimensioning any shared resource, from airline seats, to space on a bread truck, to talk time on a wireless system. The broader parent discipline is known as *queuing theory* and may appear quite unfriendly to an impatient user preoccupied with building sites and commercial in-service dates.

However, familiarity with the basic concepts of traffic engineering is necessary to the design of wireless systems, and these concepts are quite sensible if approached logically.

**1.3.5.1  Basic circuit concept.**  The basic resource to be administered is individual circuits. A *circuit* is one path over which it is possible to hold a conversation. Sometimes such a circuit is just one of a pooled group and, if so, is called a *trunk*. The whole pool might be called a *trunk group;* one trunk is a *member.*

For example, in an AMPS system, one radio carries one conversation, therefore it is considered to be one circuit. In a TDMA-3 system, one radio can carry three circuits. In a GSM system, one radio can carry eight circuits.

**1.3.5.2  Basic traffic units.**  The basic units of traffic are simply units of talk time. Normally, traffic is studied in periods of one hour. If one circuit is continuously occupied for the whole period (one hour), it is considered one *erlang* of traffic. The Erlang is named after Anger K. Erlang, a Danish telephone engineer who first developed and published much of the theory of telephone traffic engineering around 1911. In addition to Erlangs, traffic can be expressed in any other convenient units such as CCS (centum call seconds) and MOU (minutes of use). CCS is one hundred call seconds. Thus, one Erlang equals 36 CCS. Similarly, one erlang equals 60 MOU.

**1.3.5.3  Concepts of capacity, blocking, and grade of service.**  The absolute maximum traffic that can be carried on a single circuit for one hour is an Erlang. Two circuits can carry two Erlangs; twelve circuits, twelve Erlangs, and so on. If one circuit is kept busy for an entire hour, it won't be available for anybody else to use. A second prospective user would be blocked if the circuit is busy when they attempt to use it. Telecommunications systems are designed so that the probability of blocking is small. Two percent is the goal in most wireless systems. This probability of blocking is sometimes called the *grade of service* (GOS). For example, a two percent blocking probability is expressed as GOS P.02.

There is an important distinction between two types of traffic. *Carried traffic* is the traffic actually carried on a system, i.e., communications that are successfully completed. *Offered traffic* is the traffic users attempted to push through the system. If all the offered traffic gets through, the carried and offered traffic are equal. If calls are blocked, the carried traffic will be smaller than the offered traffic by an amount equal to the percentage of blocking.

**1.3.5.4  Trunking efficiency.**  There is a set of Erlang tables used by traffic engineers to determine the number of circuits needed based on blocking criteria. The traffic carried by a pool of circuits is larger than the sum of traffic that could be carried on each separate circuit. The trunking efficiency is shown in Fig. 1.16.

It demonstrates the need to organize networks in groups of trunks that are as large as possible and to avoid small groups of trunks. It is also the basis of calculated tables showing how much traffic can be accepted while meeting a given GOS on a given number of trunks.

Blocking theoretically can occur anywhere in wireless systems. The bottleneck occurs wherever the circuit passes through the smallest group of trunks along its route, which is normally at the cell site. Trunk groups from the wireless switch to the PSTN are normally large and usually are arranged to back up each other with alternate routing. Alternate routing possibilities at the cell site are rare.

There are actually several different traffic tables showing the relationship between offered traffic, number of trunks, and blocking. The tables apply to situations with different starting assumptions. The Erlang-B table assumes blocked calls are cleared or lost from the system. The Erlang-C and Poisson tables cover the situation where there is a queue and various objectives apply to the waiting time and other dynamics of the situation. The table values differ by a few percent, but are in fairly close agreement for situations where the blocking is small.

The Erlang-B table is the most widely used in wireless systems. Figure 1.17 shows the table for GOS P.02 and the number of trunks typically involved in wireless sites where the natural bottleneck of wireless systems occurs.

**1.3.5.5  The busy hour.**  *Busy hour* is the busiest hour-long period of the day on a wireless system. Circuits are planned to carry the normal everyday traffic maximum that occurs during the busy hour, while meeting the system's GOS objective. Unusual events such as storms or accidents are ignored in this planning.

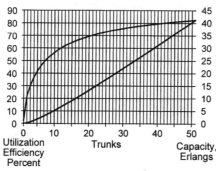

**Capacity and Trunk Utilization
Erlang-B for P.02 Grade of Service**

**Figure 1.16**  Trunking efficiency.

| #Trunks | Erlangs | #Trunks | Erlangs | #Trunks | Erlangs | #Trunks | Erlangs | #Trunks | Erlangs | #Trunks | Erlangs | #Trunks | Erlangs | #Trunks | Erlangs |
|---|---|---|---|---|---|---|---|---|---|---|---|---|---|---|---|
| 1 | 0.0204 | 26 | 18.4 | 51 | 41.2 | 76 | 64.9 | 100 | 88 | 150 | 136.8 | 200 | 186.2 | 250 | 235.8 |
| 2 | 0.223 | 27 | 19.3 | 52 | 42.1 | 77 | 65.8 | 102 | 89.9 | 152 | 138.8 | 202 | 188.1 | 300 | 285.7 |
| 3 | 0.602 | 28 | 20.2 | 53 | 43.1 | 78 | 66.8 | 104 | 91.9 | 154 | 140.7 | 204 | 190.1 | 350 | 335.7 |
| 4 | 1.09 | 29 | 21 | 54 | 44 | 79 | 67.7 | 106 | 93.8 | 156 | 142.7 | 206 | 192.1 | 400 | 385.9 |
| 5 | 1.66 | 30 | 21.9 | 55 | 44.9 | 80 | 68.7 | 108 | 95.7 | 158 | 144.7 | 208 | 194.1 | 450 | 436.1 |
| 6 | 2.28 | 31 | 22.8 | 56 | 45.9 | 81 | 69.6 | 110 | 97.7 | 160 | 146.6 | 210 | 196.1 | 500 | 486.4 |
| 7 | 2.94 | 32 | 23.7 | 57 | 46.8 | 82 | 70.6 | 112 | 99.6 | 162 | 148.6 | 212 | 198.1 | 600 | 587.2 |
| 8 | 3.63 | 33 | 24.6 | 58 | 47.8 | 83 | 71.6 | 114 | 101.6 | 164 | 150.6 | 214 | 200 | 700 | 688.2 |
| 9 | 4.34 | 34 | 25.5 | 59 | 48.7 | 84 | 72.5 | 116 | 103.5 | 166 | 152.6 | 216 | 202 | 800 | 789.3 |
| 10 | 5.08 | 35 | 26.4 | 60 | 49.6 | 85 | 73.5 | 118 | 105.5 | 168 | 154.5 | 218 | 204 | 900 | 890.6 |
| 11 | 5.84 | 36 | 27.3 | 61 | 50.6 | 86 | 74.5 | 120 | 107.4 | 170 | 156.5 | 220 | 206 | 1000 | 999.1 |
| 12 | 6.61 | 37 | 28.3 | 62 | 51.5 | 87 | 75.4 | 122 | 109.4 | 172 | 158.5 | 222 | 208 | 1100 | 1093 |
| 13 | 7.4 | 38 | 29.2 | 63 | 52.5 | 88 | 76.4 | 124 | 111.3 | 174 | 160.4 | 224 | 210 | | |
| 14 | 8.2 | 39 | 30.1 | 64 | 53.4 | 89 | 77.3 | 126 | 113.3 | 176 | 162.4 | 226 | 212 | | |
| 15 | 9.01 | 40 | 31 | 65 | 54.4 | 90 | 78.3 | 128 | 115.2 | 178 | 164.4 | 228 | 213.9 | | |
| 16 | 9.83 | 41 | 31.9 | 66 | 55.3 | 91 | 79.3 | 130 | 117.2 | 180 | 166.4 | 230 | 215.9 | | |
| 17 | 10.7 | 42 | 32.8 | 67 | 56.3 | 92 | 80.2 | 132 | 119.1 | 182 | 168.3 | 232 | 217.9 | | |
| 18 | 11.5 | 43 | 33.8 | 68 | 57.2 | 93 | 81.2 | 134 | 121.1 | 184 | 170.3 | 234 | 219.9 | | |
| 19 | 12.3 | 44 | 34.7 | 69 | 58.2 | 94 | 82.2 | 136 | 123.1 | 186 | 172.4 | 236 | 221.9 | | |
| 20 | 13.2 | 45 | 35.6 | 70 | 59.1 | 95 | 83.1 | 138 | 125 | 188 | 174.3 | 238 | 223.9 | | |
| 21 | 14 | 46 | 36.5 | 71 | 60.1 | 96 | 84.1 | 140 | 127 | 190 | 176.3 | 240 | 225.9 | | |
| 22 | 14.9 | 47 | 37.5 | 72 | 61 | 97 | 85.1 | 142 | 128.9 | 192 | 178.2 | 242 | 227.9 | | |
| 23 | 15.8 | 48 | 38.4 | 73 | 62 | 98 | 86 | 144 | 130.9 | 194 | 180.2 | 244 | 229.9 | | |
| 24 | 16.6 | 49 | 39.3 | 74 | 62.9 | 99 | 87 | 146 | 132.9 | 196 | 182.2 | 246 | 231.8 | | |
| 25 | 17.5 | 50 | 40.3 | 75 | 63.9 | 100 | 88 | 148 | 134.8 | 198 | 184.2 | 248 | 233.8 | | |

**Figure 1.17**    Erlang-B table for P.02 GOS.

Most wireless systems select the "floating" or "bouncing" busy hour for their analysis of each cell, the switch, and other common resources. Whichever hour happens to be the busiest of a given day is used as the busy hour for that day; that is, it is not a fixed hour for all days based on long-term trends but is free to float or bounce from day to day.

**1.3.5.6   Clues to geographic traffic distribution.**   Wire-line telephone systems have one degree more knowledge about their customers than wireless systems. Wire-line operators know the location of the customers, so they can plan accordingly. Wireless operators, on the other hand, can only estimate the geographic distribution of traffic on the new system and track trends once the system is in operation.

Determining the required number of circuits at each cell site is a difficult guessing game for a new system. It requires knowledge of the total number of customers, their calling habits, and distribution of traffic throughout the market. The stakes are high since network equipment capital cost is high and customers demand acceptable service.

From the overall traffic perspective, a spreadsheet is created with the collaboration of business planners and the marketing group within the operating company. The market population, customer penetration level, and resulting number of subscribers are estimated over a several-year period. The calling behavior of the subscribers is anticipated and a total value of expected traffic is computed in Erlangs.

For initial business planning, the total traffic figure is simply divided by the number of cells and the resulting average traffic-per-cell is contemplated to see if it is reasonable. The detailed cell planning attempts to distribute the traffic geographically in some form of model and allow the propagation prediction tool to estimate the traffic intercepted by each cell. Cells that appear overloaded are divided and made smaller, while cells that have little traffic may be relocated, raised, and consolidated to reach a more practical economic efficiency.

Estimating traffic geographic distribution requires: consideration of population density available from census tract maps and TIGER files; office-space totals and other business-activity benchmarks; vehicular-traffic peg counts from government agencies and highway departments; local area knowledge of dynamic, new construction projects; and development trends. In dense areas, a straight-down, aerial photograph at the peak of rush hour is a good indicator of traffic distribution.

## 1.4 Summary

This chapter provides an overview of the market and technology of wireless telecommunications. It provides information on the major vendors and operators in the different regions of the world. The motivation for the different countries to pursue wireless telecommunications depends on the existing telecommunications networks' capabilities. Developed countries are interested in providing mobility and its associated independence and efficiency to its citizens. Developing countries are primarily using wireless telecommunications as a substitute for landline telephony to meet pent-up demand. Many of these developing countries view wireless telecommunications as a quick and economical way to provide basic telecommunication services to its rural areas and instant service in urban areas. In addition to the wireless telecommunications market, the different technologies that are used in wireless telephony are briefly discussed. The details of these technologies are discussed in subsequent chapters. However, preceding the chapters on PCS digital technologies are chapters on PCS substitutes such as North American AMPS, Paging, and Personal Communications Satellite Systems. The chapters on PCS substitutes are followed by chapters on PCS market structure, digital technologies for PCS, PCS network design, and the economics of PCS.

# PCS Substitutes and Complements

# 2

# Cellular Services

## 2.1 Introduction

Due to the arrival of new PCS operators in the competitive landscape, cellular operators have accelerated their push to introduce PCS-type services. Many cellular operators, especially the large ones, have either offered or announced plans for commercial "one-number" solutions in various forms. The cellular operators hope to have a head start over the new PCS operators for PCS services. Many cellular operators rightfully believe that cellular and PCS are essentially the same, but separated by spectrum frequency. Services that are planned for PCS can and will be offered by cellular operators. In fact, it is a matter of price, packaging, distribution channels, and product/network enhancements.

Many cellular operators have lowered their prices, making them attractive in the consumer market. Cellular operators have initiated shrink-wrapped packages that allow over-the-air activation. This concept allows the phones to be displayed and sold through traditional consumer outlets such as Wal-Mart, thereby expanding and enhancing their current distribution channels. These distribution channels, when available, will make cellular phones easily available as landline phones. Again, the intention is to make the product easily available to consumers.

The cellular industry, as a preemptive measure against the PCS operators, has started investing in network enhancements. Cellular operators will have the necessary product capabilities to provide PCS-like services with greater microcell coverage, deployment of digital, seamless roaming, and intelligent network components. The cellular air interface standard has been enhanced by another standard, IS-94, which was recently incorporated into IS-91A. IS-94 is the analog air interface standard for cellular-based personal communication systems.

IS-94 allows cellular operators to provide PBX/Centrex-like features. On the consumer market, Motorola has introduced its InReach product based on these standards. InReach allows a cellular phone to act as a cordless phone within the confines of the home, thus avoiding airtime charges, and as a cellular phone outside the home. Similarly, on the commercial side, cellular phones have been introduced by Motorola, Panasonic, and others that can be used as an extension to a PBX within the office, and as a regular cellular phone outside the office.

Cellular operators are accelerating their plans to go digital by making the networks digital-ready in preparation for PCS competition. In addition, they are introducing intelligent network components to ultimately evolve the cellular network to an Advanced Intelligent Network (AIN). These enhancements will allow the operators to provide economical features such as calling-line identification, authentication, voice privacy, integrated paging, and enhanced-message services.

Cellular operators are repositioning themselves from the marketing and technical aspects of the business in order to compete with the new PCS operators. The cellular operators, unlike the PCS operators with multiple digital technology options (GSM, TDMA, CDMA, PACS, etc.), have two digital options—TDMA and CDMA. The remainder of this chapter will provide details of the AMPS market structure and technology. Future chapters will detail the digital technology of TDMA and CDMA.

## 2.2   Market Structure

The market structure of the cellular industry is determined by forces that are very broad. It encompasses social, regulatory, and economic forces. The external market forces, such as regulation, shape the industry structure. These external forces are the catalyst for strategic moves by the industry players. They affect all firms in the industry and determine their ultimate profit potential. These forces are different for different industries: intense for the tire and steel industries, but mild for industries such as cosmetics and toiletries. Also, these forces explain the low rate of return on invested capital in the tire and steel industry and the high rate of return in the cosmetics and toiletries industry. This section is concerned with identifying and describing the key structural elements that influence the cellular industry structure. Knowledge of these forces will allow the reader to better appreciate the tactical, strategic, and competitive moves of the industry participants.

The extent of competition within the industry is determined by these forces. Competition in the industry continually works to drive down the rate of return on invested capital toward the competitive floor rate of return (the return that would be earned by the economist's "perfectly competitive" industry). The presence of higher rates of return will attract capital either through new entry or through additional investments by existing competitors, leading it to move toward the competitive floor rate of return.

There are many internal and external forces that influence the cellular industry. This section will not attempt to identify or describe all of these forces. However, the important ones that shape the cellular industry will be identified and described. The main forces that have influenced the cellular industry in the last decade are:

1. Regulation

2. Subscriber demand

3. Equipment vendors

4. Substitution products such as PCS and SMR

5. Operators and competition among them

These forces are shown in Fig. 2.1.

Each force by itself does not define the cellular industry, but the combination and collective strengths of these forces along with the rivalry among the existing operators decides the cellular industry structure. Changes in these forces lead to either strategic or reactionary moves by

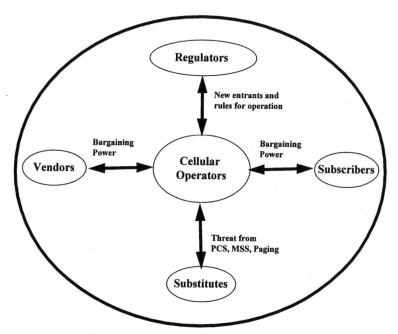

**Figure 2.1**   Forces affecting the cellular industry.

the industry participants. The change in the regulatory landscape with the FCC's announcement of the release of 120 MHz of spectrum for the PCS market led to several alliances, mergers, and consolidation by the industry participants. A notable alliance was the one among the four large cellular companies—Airtouch, US West, Bell Atlantic, and Nynex—to form PCS Primeco. Another occurred between Sprint and the cable operators Cox, TCI, and Comcast to form Wireless Company. Also, this has accelerated the cellular industry's transition to digital and the introduction of intelligent components into the network.

## 2.2.1   Regulation

For licensing purposes, the FCC divided the entire United States into 306 urban areas called Metropolitan Statistical Areas (MSAs) and 428 rural areas called Rural Service Areas (RSAs). MSAs cover approximately 80 percent of the population, but only 20 percent of the landmass, while RSAs cover only 20 percent of the population, but 80 percent of the landmass. The FCC decided to have two operators in each of the MSAs and RSAs. Each operator was allocated a spectrum in the 800-MHz frequency. One license was awarded to the local telephone (wire-line) company called the B-side carrier, while the other license was awarded to a non-wire-line company called the A-side carrier. This was to prevent the strong wire-line companies from grabbing the licenses in a head-to-head competition in the comparative-hearing process for the allocation of the licenses. This distinction between the wire-line and non-wire-line carrier was valid only in the initial licensing process. The regulation does not prevent subsequent buying, selling, consolidation, or mergers. Such postlicensing activities have blurred the distinction between the wire-line and non-wire-line carriers. For example, BellSouth owns several non-wire-line licenses in addition to its wire-line licenses. BellSouth's non-wire-line licenses, where it has more than 50 percent ownership in the top 50 MSA market, include Los Angeles, Houston, Milwaukee, Indianapolis, and Honolulu. Among Southwestern Bell Communications' top five markets, three (Chicago, Boston, and Washington) are on the non-wire-line side.

The spectrum allocated to the A- and B-side carriers are shown in Table 2.1.

TABLE 2.1   **Spectrum Allocated to Cellular Carriers**

|  | Band size (MHz) | Mobile transmit (MHz) | Mobile receive (MHz) |
|---|---|---|---|
| A-side carrier | 25 | 824–835; 845–846.5 | 869–880; 890–891.5 |
| B-side carrier | 25 | 835–845; 846.5–849 | 880–890; 891.5–894 |
| Total for cellular | 50 | 824–849 | 869–894 |

The noncontiguous frequency allocation in each transmit and receive band is due to the fact that, originally, the FCC had allocated a total of 40 MHz of spectrum for the two carriers. The FCC had minimized the risks of future spectrum shortage by setting aside additional spectrum totaling 10 MHz in proximity of the cellular allocation because of the uncertainty of the future cellular demand with estimates from reputable research companies differing widely. This spectrum was released to the existing carriers in July 1986. Each carrier now has 25 MHz of spectrum. In analog systems, voice transmission requires 30 kHz of bandwidth. Thus, each carrier has 416 channels. In the digital technology such as TDMA, each 30-KHz channel is further divided into three or six channels.

The regulation that created the cellular industry was the result of entrenched interests of AT&T, Motorola, Radio Common Carriers (RCCs), and others who benefited from the FCC deciding one way or the other. In the 1970s, the FCC believed telephone monopoly had led to universal service; telephone rates and services were much better than in other countries; the telephone network was a natural monopoly and should be treated like a national, public utility; and competition would not benefit the consumers. Throughout this era, the FCC favored and worked with the Bell System on technical matters and acknowledged the Bell System's views in matters concerning telecommunications. However, external competitive ideas from the Justice Department and the Office of Telecommunications Policy (OTP), as well as the growing political clout of Motorola and the other participants, started changing the views within the commission in the 1970s. In addition, the FCC lost several lawsuits filed by the RCC, thereby being forced to allow competition in the mobile telephony area. In the end, the FCC rejected the Bell System's idea of single-operator monopoly, which would result in spectrum efficiency and lower costs to consumers, in favor of two operators per market. The FCC believed that the gains from competition would far outweigh the gains from the traditional monopoly. The commission wrote that "the introduction of a marginal amount of competition into the cellular market would foster important public benefits of diversity of technology, service, and price that would outweigh the benefits associated with the increased efficiency of a single 40-MHz system over that of two 20-MHz systems."

The FCC initially decided to use comparative hearings for determining licenses for each market, and licenses would be awarded in several rounds. The first was for the top 30 markets. Each applicant had to submit a detailed technical and business plan. The commission would evaluate each plan, and award the license to the best one. The first-round applications for the top 30 markets were accepted in June 1982. The licenses were submitted by the telephone-operating companies as well as other large companies such as MCI. About 200 applications were received for the top 30 markets. Some of these applications contained more than a thousand pages. The FCC did not have the support structure for a comparative hearing of these applications. Furthermore, there was a strong possibility of appeal and delay in the award of license until judicial review. The FCC started encouraging applicants to form alliances and consolidate. This would reduce the number of applications and the need for comparative hearings. The wire-line applicants quickly consolidated because of their similar business past. However, the non-wire-line companies did not significantly consolidate because of their different pasts and interests, thereby leading to lengthy comparative hearings. Based on these difficulties, the FCC decided to go with the lottery procedure for the future round of licensing. This lottery process for deciding license awards opened the cellular market for small- and medium-sized investors. The playing field was now level. All investors, irrespective of their size, had an equal chance of being granted a cellular license. MCI and Metromedia had spent hundreds of dollars preparing marketing and technical plans to impress the FCC. This effort was fruitless since the FCC no longer was concerned about the potential carrier's plan. This led to several engineering houses preparing mass FCC applications for their clients who in many cases had no background in telecommunication, let alone in wireless. The sudden injection of chance into their dealings frustrated the non-wire-line firms such as Metromedia and MCI. As result of the uncertainty, companies such as MCI left the cellular market. This allowed the smaller and unknown players to take part in wireless telecommunications.

The FCC regulation led not only to the creation of a duopoly industry structure in each market but for the first time in the history of telecommunications in the United States led to the creation of 100-plus cellular operators—from the giants (e.g., AT&T) that provide service to most of the United States to small operators (e.g., Saipan Cellular Partnership) that provide services in the Marianna Islands. In Texas alone, there are more than 40 operators providing cellular services. Of course, there are only two per area. The market forces are leading to enhanced activity in mergers, acquisitions, joint ventures, and alliances among competitors because of the lack of constraints placed by FCC on the cellular licenses.

### 2.2.2 Cellular subscribers

In 1994, there were 24 million cellular subscribers in the United States. The number of cellular subscribers from 1984 to 1994 is shown in Fig. 2.2.

The population of the United States in 1994 was 259 million. Therefore, 24 million cellular subscribers represent a penetration of 9 percent. The penetration of cellular since 1984 is shown in Fig. 2.3.

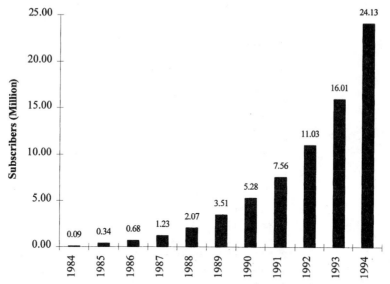

**Figure 2.2**    Cellular subscribers.

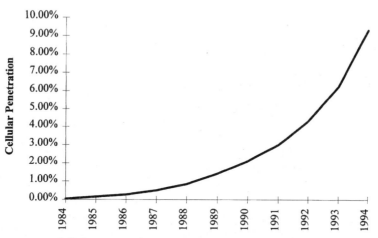

**Figure 2.3**    Cellular penetration.

This level of penetration is high compared to most countries. Countries with penetration levels higher than the United States are Australia (10 percent), Sweden (16 percent), Finland (13 percent), Norway (14 percent), and Denmark (10 percent).

The growth rate of cellular subscribers has been tremendous and beyond industry expectations in the last decade. The growth rate from 1984 to 1994 is shown in Fig. 2.4.

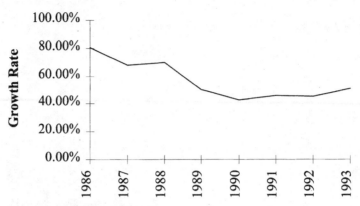

**Figure 2.4**    Cellular growth rate.

The growth rate in 1994 was 51 percent, the highest since 1989. The cellular industry is passing through the growth phase of a typical product life-cycle model. The market buyers of cellular service appear to be moving from those that initially subscribed for "high-tech/early adopter" reasons or professional needs to buyers who are subscribing not because a cellular phone is needed but because it would be marginally beneficial or comfortable to own one. A recent survey conducted by Market Research Institute for Cellular One showed that 70 percent of survey respondents use their cellular phones primarily for personal reasons. According to Cellular One, this was a 50 percent increase in personal use compared to a few years ago. This is consistent with the data provided by the CTIA as the primary reasons for purchasing a cellular phone. About 61 percent of new users purchase a cellular phone for safety or for calling friends and family. The number of new users who purchased cellular phones for business reasons was 36 percent. Some of the growth in cellular subscribers could be attributed to the increase in the population covered from 84 percent in 1989 to 97 percent in 1994. However, most of the cellular growth could be attributed to increased awareness about the utility of cellular phones. Cellular phones are not bought just for business reasons but also for personal reasons. The average monthly revenue per subscriber is declining due to the increased use of cellular phones for personal use, as shown in Fig. 2.5.

The monthly revenue per cellular subscriber is decreasing by approximately 9 percent per year. This decrease is due largely to a shift in the profile of the subscribers from those using the service for business reasons to those using it for personal reasons. The average price of the cellular phone has dropped from $2000 in 1984 to less than $200 in 1994. The $200 price tag for a cellular phone is similar to that of a high-

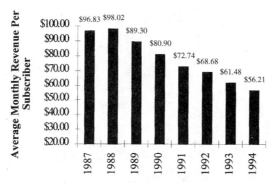

**Figure 2.5** Average monthly revenue per subscriber.

end residential cordless phone, thereby appealing to the mass market—i.e., those segments who would buy a cellular phone because it would be convenient and useful but not necessarily needed. In many markets, the cellular phone is provided free or practically at no cost for a 1- or 2-year service commitment as shown in an advertisement from the *Dallas Morning News* (see Fig. 2.6).

The buyers of cellular phones do not have a trade association to increase their bargaining power. However, they still have significant clout because cellular services offered by the carriers are standard or undifferentiated. Buyers can play one company against another for a better price just as with the long-distance telephone market. If a user is unhappy with one carrier a switch can be made to the other carrier since the cost to switch is very low. The incumbent carrier can no longer dictate terms to a buyer. In addition, most cellular networks have excess capacity, except those in large cities. Price erosion is typically found in industries with excess product or service capacity. Cellular prices are lowered to attract more subscribers who in turn utilize excess network capacity.

## 2.2.3 Equipment vendors

Equipment vendors supply the Mobile Telephone Switching Office (MTSO) and cell sites. The MTSO is the switching unit that connects the cell sites to the public network. The cell sites handle the radio portion and connect the subscribers to the rest of the cellular network. The main suppliers are AT&T, Ericsson, Hughes Network Systems, Motorola, and Northern Telecom. These vendors supply equipment for both large and small systems. However, the economics for small systems are not very attractive due to the initial capital cost of the large switch which, basically, was adapted from each manufacturer's central office product. As a result the MTSO switches are not cost-optimized

**Figure 2.6**    Advertisement for free cellular phone (*Dallas Morning News*).

for less dense, small RSA markets. These small RSA markets can either backhaul to a large MSA switch or share a large switch among multiple RSA operators. Either way, the small operators incur huge backhauling costs. Alternatively, the small operators could buy equipment from a small-equipment provider such as Celcore, NovAtel, or Plexys. These manufacturers have designed a switch based on the requirements of the small market and are cost-optimized for these markets. This system equipment, however, is not very effective for a dense, urban market. The smaller switching platforms do not have all the features that were historically transferred from the large, central, office switches to the cellular switches. It appears the trend toward a small distributed architecture may be a repeat of the computer market in 1980. In the 1980s, the centralized architecture, with its mainframe as the brain for all activities, lost its dominance to distributed intelligence of personal computers. Time will tell whether the wireless telephony market is headed for a repeat of the computer market.

The cellular network has grown since 1982, primarily due to coverage reasons. As cellular operators increased their footprints, they had to buy more switches and cell sites. The number of switches and cell sites deployed from 1984 through 1994 is shown in Figs. 2.7 and 2.8.

CTIA estimates that operators spent $5 billion in 1994. The 1995 capital expenditure was about 85 percent higher than the capital expenditure in 1993. Since 97 percent of the population was covered in 1993, the growth is due primarily to capacity increases because of increased subscribers and growth in portable terminals, which led to an increased number of smaller cell sites, transition to digital technology such as TDMA, system upgrades, and network expansion.

Among the suppliers of cellular systems, AT&T has the largest market share with 39 percent of the sales in 1994. AT&T is followed by Ericsson (24 percent), Motorola (19 percent), Northern Telecom (12 percent) and Hughes Network Systems (5 percent). The remaining one

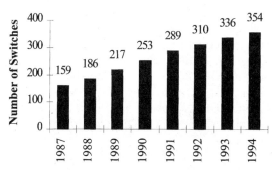

**Figure 2.7** Number of switches (Northern Business Information Estimates, 1995).

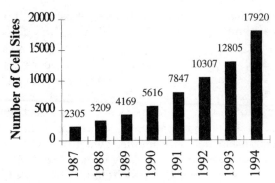

**Figure 2.8**    Number of cell sites (CTIA, 1995).

percent is distributed among Celcore, Plexys, Astronet, NovAtel, and other smaller-equipment providers.

The equipment market is essentially oligopoly. The equipment supply market has a high-entry barrier due to the high-development cost and R&D. In addition, the carriers will encounter high costs in replacing or switching between vendors. However, intense competition among the vendors has led to operators negotiating lower prices, as well as vendor financing.

### 2.2.4   Operators

The cellular industry is highly fragmented due primarily to the manner in which licenses were awarded and the lack of regulation or requirements for the FCC lottery winners. The FCC lottery system led to winners who had no knowledge or understanding of building or operating a telephone network, nor did they understand the complexity of a cellular network. Many of the lottery winners were doctors, lawyers, and other professionals. Their sole interest in entering into the lottery process was purely as a speculative investment. A cellular license was another tradable security as in the commodities market, i.e., investing in the lottery process without having any intention of building or operating the cellular network. Though FCC had rules to screen out and prevent such speculation, it did not have the resources to police it, so it was left to the losers in the lottery to police the winners through litigation and pressure on legislators and the FCC. The speculative nature, or viewing of a cellular license as another tradable security, was indicated in September 1984 when the cellular license swap meet was held in New York City to allow investors to engage in the trading of licenses. The end result of the licensing process and lack of oversight by the winners led to a industry that was highly fragmented.

Even though the industry was highly fragmented initially, some of the large players, especially the Bell operating companies, saw opportunity in the fragmented scenario. They started the process of consolidation by aggressive acquisition of cellular licenses, both wire-line as well as non-wire-line, and the distinction between the non-wire-line and wire-line carriers disappeared. Some of these players, such as BellSouth, have regional-clustering strategy in which they acquire a cellular license surrounding their current licenses. Visionaries, such as Craig McCaw, started consolidation of the non-wire-line cellular license to form the McCaw Cellular Communications Corporation, which was acquired by AT&T in 1994. McCaw Cellular, now AT&T Wireless Systems, covers about 64 million people (also referred as POPs) in 103 licenses, the largest number of POPs among all cellular operators. The next largest network belongs to GTE. GTE cellular network covers 52 million POPs with 127 licenses. The large carriers dominate the cellular market, of which the top carriers are shown in Table 2.2. The top ten cellular operators have 75 percent of the total cellular subscribers and cover 70 percent of the U.S. population.

The trend toward consolidation improves the operational efficiency and positions the operators for PCS competition. Many operators have formed joint ventures and alliances. AirTouch and US West have formed a joint venture that is 70 percent owned by AirTouch and 30 percent owned by US West. The consolidated operation provides seamless operation in most of the western United States. Similarly, Bell Atlantic and NYNEX have formed a joint venture of which 62.35 percent is owned by Bell Atlantic and 37.65 percent by NYNEX. This provides operational efficiency and seamless coverage over the northeastern United States. These mergers are the BOCs' approach to overcoming the national footprint developed by the AT&T/McCaw acquisition over the last five years.

**TABLE 2.2  Top Ten Cellular Operators**

| Top ten cellular operators | POPs covered (1994) × 1000 | Subscribers (1994) × 1000 | % of Total subscribers |
|---|---|---|---|
| AirTouch | 38,994 | 1560 | 6.5 |
| Ameritech | 24,575 | 1299 | 5.4 |
| AT&T | 63,963 | 3304 | 13.7 |
| Bell Atlantic | 30,893 | 1641 | 6.8 |
| BellSouth | 45,092 | 2155 | 8.9 |
| GTE | 49,275 | 2339 | 9.7 |
| NYNEX | 27,407 | 905 | 3.7 |
| SBC | 39,990 | 2992 | 12.4 |
| Sprint Cellular | 18,642 | 1040 | 4.3 |
| US West | 20,285 | 968 | 4.1 |
| Total of 10 operators | | 18,203 | 75 |
| Total number of subscribers | | 24,134 | 100 |

The revenue per subscriber is decreasing every year and the advent of PCS will lead to extreme price erosion, impacting profitability. In order to improve profitability, cellular carriers are banding together to allow them the scope and scale to reduce their marketing and operation costs. Equipment vendors are more likely to provide favorable prices and financing to reduce their capital costs. The banded cellular operators will put themselves into a better bargaining position than the equipment vendors.

## 2.3   Technology

The technology section will describe the AMPS cellular technology. The channel structure, system architecture, air interface, and enhancement to the air interface for PCS-type services for AMPS are discussed in this section. Digital technology is discussed in detail in subsequent chapters. This discussion is followed by an explanation of the interaction of the individual elements of the system to achieve different call-processing tasks such as call origination, call termination, and handoff. This illustrative approach will simplify the complexities of the AMPS cellular system.

### 2.3.1   Channel structure

The spectrum allocated to AMPS is 50 MHz with each carrier having 25 MHz. This is shown in Fig. 2.9.

The 800-MHz cellular band includes two identical, mirror-image range of frequencies. The lower range of frequency is used for transmission by the mobile. The upper range of frequency is used for transmission by the cell sites. The difference between the mobile transmit and mobile receive for each carrier is always 45 MHz, as shown in Fig. 2.10.

The mobile transmit is also called the *reverse channel,* while the mobile receive, or base station transmit, is called *forward channel.*

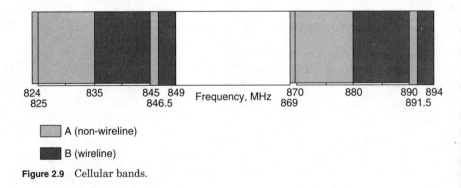

824   825   835   845   849   846.5   Frequency, MHz   870   869   880   890   894   891.5

☐ A (non-wireline)

■ B (wireline)

**Figure 2.9**   Cellular bands.

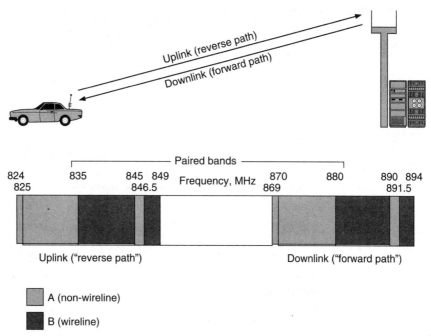

**Figure 2.10**   Mobile transmit and receive channels.

Each voice channel has a 30-kHz bandwidth. The mobile transmit channel at 825.030 MHz and its corresponding receive channel at 870.030 (separated by 45 MHz) is channel number 1. Notice that the channel number 1 does not start from 824 due to the fact that the original allocation of 40 MHz started from 825 MHz. The FCC allocated additional frequencies, enlarging each range identically. Some of the new frequencies are above the original range while others are below the previously allocated frequency. The added frequencies are called *expanded spectrum*. The original allocation included 1–666 channels. The A license has 1–333 channels and B license has 334–666 channels. The expanded spectrum added channels 667–799 and 991–1023 channels. The mobile-transmit center frequency associated with a channel number can be computed as follows:

$$0.03 \, N + 825.000 \quad \text{for } 1 \leq N \leq 866$$

$$0.03 \, (N - 1023) + 825.000 \quad \text{for } 990 \leq N \leq 1023$$

The mobile-receive center frequency can be obtained by adding 45 MHz to the mobile-transmit center frequency. The channel numbers and associated center frequencies are shown in Table 2.3.

**TABLE 2.3    Channel Numbers and Associated Frequencies**

| System | Channel number range | Center frequency mobile transmit (MHz) | Center frequency mobile receive (MHz) |
|---|---|---|---|
| A (expansion) | 991 | 824.040 | 869.040 |
| | 1023 | 825.000 | 870.000 |
| A | 1 | 825.030 | 870.030 |
| | 333 | 834.990 | 879.990 |
| B | 334 | 835.020 | 880.020 |
| | 666 | 844.980 | 889.980 |
| A (expansion) | 667 | 845.010 | 890.010 |
| | 716 | 846.480 | 891.480 |
| B (expansion) | 717 | 846.510 | 891.510 |
| | 799 | 848.970 | 893.970 |

Each cellular carrier has 832 channels or 416 duplex channels. The channels could be of different types. The channels could be control channels or voice channels as shown in Fig. 2.11.

Control channels allow the mobile to communicate with the network for setting up calls and to obtain system information. There are 21 control channels for each operator. These channels cannot be used to carry voice. They are used to obtain system information, and to obtain service—call origination, call termination, registration—from the network. Control channels do not transfer voice conversations. They are used for the transmission of digital control information from a network to a mobile, or from a mobile to the network. Unlike GSM, these channels are dedicated so the mobile does not have to scan all 416 channels within a short time to request service from the network. The control channels are located at the center where the A and B operators have their boundaries. The first 21 channels on each side of the border are

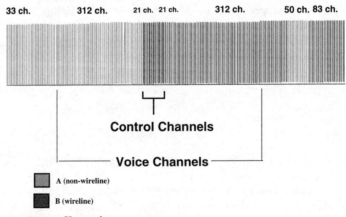

**Figure 2.11**   Channel types.

designated as control channels. The control channels for the A operator are within channel number 313 and channel number 333. The control channels for the B operator are within channel number 334 and channel number 354.

The voice channels are used to transmit voice information. They can also transfer control messages particular to the mobile station when it is operating as a voice channel, e.g., when the mobile requests a feature such as call waiting. Since 21 of the 416 channels are used as control channels, a total of 395 channels are available for voice.

### 2.3.2   System architecture

The generic system architecture, as provided by the Telephone Industry Association (TIA) standards body, is shown in Fig. 2.12.

Many of the functions of the HLR and VLR are often implemented in the switch from a practical point. Thus, the essential elements in an AMPS system are the network components comprising the MTSO, the base station, and the subscriber components. The cellular system components are shown in Fig. 2.13.

**2.3.2.1   MTSO.**   The MTSO interconnects the cellular network to the PSTN network for connection to the local central office (CO) and long-distance toll centers. It provides all the CO type functions, such as

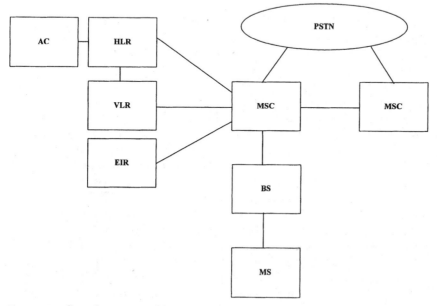

**Figure 2.12**   Generic system architecture.

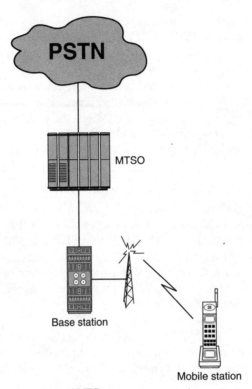

**Figure 2.13**   AMPS system components.

switching, networking, call processing, call statistics, and billing for the cellular network. Most MTSOs from major vendors are a modification of their existing PBXs, COs, or toll switches. The MTSO is essentially a trunk switch with mobility functions added to it. Figure 2.14 is a cellular switch from Northern Telecom.

The MTSO coordinates all of the base-station activities such as channel assignments for users in each cell. It controls handoff decision and timing and power control; routes signals and voice to and from the PSTN; and performs administrative functions such as billing and maintenance, diagnostic testing, and alarm monitoring. The MTSO consists of four major subsystems: central control, switching assembly, input/output controllers, and communications link processors.

The central control subsystem is the brain of the MTSO. It receives, interprets, and sends commands and responses to and from the base station, and coordinates the communications links to the landline network. These components are fully redundant and fault tolerant. The components run parallel, with one of the components being active while the other is on standby. The standby executes all the same

**Figure 2.14**　MTSO (*Courtesy Northern Telecom*).

instructions as the active. The two components exchange synchronization with each other. Network and subscriber information is stored in the central control which validates subscriber requests for access and the tracking of the subscriber's call. Most vendors, at least for smaller systems, have the VLR and HLR functionality integrated into the MTSO functions to reduce time for call processing. The central control also has a view of the network and makes decisions regarding handoff to neighboring cells based on signal strength measurement received from the base stations.

The switching-assembly subsystem is similar to the landline's CO and tandem switches. It allows nonblocking connections between base-station communications channels and the PSTN communications channels. The telephone switching function was originally manually performed by a switchboard operator. Later on, it was *electromechanical*—electric switches driven by electromagnets, which by mechanical movement performed the switching function. The electromechanical switching evolved into space switches which connected one circuit to another circuit with a specific physical connection in space. The space switches need $n \times n$ connection to connect $n$ inputs to $n$ outputs. This limited their capacity. Later on, time switches were introduced with the availability of digital techniques. In time switches, the input signal is rearranged or switched under the direction of the control store so that each incoming time slot is connected to the desired outgoing time slot. Modern switches use a combination of time and space switches. The switching assembly subsystem, like the central control subsystem, is fully redundant, with designated active and inactive processors.

The input/output controller provides the interface to input/output devices, such as system tape, statistics tape, billing tape, alarms, and visual-display terminals used for human–machine interface for maintenance and administration. These controllers were also used for handling signaling messages between the MTSO and the cell sites through X.25 HDLC in the early cellular switches.

The communication link processors connect the MTSO to the external world and provide the necessary signaling so that the MTSO can communicate with other external switches. Cellular switches have trunk interfaces. Trunk interfaces are T1 in ANSI countries and E1 in CCITT countries. The signaling between the MTSO and the PSTN can take many forms. They could be either ANSI or CCITT standard MF, DTMF, ISUP, TUP, R1, and R2. This is further complicated by the fact that the standard signaling has variants for each country. These variants for each country require additional software changes. ISUP signaling in the United States is different from the ISUP in France. Similarly, each country has it own variant of R2. The communications-link processors help to localize the differences and reduce the impact of the differences on other subsystems at the MTSO.

The MTSO can be connected to the PSTN by two basic types of direct trunk connection. These trunk interconnection arrangements are type 1 and type 2 as shown in Fig. 2.15.

Type 1 trunk interconnection is similar to the direct trunk connection that exists between a private branch exchange (PBX) and the central office. Charges are rendered on each call to the cellular carrier and billed by the local exchange company (LEC). In most cases the telephone company bills the cellular subscriber directly. This method is useful for fast market entry and to lower capital expenses, but it reduces the profit potential for the cellular operators.

Type 2 trunk interconnection is equivalent to that of an interexchange carrier's connection to a local exchange carrier. Minutes of use are billed but not identified on a call-by-call basis. These minutes of use are billed on a bulked, monthly basis sorted by the office code. The cel-

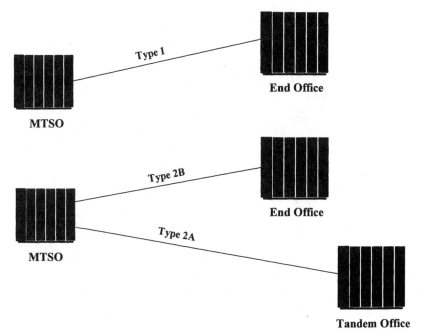

**Figure 2.15**    Cellular interconnection types.

lular operator bills its mobile customers directly, passing along its rate. This provides the cellular operator better marketing and billing control, therefore more profit. There are two types of type 2 interconnection, called type 2A and type 2B. Type 2A connects the MTSO to a class 4 tandem switch while type 2B connects to a class 5 end office.

**2.3.2.2  Base station.**    A base station is the network element that interfaces the mobile station to the network via the air interface. Each cell in the network has a base station associated with it. One of the factors determining the size of the cell is the base-station transmit power. The greater the transmit power of the base station, the larger the cell, and vice versa. The transmit power of the base station cannot be arbitrarily increased as it effects frequency reuse. The cellular base station from Northern Telecom is shown in Fig. 2.16.

The primary function of the base station is to maintain the air interface, or medium, for communication to any mobile station within its cell. Other functions of the base station are call processing, signaling, maintenance, and diagnostics. The base station communicates to the mobile station via the air interface, and to the MTSO by dedicated communication links, such as T1 trunks. Communication links on the base station to the MTSO interface are also classified into voice links and

**Figure 2.16**   Base station (*Courtesy Northern Telecom*).

signaling links. Each T1 contains 24 slots called DS0s. Typically 23 DS0s of the T1 are used for voice and 1 DS0 is used for signaling. There is a direct one-to-one correspondence between a voice channel and a DS0. Thus, if a base station supports 20 voice channels, then 20 DS0s to the MTSO will be used for voice. The single DS0 used for signaling is shared among all the voice channels. This DS0 is not dedicated to a particular voice channel or user. Traffic on the control channel is transported by a single DS0 which has a 64-kbps rate. The signaling interface between the base station and the MTSO is not standardized, unlike in GSM. The protocol between the base station and the MTSO is specific to each vendor. This means that a cellular operator has to purchase the base station and the MTSO from the same vendor. The functional elements of a base station are shown in Fig. 2.17. The functional elements consist of the antenna subsystem, multicoupler, transceivers, combiner, and base-station controller.

The antenna subsystem converts the RF energy from the transceivers into electromagnetic waves propagating in free space. It also

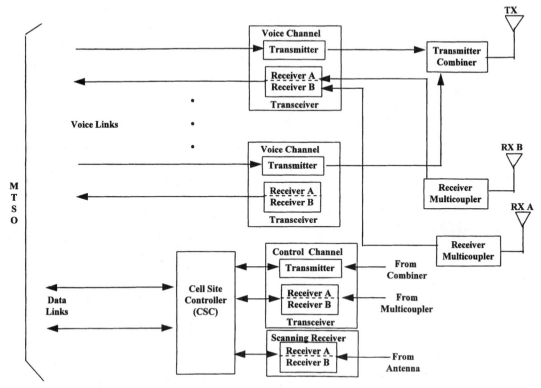

**Figure 2.17**    Functional elements of a base station.

converts received electromagnetic waves from the mobile station into electrical signals sent to other components of the base station. The type of antenna used depends on its application: (1) *omnidirectional* antennae transmit and receive in all directions; (2) *sectored antennae* transmit and receive energy in a specific direction. Typically sectored antennae are 60° and 120°. A 60°-sectored antenna will partition a cell into six sectors for the purposes of transmitting, receiving, or both. Sectorizing a cell divides the cell into sectors of desired coverage, reducing cochannel interference and increasing the ability to reuse frequencies more often. Increases in traffic are handled by cell splitting, as discussed previously. Typically, omnidirectional antennae are used when a cell is first commissioned. As the traffic in the cell increases, either a new cell is added or the cell may be sectored. Figures 2.18 and 2.19 show the omnicell and 120°-sectored cell.

Base stations use two receive antennae to provide for receive diversity, a technique which enables sampling of the received signal to more than one receiver. The purpose of diversity is to combine signals in

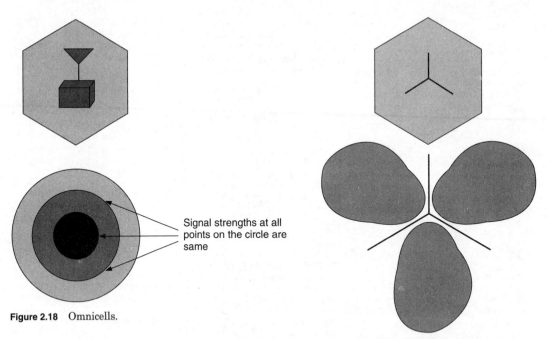

**Figure 2.18**   Omnicells.

Signal strengths at all
points on the circle are
same

**Figure 2.19**   120° sectored cells.

order to enhance the signal or to select the best-quality signal. Diversity schemes greatly enhance the received signal from the mobile station and improve the quality of the wireless link. Diversity is a very effective way to combat the deleterious effects of multipath. One way to receive the same signal twice is to use two receive antennae whose outputs are either combined digitally to enhance the received signal strength, or the best signal is chosen for further processing. These techniques are known as *diversity reception* or *receive diversity* as illustrated in Fig. 2.20.

The *multicoupler* allows a single antenna to receive multiple RF channels. It is more advantageous to use a single receive antenna to serve several RF channels than to use a single receive antenna for each RF channel. While this approach results in the efficient usage of receive antennae, it also necessitates the use of a multicoupler. The multicoupler is needed to separate the individual RF channels from the composite received signal. In the process of splitting the individual channels, the multicoupler reduces the power level of the signal. To account for loss in the multicoupler and also to account for cable loss from the antenna, several low-noise RF preamplifiers are used to boost the power levels of the signals received prior to the multicoupler. The split output from the multicoupler can then be fed to individual receivers.

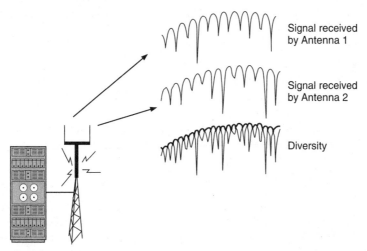

Signal received
by Antenna 1

Signal received
by Antenna 2

Diversity

**Figure 2.20**   Receive diversity concepts.

The transceiver section has a transmitter, a receiver, and logic-processing components. The transmitter has audio processing, modulation, and power-amplifier elements. The transmitter codes the audio signals received from the MTSO and transmits them onto the carrier frequency. Since AMPS uses analog technology and frequency modulation, the transmitter continuously modifies the frequency of the baseband signal based on the received audio signal. It also shifts the frequency of the signal to the desired range. The transmitter also has a power-amplifier element. The purpose of the power amplifier is to increase the transmitter's output power. A Class C, RF, power amplifier is used for high efficiency in FM modulation. Typically, the maximum effective radiated power (ERP) for urban areas is 100 watts and for rural areas is 500 watts.

The receiver processes low-power-level RF signals into their original audio components. Since the receiver has diversity reception there are two dual RF amplifiers, a demodulator, and audio-processing elements. The amplifier increases the level of the signal to the workable range. The logic-processing section of the base station inserts and extracts signaling messages from the radio channel. Typically, transceivers can be configured to operate in either a control channel mode where they can transmit and receive data, or in the voice channel mode where they are used only for voice. This is achieved with an embedded software command. The output from the transceiver to the MTSO is a four-wire audio interface connected to the channel bank. The channel bank digitizes the audio and puts it on one of the 24 T1 slots or 32 E1 slots at 64 kbps.

The combiners allow outputs from the multiple transmitter to be fed into one antenna. Without a combiner, if the transmitter outputs were mixed at a junction, then the power from one transmitter will interfere with the power from another transmitter. This would lead to intermodulation products. Two main types of combiners are used in cellular. They are the *tuned combiner* and *hybrid combiner*. The tuned combiner uses filters tuned to pass only a particular frequency. These combiners are either manually tuned or automatically tuned. Tuned combiners require a minimum 21-channel separation to achieve a 17-dB isolation. It adds 1.5 dB insertion loss for up to 16 inputs. The hybrid combiners have two input ports with special circuits that prevent a signal from exiting backward through another port. It adds about 3.5 dB insertion loss. For more than two inputs, the hybrid combiners are connected in cascade with each hybrid combiner adding 3.5 dB insertion loss. Thus, for 16 inputs, the insertion loss is 14 dB compared with 1.5 dB insertion loss for tuned combiners. The hybrid combiners are cost-effective for small numbers of inputs.

The base-station controller (BSC), sometimes called the cell-site controller (CSC), coordinates the operation of the base-station equipment through its own stored software program and acts on the commands sent to it from the MTSO. The controller manages the voice and control transceivers, provides maintenance and diagnostic functions, and handles call-processing functions.

**2.3.2.3   Mobile station (MS).**   The mobile station consists of three elements: transceiver, antenna, and user interface. Most of these elements are the same as described in the base-station section, and they are functionally similar. The user interface exists only at MS. The user interface consists of a display, a keypad for entering numbers, and an audio interface for speaking and hearing voice conversation.

There are three mobile station power classes defined for EIA-553 mobiles: classes I, II, and III. There are also eight mobile station power-level commands, 0–7, which each of these mobiles can operate. A class I mobile is capable of operating at 6 dBW, class II at 2 dBW, and class III, for portables, operates at −2 dBW. Each lower-power level changes the power from the previous higher-power level by 4 dB. The power level of the mobile can be changed on command from the BS. The mobile station nominal power levels are shown in Table 2.4.

The mobile station also stores several parameters in its permanent memory. Some of the important ones are the mobile identification number (MIN), electronic serial number (ESN), and station class mark (SCM). These are transmitted upon power on, cell-initiated sampling, and call origination.

TABLE 2.4 Nominal ERP for Different Classes of Mobile Stations

| Mobile station power level | Nominal ERP (dBW) for mobile stations | | |
|---|---|---|---|
| | Power class I | Power class II | Power class III |
| 0 | 6 | 2 | −2 |
| 1 | 2 | 2 | −2 |
| 2 | −2 | −2 | −2 |
| 3 | −6 | −6 | −6 |
| 4 | −10 | −10 | −10 |
| 5 | −14 | −14 | −14 |
| 6 | −18 | −18 | −18 |
| 7 | −22 | −22 | −22 |

The mobile stores its MIN in its permanent memory. The MIN uniquely identifies the mobile in the network and, in several cases, is also a network-dialable number, that is, a number that another mobile user or a landline user can dial to reach the mobile user. The MIN is 10 digits long and has a format similar to that of a directory number.

The electronic serial number is a 32-bit binary number that uniquely identifies a mobile to any cellular system. It is set in the factory and not alterable in the field. With this security, the circuitry that provides the serial number is protected from fraudulent use and tampering. Attempts to change the serial number circuitry render the mobile inoperable.

The station class mark (SCM) parameter indicates the capabilities of the mobile. It indicates the power class of the mobile, extended-frequency support, and discontinuous transmission support. The SCM is a 4-bit pattern that identifies the capabilities of the mobile station. The first two bits represent the power class of the mobile, the third bit represents continuous or discontinuous transmission support, and the last bit represents the mobile station's capability to support frequency range, i.e., 20 or 25 MHz. The SCM is permanently stored in the mobile station.

## 2.3.3 Air interface

The air interface between the mobile station and the BS for AMPS is defined by the standard EIA-553 mobile-station/land-station compatibility specification. It specifies the transmitter, receiver, security and identification, supervision, call-processing, and signaling-format specifications for the mobile station and the base station. Any EIA-553 compliant mobile station will be able to interface to any other EIA-553 compliant base station over this standard air interface. Operators are free to choose the mobile-station equipment and base-station equipment vendors. The users are not locked to the mobile station of a vendor.

As previously discussed, there are essentially two types of RF channels defined in EIA-553, control channels and voice channels. The control channels could be further classified as forward control channel signaling (FOCC) and reverse control channel signaling (RECC).

**2.3.3.1  Forward control channel signaling.**  The forward control channel signaling (FOCC) is monitored automatically by each MS. The FOCC contains general information about the system, such as system identification. The FOCC has a data transmission rate of 10 kbps and the information is sent via frequency shift keying (FSK). The control channel message is preceded by a dotting scheme which is an alternating scheme of 1s and 0s. A synchronization word follows the dotting scheme. The synchronization word has a unique pattern of 1s and 0s which enables a receiver to instantly gain synchronization. The synchronization word is followed by the actual message that is repeated five times as a counter to the deleterious effects of fading. The receiver will assume that three out of five identical words constitute the message.

Each forward-control channel consists of three discrete information streams: stream A, stream B, and the busy-idle stream. All these streams are multiplexed together. Messages to the mobile with the least significant bit of their MIN equal to 0 are sent on stream A and those with the least significant bit of their MIN equal to 1 are sent on stream B. The busy-idle stream contains busy-idle bits, which are used to indicate the current status of the reverse-control channel. The reverse-control channel is busy if the busy-idle bit is equal to 0 and idle if the bit is 1. A busy-idle bit is located at the beginning of each dotting sequence, at the beginning of each word sync sequence, at the beginning of the first repeat of word A, and after every 10-message bit thereafter. Each word contains 40 bits. The format of the FOCC is shown in Fig. 2.21.

Each FOCC message can consist of one or more words. The types of messages transmitted over the FOCC are the mobile station control message and overhead message train (OMT).

The mobile-station control messages are used to control or command the mobile to do a particular task when the mobile has not been assigned a voice channel. Typical mobile-station control messages are the following:

1. *Initial voice channel designation message.*  This message directs a mobile station to an assigned voice channel. It identifies the mobile for which the message is targeted by its MIN, indicates the voice channel number, and so on.

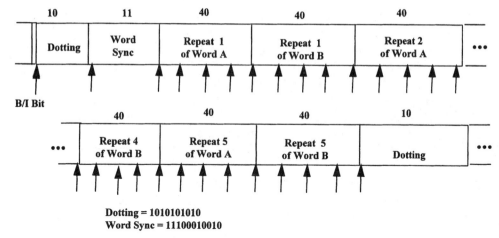

Dotting = 1010101010
Word Sync = 11100010010

**A mobile reads only one of the two interleaved messages (A or B).**
**Busy-Idle bits are inserted at each arrow.**

**Figure 2.21**  Format of forward control channel signaling (FOCC).

2. *Directed retry message.*  This message indicates to the mobile station that this particular cell is overloaded and does not have any available voice channels. The mobile is directed to neighboring cells.

3. *Alert.*  This message commands the mobile station to turn on its signaling tone, wait for 500 ms, and then wait for an answer. This command is sent when the mobile is required to notify the mobile user of an incoming call by physical ringing.

4. *Change power.*  This message commands the mobile to adjust its transmitter power level as indicated in the message.

Overhead messages are sent in a group called *overhead message train* (OMT). The first message of the train must be the system parameter overhead message. The desired global action messages and/or a registration ID message must be appended to the end of the system parameter overhead message. The overhead messages are grouped into the following classes:

1. *System parameter overhead message.*  This message is used to control mobile stations monitoring a control channel. It contains information about the system identification (SID), number of paging channels, number of access channels, indication if discontinuous transmission is supported or not, parameters used to support registration, and so on.

2. *Global action overhead message.*  Each global action overhead message consists of one word. These messages provide information

about access type parameters, access attempt parameters, new access channel set, registration increments, and so on. These messages, along with other messages in the OMT, indicate the digital color code (DCC) to the mobile. The DCC is transmitted from the BS to the mobile. The MS then uses the DCC to identify to the BS which BS transmitter the MS is receiving.

3. *Control filler message.*   The control filler message consists of one word. It is sent whenever there is no other message to be sent on the forward control channel. It may be inserted between messages as well as between word blocks of a multiword message. The control filler message is also used to specify a control mobile-attenuation code (CMAC) for use by mobile stations accessing the system on the reverse control channel.

**2.3.3.2   Reverse control channel signaling.**   The reverse control channel (RECC) is the control data sent from the MS to the BS. The control data includes page responses, access requests, and registration requests. The RECC data stream is generated at a rate of 10 kbps. The format of the RECC is shown in Fig. 2.22.

All messages begin with the RECC seizure precursor that is composed of a 30-bit dotting sequence, an 11-bit word sync sequence, and the coded digital color code (DCC). Each word contains 48 bits and is repeated five times. The RECC messages are described in the section on system access.

**2.3.3.3   Voice channel signaling.**   The analog voice channel transports both voice and digital signaling information. By inhibiting the voice, the voice information is carried by FM voice; the digital messages are carried by FSK. When signaling data is to be sent on the voice channel the audio is suppressed and replaced with digital messages. This technique is known as *blank-and-burst*. The bit rate for the digital messages is 10 kbps. To inform the receiver that a digital signal is going to be sent, a 101-bit dotting sequence precedes the message. The BS

| 30 | 11 | 7 | 240 | 240 | 240 | |
|----|----|----|----|----|----|----|
| Dotting | Word Sync | Coded DCC | First Word Repeated 5 Times | Second Word Repeated 5 Times | Third Word Repeated 5 Times | ••• |

Dotting = 1010...010
Word Sync = 11100010010

**Figure 2.22**   Format of reverse control channel signaling (RECC).

receiver on detecting this pattern readies itself to receive a digital message. After the dotting sequence, a synchronization word is sent to denote the start of the message. On the forward voice channel, messages are repeated eleven times to ensure that control information is properly received by the MS and that on the reverse voice channel words are repeated only five times. Words on the forward voice channel contain 40 bits and on the reverse direction contain 48 bits. The format of the signaling on the forward and reverse voice channels is shown in Figs. 2.23 and 2.24.

Supervisory audio tones (SATs) are used to ensure reliable voice communications. An AMPS system has SAT codes (6 kHz tones at frequencies 5910, 6000, and 6030) that are transmitted on the speech channels. These tones are used to identify the local cell traffic from adjacent, interfering traffic. The SAT is generated by the base station and looped back by the mobile station. If the BS receives the same SAT that it is sending, that means the MS is tuned to the right BS. Allocation of SAT codes is done on the principle that adjacent cells must have a different SAT. The DCC received on the OMT indicates the particular SAT code to use. Signaling tones (STs) is a 10-kHz tone burst used to indicate a status change, for example, during termination. It confirms messages sent from the base station. The SAT assignment in a cellular system is shown in Fig. 2.25.

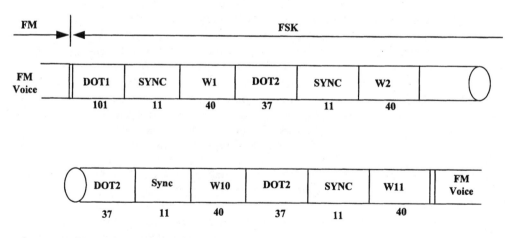

DOT1 = 101 bit dotting sequence
DOT2 = 37 bit dotting sequence
SYNC = Synchronization word
WN = Message word (N)
N = Number of repeated message words

**Figure 2.23**   Format of forward voice channel.

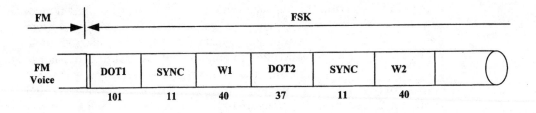

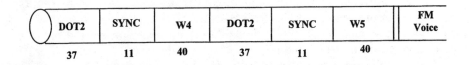

DOT1 = 101 bit dotting sequence
DOT2 = 37 bit dotting sequence
SYNC = Synchronization word
WN = Message word
N = Number of repeated message words

**Figure 2.24**   Format of reverse voice channel.

### 2.3.4   Enhancement of AMPS air interface
### for PCS services

IS-94 enhances the current AMPS air interface to support PCS services. Cellular has long recognized the need for geographically limited capability that offers affordable, lightweight, portable communications. The concept raised by this need is that certain specialized services need to be offered exclusively to a set of users called a *closed user group*. The users of such systems might be expected to be able to access PBX/Centrex features in the same way they would on wired PBX/Centrex terminals. Vendors and operators have evolved these features to support other applications, such as the use of a cellular phone as a cordless phone at home where the subscriber will not incur airtime charges.

The following are the major enhancements to the AMPS air interface:

1. There is extensive support for low-power systems. The lowest power level supported is 10 instead of 7 in the current air interface (EIA-553). A new power class IV has been added. The lowest ERP for the EIA-553 cellular phones is −22 dBW. With the addition of the new power level and power class, the lowest ERP is −34 dBW. This makes the system more suited for indoor environment and results in better battery life.

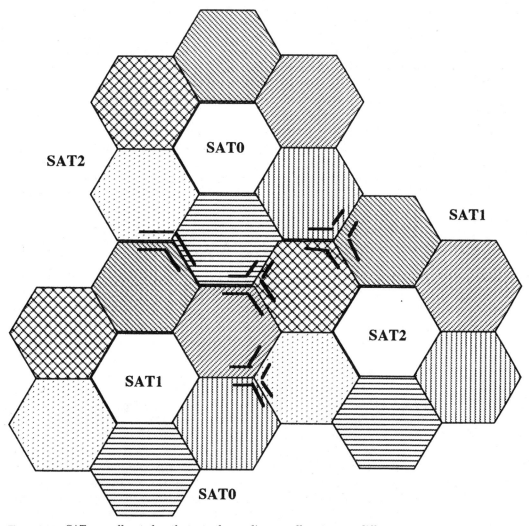

**Figure 2.25** SATs are allocated so that any three adjacent cell groups are different.

2. The private system control channels are allocated in the voice channel band. The new control channels are adjacent to the EIA-553 control channels. This prevents EIA-553 cellular phones from accidentally using a private system.

3. Additional message information is available to discriminate between private systems and public systems.

4. Mobile handsets have additional capability such as least-cost routing (private over public) and extension-number display.

### 2.3.5   Call processing

When the mobile is powered on, it checks its internal memory for several stored initialization parameters. One of the parameters stored in the mobile is the choice of the home system—A or B. If the preferred system is A, then the serving-system status is enabled; if the preferred system is B, then the serving-system status is disabled. The serving-system determination is important to the mobile because it aids in determining the RF channels to scan along with other information. The mobile then scans the dedicated control channels assigned nationwide to system A or B depending on the status of the serving system. It records the signal-strength measurements of these channels. The mobile tunes to the strongest dedicated control channel and receives a system parameter message and records information from the message. This information is used for subsequent access to the network. The mobile stores information such as the system identification (SID) and the number of paging channels and determines the range of paging channels to monitor. The mobile station tunes to the strongest paging channel and receives an overhead message train. The mobile based on the SID being transmitted on the paging channel determines if it is roaming or not. The roam indicator in the mobile is set according to this determination.

The mobile, after completing the initialization procedures, enters the idle mode. In this mode the mobile periodically receives an overhead message train and captures system access information, system control information, and system configuration information. This information is used by the mobile stations for subsequent access, for the provision of diagnostic information, and so on. It also responds to any messages received in the message train. Messages which the mobile might respond to are local control messages and registration increment messages that increment the registration timer in the mobile.

The mobile station also monitors mobile station control messages for page messages. If a page is received, the mobile attempts to match the received MIN to the stored MIN. If the MIN matches the mobile, it determines that the page is for itself and enters the system access task with page response indicator. The mobile also enters the system access mode for user-initiated call origination, registration, and so forth, with the appropriate indication.

When the mobile enters the system access mode it starts a timer associated with each individual cause indicator. For example, the timer value for call origination is set to 12 seconds, for a page response it is set to 6 seconds, and so on. The access mechanism is an ALOHA type of access. In an ALOHA access several uncoordinated users attempt to use a shared resource. Essentially, each user is allowed to begin transmission at any time, independent of other users. When two users start their transmissions at the same time, their transmissions will collide and ALOHA specifies graceful back-off mechanisms, wherein each user waits for a random amount of time before reattempting access. EIA-553 uses an ALOHA-based protocol, specifically, CSMA/CD (carrier sense multiple access with collision detection). Prior to attempting transmission, the mobile reads the busy-idle status of the channel. If the channel is busy, it awaits a random amount of time before re-attempting access. This ensures that once a channel is captured by a mobile it will be able to complete its transmission fully. If the channel is not busy, the mobile starts transmission. It also monitors the busy-idle status of the channel to ensure that it does not become busy prior to 56 bits being transmitted. Once the beginning of a transmission is received by the BS, it sets the busy-idle bit in the control channel to busy. This process takes approximately 56-bit time intervals. If the busy-idle bit is set to busy after 56 bits have been transmitted by the mobile, it means that the channel has been captured by the mobile for its exclusive use. If the busy-idle bit is changed to busy prior to 56 bits having been transmitted, then the mobile stops transmission and waits for a random amount to time because the channel has been captured by a different mobile.

The initial message transmitted by the mobile depends on the type of access attempt. The access attempt messages are a combination of one to five words and are transmitted on the reverse control channel. The types of messages that can be transmitted over the reverse control channel are the page response, origination, order confirmation, and order messages. Every reverse channel message will contain an abbreviated address word (word A) which indicates the seven-digit NXX-XXXX of the MIN. For certain messages this word will be followed by the extended address word (word B) which indicates the NPA of the MIN. If the network requests the serial number of the mobile on access messages, the serial number word (word C) is included, which indicates the serial number of the mobile. If the access attempt is an origination message then the called address number is indicated in words four and five as shown in Fig. 2.26.

**2.3.5.1 Mobile-to-land call.** A mobile user dials the phone number of the landline user. Unlike a landline phone the mobile phone stores the dialed digits in memory until the user presses the SEND button. The mobile station, prior to the SEND key being pressed, is in the IDLE mode. Once the SEND key is pressed, the mobile enters the system access mode with the indicator as origination. The mobile station tunes to the strongest control channel and attempts to send an origination message. The ORDQ, ORDER, and T fields are set to indicate that this is an origination attempt. The first and second words of the called address will contain the user-dialed digits. The base station ensures the

**Word A - Abbreviated Address Word**

| F = 1 | NAWC | T | S | E | RSVS = 0 | S C M | MINI $_{23-0}$ | P |
|---|---|---|---|---|---|---|---|---|
| 1 | 3 | 1 | 1 | 1 | 1 | 4 | 24 | 12 |

**Word B - Extended Address Word**

| F = 1 | NAWC | LOCAL | ORDQ | ORDER | LT | RSVD = 000...0 | MINI $_{33-24}$ | P |
|---|---|---|---|---|---|---|---|---|
| 1 | 3 | 5 | 3 | 5 | 1 | | 10 | 12 |

**Word C - Serial Number Word**

| F = 0 | NAWC | SERIAL | P |
|---|---|---|---|
| 1 | 3 | 32 | 12 |

**Word D - First Word on the Called Address**

| F = 0 | NAWC | | 2nd DIGIT | ... | ... | ... | 7th DIGIT | 8th DIGIT | P |
|---|---|---|---|---|---|---|---|---|---|
| 1 | 3 | 4 | 4 | 4 | 4 | 4 | 4 | 4 | 12 |

**Word E - Second Word on the Called Address**

| F = 0 | NAWC = 000 | 9th DIGIT | 10th DIGIT | ... | ... | ... | ... | 15th DIGIT | 16th DIGIT | P |
|---|---|---|---|---|---|---|---|---|---|---|
| 1 | 3 | 4 | 4 | 4 | 4 | 4 | 4 | 4 | 4 | 12 |

**Figure 2.26** Messages made up of a combination of five words.

integrity of the message by checking the parity and forwards it to the MTSO. The base station may send only those fields it deems are necessary for the MTSO to validate the origination request. The MTSO checks the user's subscription to ensure that the user is a legitimate user and can be provided service. It might also check the user's service profile to ensure that the user is allowed to make outgoing calls or has a long-distance service subscription (if the dialed number is a long-distance number). After the mobile successfully passes all the checks, the MTSO decides to provide service to the user. It indicates this by sending a message to the BS to assign a voice channel. The BS assigns a voice channel for the particular mobile and indicates the voice channel particulars, such as channel number, to the mobile station via an initial voice channel designation (IVCD) order. The BS transmits the SAT code. The MS tunes to the assigned voice channel and transponds the SAT on that voice channel back to the BS via the reverse voice channel. Upon detecting the SAT, the BS ensures that the mobile is on the channel and sends an SAT received message to the MTSO which dials the called number and connects the call to the landline user via the PSTN. The mobile user receives the call-progress tones, such as ringing or busy, from the PSTN. When the landline user answers the call, conversation begins. If the mobile user terminates the call, the MS transmits a signaling tone (ST) for 1.8 seconds and terminates the transmission. The BS detects the ST and generates a disconnect message which is sent to the MTSO. The MTSO, on receiving this message, releases the connection with the PSTN and the landline user. If the landline user terminates the call, the release from the PSTN is converted to a release channel message by the MTSO and sent to the BS. The BS sends a release channel to the MS, which releases the connection and transmits ST for 1.8 seconds indicating to the BS that it is no longer using the channel. The BS sends a channel available message to the MTSO. These message flows for mobile-to-land call are shown in Fig. 2.27.

### 2.3.5.2  Land-to-mobile call.

The landline user dials the telephone number (MIN) of the mobile station. The MIN provides routing information to the PSTN to enable the call to be routed to the correct

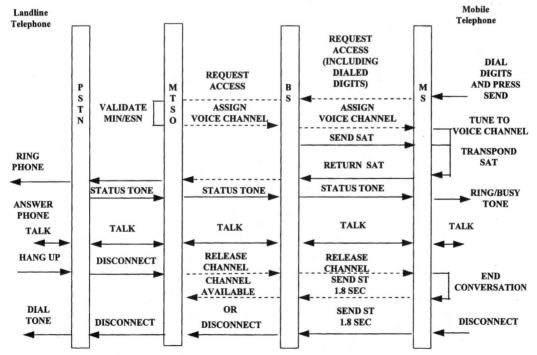

**Figure 2.27**  Mobile-to-land call message flows.

MTSO. The MTSO on receiving the call has to contact the MS to indicate that there is an incoming call. The MTSO also ensures that the MS is not already involved in a call. Also, the MTSO might have a prior knowledge of the general area where the MS is, due to registration. The MTSO can optionally try to contact the MS only in the registration area or in the entire system. The MTSO sends a page request to the base stations to page the MS. The BS pages the mobile by sending a page message on the paging channel. The mobile station has to be powered up and idle to respond to the page. The mobile station on receiving the page sends a page response on the access channel to the BS. This message will contain fields indicating that this is a page origination. The BS forwards this message to the MTSO. The MTSO, in turn, after verification of the MS's identity, commands the BS to assign a voice channel. The BS sends an IVCD order to the MS commanding it to tune to the indicated voice channel. The MS tunes to the voice channel and transponds the SAT back to the MTSO. The BS on detecting the SAT sends an alert order to the MS. On receiving this order, the MS locally generates power ringing to alert the user that there is an incoming call. After the MS has alerted the user, it generates the signaling tone on the reverse voice channel. The BS on detecting the ST generates a message to the MTSO indicating that the MS has started alerting. On receipt of this message the MTSO generates ring-back tones for the landline user. When the mobile user answers the phone, the MS stops generating ST and connects the voice path. The BS detects the absence of ST and indicates that via a message to the MTSO. The MTSO terminates ring-back toward the landline user and connects the voice path. The conversation between the landline user and the mobile user now begins. The call-termination phase is identical to the mobile-to-land call flow. The message flow for land-to-mobile call is shown in Fig. 2.28.

**2.3.5.3 Handoff.** Handoff is initiated when the quality of the received signal at the BS is below a predetermined threshold. The quality usually deteriorates due to movement of the MS away from the BS. The handoff threshold is determined by the operator and may differ from cell to cell. The quality of the received signal is measured by a dedicated receiver at the BS, known as a *locate receiver*. The function of this receiver is to scan all the voice channels in use and to monitor the signal of each channel. The time taken to scan is usually 50 ms per channel. The measurements taken are also averaged over a period of time because of variance caused by multipath fading. When the quality of the signal falls below a given threshold the BS initiates the handoff procedure. It requests the MTSO to execute the handoff procedure for the particular mobile. The MTSO has a database which stores the identity of the cell and its neighboring cells and will send a request to all

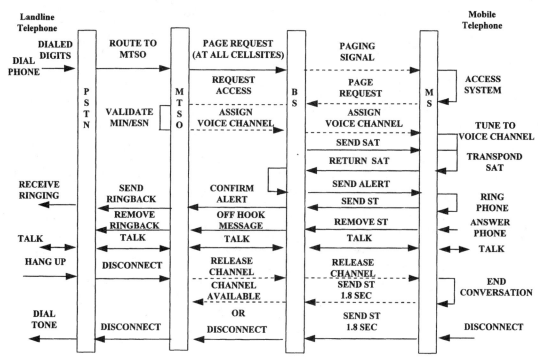

**Figure 2.28**   Land-to-mobile call message flows.

the neighboring base stations to scan the reverse voice channel frequency of the MS, take quality measurements, and report back to the MTSO. Reported values are compared with each other and against a set threshold. The best value is chosen and the BS that reported it is directed to assign a voice channel to the MS. The characteristics of this new voice channel are then transmitted to the MS via the handoff command. This command also directs the MS to tune to the new channel and begin transmission. Once the MS successfully transponds SAT on the newly assigned voice channel, the voice path is set up and conversation begins. The MTSO will then release the old voice channel. Note that there is a small break in conversation when the MS stops transmitting on the old voice channel and before it begins transmission on the new voice channel in a different cell. However, this break is a few milliseconds and hardly discernible to the user. The message flow for handoff is shown in Fig. 2.29.

## 2.4   Summary

Cellular operators are preparing for the onslaught of competition from the PCS operators by lowering prices, offering shrink-wrapped pack-

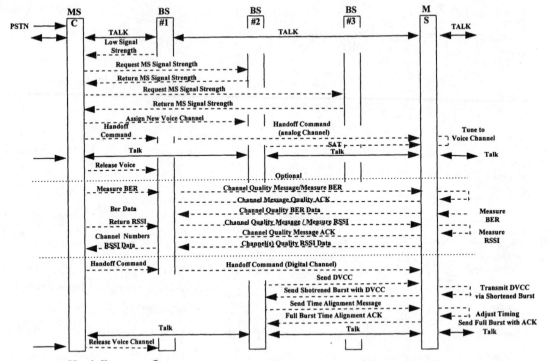

**Figure 2.29**  Handoff message flows.

ing, enhancing the distribution channels and the product through digital transition, and adding intelligent network components. This chapter provided the market structure of cellular industry and the details of AMPS technology that are predominantly deployed in North and South American cellular network. The cellular network will evolve to digital due to capacity and new-feature reasons, but more so due to competition from PCS. The digital technology available for cellular is TDMA or CDMA. However, AMPS will continue to be a prominent technology in cellular for several years because of its legacy and historical weight. The next few chapters deal with the other PCS substitutes, namely, paging and mobile satellite services. Later chapters detail the PCS market, digital technologies, network design, and economics for a PCS operator within the new, competitive environment.

# Personal Communications Satellite System

## 3.1 Mobile Satellite Overview

This chapter is a discussion on the role of mobile satellite system (MSS) as the new generation of wireless telephony: its history, technology, market, industry profile, advantages, disadvantages, the players, applications, and legal and regulatory issues. Although there are many names for the application of satellites to wireless telephony, personal communications satellite system (PCSS) will be the only one referred to in this chapter; MSS refers to just a satellite as a transponder of communications, mobility is limited within the footprint of the satellite, and it does not provide any of the PCS-enhanced services. Traditional MSS provides voice, data, and fax capabilities through very large, expensive, and high-power terminals via geosynchronous orbit (GEO) satellites. PCSS is similar to PCS, but uses low-earth orbit (LEO) and midearth orbit (MEO) satellite technology transmitted to small handheld portables which are low-powered (0.5 to 1 watt maximum). The service and feature sets will be the same as terrestrial PCS service, but will be universally available anywhere in the world. A person will be able to make or receive a call anytime, anywhere in the world, as shown in Fig. 3.1.

## 3.2 History

Satellite communications started with the advent of the cold war. During the cold war, satellite communications technology made great strides because the cold war participants wanted to dominate space and to provide covert communications that could be transmitted and

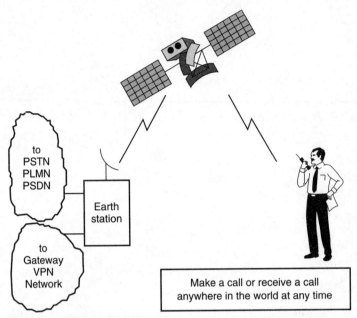

to
PSTN
PLMN
PSDN

Earth
station

to
Gateway
VPN
Network

Make a call or receive a call
anywhere in the world at any time

**Figure 3.1**  PCSS application.

received anywhere in the world for secret activities and military reporting. On October 11, 1958, TRW, a major defense contractor and pioneer in space technology, sent up Pioneer 1. It was the first satellite to be constructed and launched by a private contractor. Its purpose was scientific—to measure the Van Allen radiation belts and interplanetary magnetic field. Pioneer 1 was the first NASA spacecraft. The Department of Defense funded further development of satellite communication. This led to the development of the Defense Communications System (DSCS II). DSCS II became the first satellite that provided secure two-way voice, teletype, and digitized data to government users. DSCS II was the first truly global military command and control network. Satellite communication was born and soon became reliable technology for commercial voice, data, and video applications.

In the private commercial market, MSS has its roots in the transportation industry, where a dispatching station or a central location needs effective communications with its fleet. There are four major industry applications of MSS that are widely deployed throughout the world today. These applications are maritime, aeronautical, telemetry, and land transportation industries.

In 1976, Comsat Corporation launched three satellites for commercial communications service serving transoceanic ships. It was the first commercial satellite system in the world. A global consortium of

PTTs formed The International Maritime Satellite Organization (INMARSAT) to govern and market universal MSS telephony internationally. INMARSAT consists of signatories of 64 countries, each sharing in its profits. It is recognized as the world leader with multinational agreements in place, with 18,000 ships equipped with MSS, and with 22,000 international telecommunications users on land, sea, and air.

Satellites will replace air-to-ground communications when they are launched for aeronautical application. Currently, the aeronautical industry uses long-haul high-frequency (HF) voice radio. HF has quality problems with noise and long-range fading in air-to-air, air-to-ground, and ground-to-air voice communications.

Telemetry applications are for oil and gas wells, water and flow levels for dams and rivers, pressure and flow levels of pipelines, oil platforms, underground instruments, geologic and seismic readers, and the presence of critical situations and disaster. MSS would significantly reduce the cost of providing telemetry in remote locations where there are neither telephone, wireless, nor electric utilities available.

The land transportation industry will offer position-location services called Radio Determination Satellite Service (RDSS) for vehicle and rail tracking. This technology will compete with terrestrial wireless carriers such as cellular, PCS, and especially ESMR operators, to track truck, car and rail fleets, and report vehicle location and status reports, such as fuel level or cargo temperature, to central dispatch and control centers. Other service offerings will be smart-car technologies, highway management monitoring and control, and traffic management systems.

Universal wireless coverage is the developing MSS application for voice and data. It promises to provide the user global coverage anywhere in the world, anytime. Satellite transmission will provide the most widespread application of cellular and paging services with large-coverage footprint. It will be complimentary to domestic cellular, paging, mobile data, and PCS. Typical users will be frequent travelers, thin route users, government agencies, and land and marine fleet operators.

The acceptance of cellular as an economical and reliable form of communications has accelerated the market penetration for mobility, and has provided an answer for the pent-up demand for telecommunications in rural and international markets. This resulted in technological breakthroughs in handset design, weight reductions, low-power consumption, and low-power transmission from handsets. With increased enthusiasm for cellular, ESMR, PCS paging and wireless data, coupled with technological breakthroughs in satellite technology, PCSS promises additional opportunity for the satellite players to provide personal voice and data telephony services all within their own domain; that is, a PCSS system can be an independent, stand-alone system without ever connecting to the PSTN or PLMN.

The landscape for PCSS looks exciting but the climate is changing daily. There is a great deal of demand rising from terrestrial wireless, and pent-up demand in uncovered regions. PTTs are privatizing and easing their monopolies. There are exciting and new satellite technologies that can deliver PCSS. However, the market is being served with inexpensive wireless service, low-cost terminals, and expanding area coverage. With the emergence of PCS, SMR deployment, cellular build-out on a global basis, and with over 15 want-to-be PCSS providers, this market will have to be viewed carefully.

The revenue potential of PCSS is great; however, large investment, near and long-term returns, international spectrum allocation issues, unproven technology, many contenders, and a cautious high-tech investors climate will cause many PCSS players to rethink their deployment strategies, consolidate with others, or give up their dream. Only a handful of PCSS hopefuls will survive to share the available global market, of which 60 percent is anticipated usage in the United States. It is very probable that PCSS will survive and will be deployed. However, its rollout will not be as ambitious as currently touted and coverage will be regional and piecemeal to fill gaps in the terrestrial wireless and landline service coverage.

The most viable market will be the United States, followed by the rest of North America, Europe, and Asia. PCSS will assimilate the maritime, aeronautical, telemetry, and land transportation markets. However, the greatest potential is for PCSS to fill in the need for global roaming from the existing cellular and PCS operators. As roamers travel out of their home location register (HLR) area, they could continue a call by handoff to PCSS rather than to a rural service area (RSA) provider or other adjacent terrestrial operators; thus, an operator would keep a percentage of the roaming revenue rather than passing it to the RSA operator.

This would happen over a dual-mode (wireless of any air protocol and PCSS). A dual-mode terminal will allow a traveler to carry one handset of the technology (AMPS, CDMA, GSM, etc.) of his home carrier anywhere in the world for call origination and call delivery over the PCSS mode. This might be particularly important to PCS carriers who will build out their networks in dense, high-traffic areas without sacrificing total coverage of their licensed area. PCS fill-in and niche markets will be successful; for example, PCSS services will include rural and remote phone service, air-to-ground passenger telephone on commercial and private aircraft, inland and coastal maritime telephone, telephone for rail passengers, fleet management, telemetry, surveillance, inventory control, dispatch radio for delivery and industrial field operations, supervisory control and data acquisition, and remote security systems. Other niche applications are emergency services, search and rescue, law and regulatory enforcement services, aeronautical flight safety services. Each law enforcement agency of the federal government is planning its own satellite PCSS system, as are the military authorities of several countries. Although PCSS is exciting and holds promises of ubiquitous, seamless, global coverage, another telephone network may be too much for total world demand, and many players will fall out, consolidate, or fail.

## 3.3  The PCSS Market

PCSS will require the development of a low-powered handset for use with MEO and LEO technology, similar in size and power output to a cellular handset. The handsets will need to be dual-mode with a terrestrial carrier technology (AMPS/GSM/PCS/DECT/TACS, etc.). Even though PCSS could provide independent service from the PLMN or PSTN, most business cases of the PCSS operators will require partnerships with a terrestrial wireless operator. Therefore, a mobile station (MS) would be connected to their terrestrial wireless carrier while they were in their service area. As the MS would roam beyond their service area, or in a noncoverage area, the call would be handed off to the PCSS satellite. A call would not be handed off from PCSS to terrestrial wireless because it would be difficult to determine which operator and which cell to hand off.

Based on industry surveys, a PCSS MS requires both regional and global coverage. Subscribers want to use just one wireless MS as they travel to countries or regions with noncompatible operating technologies, such as AMPS or GSM. Today, an AMPS subscriber would have to carry a second or third MS when traveling to another country or region utilizing another access technology, such as GSM, NMT, or TACS.

Subscribers want coverage in rural areas, isolated areas, or where there is limited or no PSTN or PLMN: waterways, highways, airways, unpopulated areas, remote areas, and wilderness. Subscribers want lower cost than previous MSS service offerings, but with comparable features, service, and terminals to traditional terrestrial wireless service. Users want handheld MS with the following feature sets and features: universal ubiquitousness; integrated messaging, voice, data, fax, tracking, etc.; vehicle/craft location; telemetry; and universal roaming. There are four PCSS subscriber segments: frequent travelers, thin-route users, government agencies, and land/marine fleet operators. Figure 3.2 shows forecasted worldwide PCSS subscriber growth, from 25,000 in 1991 to 12.3 million subscribers in the year 2000. Rapid growth starts in 1995 with the launch of additional GEO service, and MEO/LEO service available in 1997.

Suscribers × 1,000

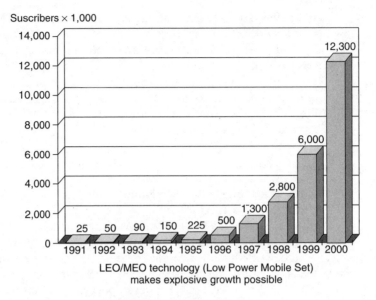

LEO/MEO technology (Low Power Mobile Set)
makes explosive growth possible

Note: PCSS subscribers only, does not include fleet subscribers
Source: Frost & Sullivan, TRW Space and Electronics Group

**Figure 3.2**    Forecasted PCSS subscriber growth.

### 3.3.1  PCSS industry needs

The PCSS industry requires seamless interfaces and feature sets
between and among networks that are totally transparent for the
PCSS user. The PSTN interfaces must be able to operate with both
ANSI and CCITT networks with national variants. The PCSS system
must be able to provide interpretability with AMPS/IS-41, GSM, TACS,
PCS, and so on; and the MS must be able to have dual-mode capability
with a terrestrial wireless air-access protocol, such as AMPS, TDMA, or
CDMA. The infrastructure cost of a PCSS system should be cost comparable to land-based networks in order to provide affordable cost to
subscribers and to provide reasonable payback and adequate return to
the investors. PCSS has unique feature sets and characteristics as follows: inter-/intrasatellite handoff, land-based to satellite handoff, land-
base and PCSS dual registration, and the Doppler effect of a moving
satellite. PCSS is the missing link for a truly global intelligent digital
wireless network. Figure 3.3 is a complete portfolio for enhanced wireless services with which an operator can interface in order to generate
increased revenue and provide more services to subscribers.

In the intelligent digital wireless network, interfaces could be made
to and from AMPS/PCS/PCSS switching platforms with access to HLRs
and VLRs; integrated signaling conversions would be done within the

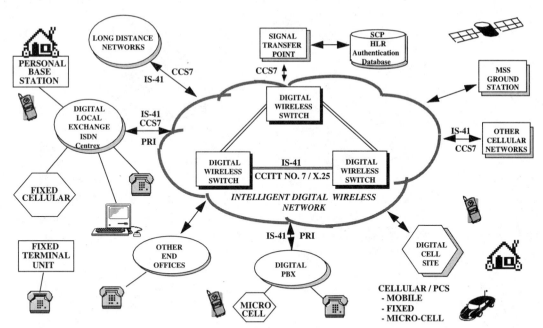

**Figure 3.3**   Intelligent digital network.

switch itself. This would allow for universal call delivery (IS41/GSM-MAP), internetwork (protocol) roaming, and integrated land/satellite-based services as shown in Fig. 3.4.

The PCSS network configuration options must consider minimum or no cellular functionality at ground station and satellite. Both GSM and AMPS are viable options to be the base-switching platform for PCSS. A manufacturer could lock up the opportunity of being a major provider of PCSS switching systems by being instrumental in defining the solution to the PCSS planners. Figure 3.5 shows the network configuration options.

### 3.3.2   PCSS advantages and obstacles

The advantages of PCSS is that it provides coverage and integrated services virtually everywhere. PCSS can fill in the gaps between land-cellular operators and PCS operators and provide wide-area coverage. PCSS also provides integrated wide-area paging on a regional or global basis and multiple services to a global customer base with ubiquitous and seamless service. PCSS extends wireless services worldwide with global coverage. Since PCSS can bypass the PSTN, and because of the low cost of satellites compared to multiple cell sites, PCSS would have a lower cost for infrastructure.

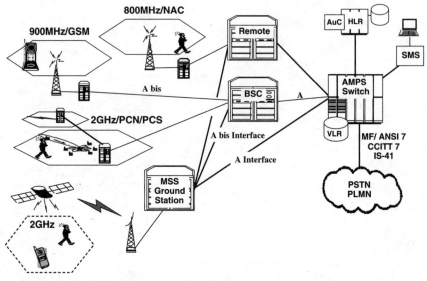

**Figure 3.4**   Universal wireless system.

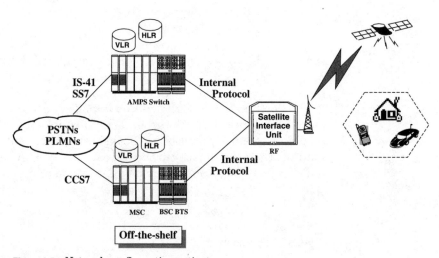

**Figure 3.5**   Network configuration options.

PCSS is ideal for fixed cellular applications. It can provide full telephone service to places that cable can never go because of cost, technical, and economic reasons, or return-on-investment concerns. PCSS could provide complementary and backup telephony services to large organizations with multiple operations, such as retail, manufacturing, finance, trucking, government, and insurance (disaster).

The obstacles to PCSS are the high cost and risk of launching a satellite. There is a high cost for the terrestrial networking and land infrastructure (estimated at $600 million to $3 billion) and a current high price for small, low-powered, dual-mode handsets that would have to be developed. Current MS technology provides a large, high-powered MS with cumbersome antennae. These MSs are expensive (recent cost down to about $2,200) with a high-installation cost provided by a local, wireless carrier.

The worldwide proliferation of cellular and PCS services and associated wireless networks diminishes the need to subscribe to a PCSS operator, unless there are differential advantages and sufficient cost differences to justify a PCSS. Also, satellite service has a poor public image, with users believing that there is less quality than with terrestrial-based telecommunications. This poor image is from the delays and the echo that satellite service provided until the deployment of digital cancellation devices in the network that minimize delay and associated echo when a satellite is used. For the most part, poor quality from satellites has been improved, but the image and perception remains.

### 3.3.3  PCSS versus cellular and PCS

PCSS can offer universal service across LATA boundaries, MTAs, BTAs, and any other geographic and international boundary. PCS, LEC, and IXC cellular operators are restricted to coverage areas. Also, PTTs have restrictions limited within their sovereign territory. PCS services have to contend with a plethora of compatibility issues from air protocol to switching platforms, interexchange, and international exchange issues.

PCSS can share the market with wireless operators (cellular, PCS, and private) to offer dual-mode phones (satellite/land-based) to its subscriber base. This can provide wireless operators a strategy for coverage issues while maintaining roaming revenues. Billing could be handled on a per-operator basis rather than multiple bills and late bills from clearinghouses for roaming and long distance while roaming. PCSSs can provide basic service extension at less cost to the wireless operators rather than constructing new terrestrial cellular systems and cell sites.

Cellular will remain the prevalent service and buy-in PCSS to fill voids in coverage. PCS, while providing coverage in dense, urban areas, could use PCSS for rural coverage, roaming capability, and call overflow. While PCS is primarily for local coverage, PCSS provides global land, sea, and air coverage, with universal features and seamless worldwide roaming. The PCS and PCSS can coexist and, in fact, can use PCSS as a building block of its infrastructure.

The biggest barrier to PCSS is price. Price is high because the would-be PCSS operators are looking at what INMARSAT is currently charging for service rather than economies of scale due to explosive deployment of wireless systems on a global basis. The price for PCSS service will eventually equalize as the PCSS survivors struggle, competing with terrestrial wireless operators.

AMSC complemented cellular offerings with their "Skycell" PCSS offering. Service is available only through existing cellular operators, exclusive with AT&T Wireless Systems. In order for the PCS operators to compete with the incumbent cellular operators, PCS could become a marketing channel for PCSS services. Also, PCSS is ideal for in-building, macrocellular, and PCS applications where, as callers leave their in-building, a call could continue by being handed off to the satellite rather than a macro cell site.

### 3.3.4   The players

The top voice PCSS contenders are listed in Table 3.1. This table describes who the top PCSS contenders are, when the company/consortium was started, the number of satellites planned to be launched, projected launch date, the total number of satellites comprising the PCSS system, type of satellite to be deployed, projected cost of implementation, and the estimated number of switches required for the earth stations or gateways to the PSTN/PLMN.

Due to the dynamic nature of the PCSS industry, these system configurations in Table 3.1 are changing as the economic realities are being realized. Because the PCSS operators have strong satellite backgrounds but relatively weak telephony backgrounds, as they design the PCSS network, the complexities, technical parameters, and business structure are constantly updated to accommodate the new realization (e.g., Iridium and Odyssey). Funding is difficult in today's investment climate. There appears to be too much PCSS competition, and there are concerns of competition with terrestrial wire-line carriers. The following sections describe the major PCSS players in alphabetical order.

#### 3.3.4.1   American Mobile Satellite Corporation (AMSC).   AMSC is owned by AT&T Wireless, formerly McCaw (32 percent), GM/Hughes (29 percent), Mtel (17 percent), Singapore Telecom (12 percent), and private investors (10 percent). They are the only FCC license holders to date who launched three satellites to provide mobile telephone and mobile data services in the United States and 200 miles of coastal waters. Service began in April 1995 with a single satellite containing 2500 circuits utilizing FDMA access technology. Three more GEOs are planned for the future. The estimated cost of the system is expected to be $3 billion.

**TABLE 3.1    PCSS Operator Profiles**

| | Comp. start | Initial sats. | Launch date | Total sats. | Type | Cost |
|---|---|---|---|---|---|---|
| AMSC | 1992 | 1 | 1995 | 3 | GEO | $3.0 b |
| Calling Communications Corp. | 1990 | 10 | 1998 | 840 | LEO | $6.5 b |
| Celsat | 1990 | 1 | 1997 | 4 | GEO | $0.7 b |
| Constellation Comm. (Aries) | N/A | 3 | 1998 | 54 | LEO | $1.7 b |
| Comsat | 1962 | * | | | | |
| CruisePhone, Inc. | N/A | * | | | | |
| Ellipso | 1990 | 6 | 1996 | 24 | E-LEO† | $500 M |
| Eutelsat | 1991 | 8 | 1991 | N/A | GEO | N/A |
| Hughes Spaceway | N/A | 1 | 1998 | 8 | GEO | N/A |
| IDB Mobile Communications | 1989 | * | | | | |
| INMARSAT-P | 1982 | 1 | 1999 | 10 | MEO | $2.6 b |
| Intl. Mobile Satellite Corp. | 1979 | N/A | N/A | N/A | GEO | N/A |
| LEO One USA | N/A | N/A | 1997 | 48 | LEO | $250 M |
| Loral/Qualcomm (Globalstar) | 1991 | 4 | 1997 | 52 | LEO | $1.9 b |
| Maritime Cellular Networks | N/A | * | | | | |
| Mobile Comm. Holdings, Inc. | 1990 | 6 | 1998 | 14 | E-MEO‡ | $1.2 b |
| Motorola Iridium | 1990 | 2 | 1996 | 66 | LEO | $3.7 b |
| Norcross Networks, Inc. | N/A | * | | | | |
| Orbital Communications Corp. ORBCOMM ** | N/A | 2 | 1995 | 36 | LEO | $190 M |
| Starsys Global Positioning ** | N/A | 2 | 1997 | 24 | LEO | $196 M |
| TMI | 1988 | 1 | 1995 | 1 | GEO | $450 M |
| TRW (Odyssey) | 1958 | 1 | 1999 | 12 | MEO | $2.5 b |
| Teledesic Corp. | N/A | 840 | 2001 | 840 | LEO | $9.0 b |
| VITA Sat | 1959 | 1 | 1995 | 3 | LEO | $25 M |

\* = Reseller
\*\* = Data-only service
†E-LEO = elliptical LEO orbit
‡E-MEO = elliptical MEO orbit

AMSC offers voice, data, messaging, and satellite tracking services. Handsets will cost approximately $1800 for a dual-mode satellite/cellular unit when they are available from Westinghouse and Mitsubishi; however, current MS units, depending on the antenna, cost from $3000 to $12,000 plus installation. Monthly subscription is $30 plus airtime charges of $1.49 per minute to or from anywhere in the coverage area.

AMSC has reached agreement with 60 U.S. cellular operators to provide adjunct service for 152 million POPs/633 markets/service providers/distributors under the name Skycell. An integration agreement with TMI was made to provide redundant coverage throughout North America. Skycell will be integrated with land-based cellular operators for call origination and call delivery. According to AMSC, revenues are expected to reach $473 million by 1996.

**3.3.4.2 Celsat, Incorporated.** Celsat is an affiliate of the Triton Corporation. It plans to launch GEO satellites with CDMA technology. They are scheduled to begin in 1997 by launching four GEOs that will provide voice, high-speed fax and data, global, positioning determination, picturephone, and nationwide paging service. Projected expenditure is $700 million. Estimated airtime is $.25 per minute.

**3.3.4.3 Comsat, Incorporated.** Comsat offers maritime and land mobile satellite communications through its membership in INMARSAT. It will provide voice, data, high-speed fax, position reporting, remote monitoring and messaging for maritime, and land and aeronautical applications.

**3.3.4.4 Constellation Communications (Aries).** Constellation Communications is a consortium based in Herndon, Virginia. This consortium includes Defense Systems, Incorporated, International Microspace, and Pacific Communications Sciences. There are other investors, including a BOC and a large system integrator. International Microspace, Incorporated, will provide launch services; Pacific Communications Sciences, Incorporated, will provide the user terminals; and Defense Systems, Incorporated, is developing the satellites. Fifty-four LEOs are planned with launch scheduled in January 1998. The satellites will be launched in phases with a planned start in early 1998 and completion in 2000. Initially, there will be 48 satellites positioned in four planes with 12 satellites per plane for global coverage. Each will make a complete orbit in 1.75 h.

The access scheme is TDM/SDMA direct sequencing over 16 MHz. CDMA dual-mode handsets and portables are estimated to be priced between $2500 and $3500, and airtime is estimated at less than $1.00 per minute. The projected cost of implementation is estimated at $1.7 billion.

**3.3.4.5 Ellipsat (Ellipso).** Ellipsat is based in Washington, D.C. The original plans called for 24 LEOs between 1995 and 1996 that consist of two complementary constellations: Ellipso Borealis, serving the northern hemisphere with 15 satellites; and Ellipso Concordia, serving the southern hemisphere with six satellites, and three fill-ins covering specific areas. However, since they partnered with Westinghouse in January 1994 they now plan for a MEO satellite system similar to the Odyssey system. Estimated start-up cost is $500 million, and they project 600,000 subscribers worldwide.

Partnerships are arranged with Mexico (Telmex), Canada (Skylink Communications), Australia (Ellipsat Australia), Israel Aircraft Industries, Harris Corporation, IBM, Westinghouse, Barclay Bank as an investor only, and others. Satellites are to be built by Fairchild Space and Israeli Aircraft Industries. Handset estimated price is between $1000 and $2000. Airtime is anticipated at $.50 per minute.

This satellite-based network is intended to complement their existing commercial terrestrial mobile telecommunications service. It is intended to provide global service with voice, real-time and store-and-forward data, paging services, and geopositioning.

**3.3.4.6  INMARSAT.**  INMARSAT is recognized as the world leader in MSS with multinational agreements in place to provide service currently to 22,000 users on land, sea, and air using very small hand portables. "Project 21" is the upcoming generation of satellite-based global pocket-phone service called INMARSAT-P for Europe, Asia, Latin America, North America, and other Comsat signatories. The technical aspects are being studied by Marconi Space Systems. INMARSAT will use MEO satellites and LME's (Ericsson) MSC switches.

**3.3.4.7  Iridium.**  Motorola is the prime for Iridium. Partners include General Electric, Lockheed, Raytheon, McDonnell Douglas, Scientific Atlanta, Sony, Kyocera, Mitsubishi, DDI, Kruchinev Enterprises, Mawarid Group of Saudi Arabia, STET of Italy, Nippon Iridium Corporation of Japan, the government of Brazil, Muidiri Investments BVI, LTD of Venezuela, China Great Wall Industry of China, United Communications of Thailand, the U.S. Department of Defense, Sprint, and BCE. Siemens is their current switch partner.

Launch is projected in 1996 with service available in 1998. Iridium will offer voice, messaging, GPS tracking, fax, and data services. Mobile and portable handsets will be priced from $2000 to $3500 each. Monthly service is planned to range from $50 to $100, with airtime at $3.00 to $3.50 per minute.

Iridium will target the upper 1 percent of the 100 million subscribers worldwide. They expect 1.75 to 2.25 million subscribers by the year 2000.

The network will consist of a total of 66 LEOs using TDMA/FDMA modulation, each with 48 beams and up to 100 channels per beam, for a total of 2000 channels per satellite. They will provide low-power dual-mode phones for land-based cellular. The project cost is estimated at $3.7 billion. Iridium's "Backbone in the Sky" network concept may encounter political dissection because it violates WARC 1992's mandate against bypass. Potential competition is INMARSAT (67-member-country global satellite cooperative).

An Iridium LEO is planned to have 48 beams with a diameter of 2500 miles. The beams can be directional for high-usage or desired-coverage areas. The satellite will provide beam-to-beam handoff and the earth station will coordinate satellite-to-satellite handoff. The Iridium LEO will traverse from horizon to horizon in 10 minutes. Embedded switching will allow a call to pass from satellite to satellite without ever going through the PSTN or PLMN, thus having the ability of call delivery from one Iridium MS to another Iridium MS with total bypass.

Each earth station provides Ka-band communications to the satellites as well as tracking, telemetry and control, and connection to the PSTN and PLMN. A summary of the technical parameters of the Iridium earth station is shown in Table 3.2.

Iridium plans to provide GSM service with 4.8 kbits/s for encoded voice and 2.4 kbits/s for data. In order to comply with WARC and not interfere with other LEO operators, the FCC proposed in 1994 that the frequency range for LEOs be segmented, allocating 5.15 MHz to Iridium and 11.35 MHz to other big LEO systems. There needs to be additional coordination with the Russian Glonass system for this proposal to become a reality.

**3.3.4.8    LEOSAT.**    LEOSAT is a company based in Silver Spring, Maryland. It is owned by MARCOR Corporation and several private investors. The LEOSAT system would be comprised of 24 LEOs that will provide worldwide mobile data services primarily to automobiles with very small inboard terminals. The builder of the satellites is Defense Systems, Incorporated, at a cost of $98 million. The data service is planned to be low-cost. This system is considered to be a medium risk with a low return to its investors. LEOSAT's application has been pending with the FCC since 1992.

**3.3.4.9    Loral/Qualcomm (Globalstar).**    Globalstar was formed by Loral Aerospace Corporation and Qualcomm. Other partners include Air-Touch, Aerospatiale, Alcatel, Alenis Spazio, Hyundai, Datcom, Loral Corporation, and Vodaphone. They plan to offer wholesale voice, data, and satellite GPS tracking services at tariffs similar to current cellular rates through U.S. cellular operators and other cellular operators throughout the world. Each operator would have to invest about $350,000 for an earth station for PLMN interconnection, plus purchase satellite capacity. Globalstar's network will consist of 52 LEOs, 24 of which will be for U.S. coverage. The first 4 are planned for launch in mid-1997 with 48 additional satellites phased in at eight batches of 6 satellites. The system is estimated to cost $1.9 billion.

This system is planned to have 125 ground stations linked to the PSTN and PLMN, and would use CDMA technology. Globalstar will pick up from Qualcomm the "Omnitracks" vehicle positioning system in North America, Mexico, Brazil, Japan, and 11 European countries with 44,700 customers in the United States.

**TABLE 3.2   Iridium Earth Station Technical Characteristics**

| | |
|---|---|
| Data rate | 12.5 Mbits/s |
| Error-correcting coding | Convolution, rate = 1/2, K = 7 |
| Modulation | QPSK |
| Frequency bands: | |
| —Transmit (uplink) | 27.5–30.0 GHz |
| —Receive (downlink) | 18.8–20.2 GHz |
| Ground tracking antenna: | |
| —Dish diameter | 11 feet |
| —Gain | 53.7 dB @ 20 GHz, 57.3 dB @ 30 GHz |
| —Side lobe level | 47 CFR criteria met § 25.209(a)(2) |
| —3-dB beam width | 0.36° @ 20 GHz, 0.24° @ 30 GHz |
| —Point angle range | 360° azimuth, +5–90° elevation |
| Ground acquisition antenna | Passive array configuration |
| Transmitter EIRP | |
| —Clear weather | 51.6 dBW (3 dB) |
| —Heavy rain | ≤77.6 dBW maximum (±3 dB) |
| Receiver GT | 22.9 dB/K |
| Required Eb/No | 6.7 dB @ BER $10^{-6}$ |

**3.3.4.10   TMI Communications.**   TMI is a Canadian company and is controlled by BCE (Bell Canada Enterprises). They are the only Canadian licensee to date. TMI plans to provide network redundancy for AMSC with only one GEO satellite.

**3.3.4.11   TRW (Odyssey).**   Odyssey is owned by TRW based out of Cleveland, Ohio ($7.9 billion in 1992). TRW Space and Defense is based in Redondo Beach, California, and is leading the development of Odyssey. TRW originally intended to provide the system for INMARSAT but decided to proceed on their own with multicorporation participation. TRW plans were changed due to lack of funding so a less aggressive deployment plan was imposed. They plan to launch 1 MEO at start-up plus 12 later, with 12 to 18 earth stations and 80–100 gateway switches. Initial plans called for an initial launch of six satellites followed by another launch of six. Total system cost is estimated at $2.5 billion.

Odyssey plans to offer voice, data, and messaging services complementary to land-based cellular. It is proposed that the system would be used where connection to terrestrial-based cellular service is unavailable.

Modulation is CDMA with 0.5-watt terminals. The handsets, with a power range from 0.5–1.0 watt, are planned to be sold at less than $300 each. The planned monthly rate is $8.00, plus $.65 per minute.

**3.3.4.12   Volunteers In Technical Assistance (VITA Sat).**   VITA plans to launch one LEO in 1995, to complement its existing GEOs. They plan to provide global coverage to underdeveloped countries. Their network will cost about $25 million. Service charges have not been determined, but will be limited to those who can pay.

### 3.3.5  Economic considerations

Although the promise of PCSS and new technology seems intriguing and there has been considerable investment and effort spent, the harsh realities of return on investment, competition, and risk should not be taken lightly. Some of the cost and revenue components will be explored in order to understand the business case for PCSS. Revenue estimates vary widely since PCSS is highly speculative and most of the design is still in R&D. In any event, the revenue numbers are large.

Table 3.3 is the satellite communications market revenue for equipment and services projected by Edge Media. This chart represents the economic impact in both expenditures to the equipment vendors and the revenues expected from services rendered to subscribers. Revenue grows from $6.414 million in 1994 to $14.387 billion in 1999. Mobile subscriber terminal equipment is not included. The first column lists the category of revenues by PCSS segment equipment type (expenditures for equipment, or revenue to the equipment vendors) as well as revenue for services. The revenues are actual from 1994 with a five-year projection to 2000. The last column is the cumulative annual growth rate over the reported and projected period.

**3.3.5.1  Mobile earth station.**  The *mobile earth station* is the satellite tracking and communications point to and from the earth. In addition to the earth station, a gateway switch and cell site equipment will be

**TABLE 3.3    Satellite Communications Market Revenue ($ millions)**

| Revenue | 1994 | 1995 | 1996 | 1997 | 1998 | 1999 | CAGR |
|---|---|---|---|---|---|---|---|
| Space equipment segment | $890 | $950.0 | $1,005 | $1,065 | $1,130 | $1.192 | 7.9% |
| Fixed ground equipment | $535 | $594.0 | $655.0 | $720.0 | $785.0 | $850.0 | 10.5% |
| Mobile ground equipment | $254 | $338.6 | $445.0 | $565.5 | $740.0 | $965.0 | 32.0% |
| Mobile satellite services | $1,460 | $2,495 | $3,295 | $3,955 | $4,760 | $5,615 | 52.5% |
| Fixed satellite services | $3,275 | $3,735 | $4,220 | $4,725 | $5,245 | $5,765 | 13.0% |
| Satellite equipment | $1,679 | $1,882 | $2,105 | $2,350 | $2,655 | $3,007 | 13.0% |
| —Yearly growth | 11.2% | 12.1% | 11.8% | 11.7% | 13.0% | 13.3% | |
| Satellite services | $4,735 | $6,230 | $7,515 | $8,680 | $10,005 | $11,380 | 13.0% |
| —Yearly growth | 37.7% | 31.6% | 20.6% | 15.5% | 15.3% | 13.7% | |
| Total mobile satellite | $6,414 | $8,112 | $9,620 | $11,030 | $12,660 | $14,387 | 13.0% |
| —Yearly growth | 29.6% | 26.5% | 18.6% | 14.7% | 14.8% | 13.6% | |

required to provide transmission and interconnection to the PSTN and PLMN. The cost of each station will range from $350,000 to $5 million for a simple satellite-tracking earth station. Associated gateway switching equipment, depending on the type (MSC/BSC, ANSI base with cell site, hybrid or custom) and size, will cost from $500,000 to $4 million. Figure 3.6 estimates revenue to vendors growing to $6.5 billion in 2000. If this trend continued to the year 2002, the infrastructure cost would approach $20 billion.

A breakdown of cost associated with the ground segment is shown in Table 3.4.

**3.3.5.2 Satellite and launch.** The cost of a satellite varies by its type, its technology, orbit, and its launching vehicle. Whereas a LEO costs more than a MEO or GEO, its launch costs are much less. A LEO can be shot from an L-1011 jumbo jet over the Pacific Ocean, but a lower-cost MEO or GEO requires a Titan class or greater missile to place the satellite in it orbit. The farther out the satellite, the greater the cost to launch. Figure 3.7 shows the projected satellite cost for full-time coverage of four representative PCSS proposed systems. Figure 3.7 shows Celsat's single satellite cost of about $250 million, TRW Odyssey's satellites costing about $450 million for 12 satellites, Loral/Qualcomm's satellites costing about $1.2 billion for 52 satellites, and Iridium's costing over $2 billion for 66 satellites.

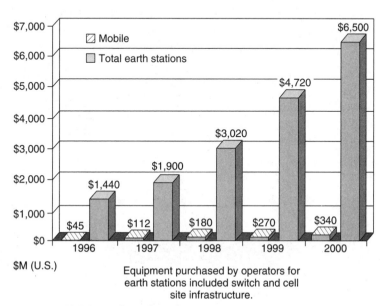

Figure 3.6  Mobile earth station cost.

**TABLE 3.4    Estimated Equipment Cost by PCSS Operator**

|  | Earth stations | Switch revenue | Ground stations | Switch revenue | Total cost |
|---|---|---|---|---|---|
| Iridium (Motorola) | 50 | $300 M | 150 | $500 M | $800 M |
| TRW (Odyssey) | 12–18 | $250 M | 80–100 | $350 M | $600 M |
| Westinghouse |  |  |  |  |  |
| —AMSC | 15 | $100 M | 146 | $730 M | $830 M |
| —TMI | 3 | $10 M | 10 | $50 M | $60 M |
| Aries | 50 | $290 M | 150 | $750 M | $1,400 M |
| Ellipso | 4 | $15 M | 20 | $80 M | $95 M |
| Globalstar | 50 | $300 M | 125 | $625 M | $925 M |
| U.S. Government | 8 | $100 M | — | — | $100 M |
| VITA | 7 | $7 | 25 | $125 M | $132 M |

Revenue derived from new mobile subscriber terminal sales from 1995 to 2000 is $710 million, as shown in Fig. 3.8.

**3.3.5.3  Cost per voice channel.**  The annual cost per voice channel, based on the earth system and satellite equipment, ranges from about $.50 for Celsat, about $4.75 for TRW Odyssey, about $7.90 for Loral/Qualcomm, and about $11.75 for Iridium. These estimates are derived by dividing the total cost of the proposed PCSS system by the total number of voice circuits over the projected life of the satellite upon deployment. These numbers vary by the number of satellites and the number of circuits designed in the satellite.

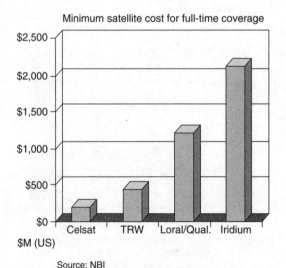

Source: NBI

**Figure 3.7**    Cost of satellite segment of PCSS network.

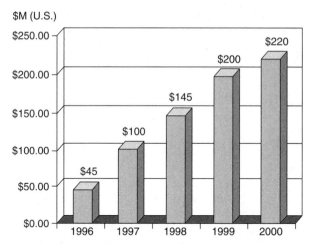

**Figure 3.8**  New subscriber terminal revenue from MSS and PCSS.

**3.3.5.4  Service offering.**  Each PCSS operator has his or her own view on what the market will bear regarding PCSS (the value of seamless and ubiquitous service). The proposed service offerings are based on the projected costs of their PCSS network, division of revenue with the land-based carriers, and PTTs. Table 3.5 shows the range for select PCSS operators. This is what the providers ultimately want to charge by the year 2000. In the case of AMSC's Skycell service, the monthly rate is $39 with airtime at $1.49 per minute; however, the terminal cost is greater than $6,000.

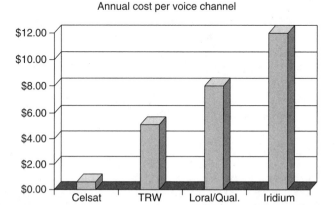

**Figure 3.9**  Annual cost for each satellite circuit, i.e., each circuit within a beam.

TABLE 3.5    Estimated Service Offerings by PCSS Operator

|            | Monthly rate   | Airtime/minute | Terminal price    |
|------------|----------------|----------------|-------------------|
| AMSC       | $25 to $45     | $1.49          | <$1,800           |
| Aries      | —              | <$1.00         | $2 k to $2.5 k    |
| Celsat     | —              | $.25           | —                 |
| Ellipso    | —              | $.50           | $2 k              |
| Globalstar | —              | —              | —                 |
| Iridium    | $50 to $100    | $3.00          | $2 k to $3.5 k    |
| Odyssey    | $8             | $65            | $300              |

AMSC was the first to deploy service in April of 1995, and has since become the de facto trendsetter. The service rate is $39.00 per month and $1.49 per minute for airtime without any long-distance charges, and a terminal charge that ranges from $3000 to $12,000 with additional installation charges from the local cellular operator. The terminal charge will vary because of newer technology but greater demand will drive down its cost. However, it will be very hard for Iridium to charge double the airtime and monthly rate without some kind of value-added feature.

The competitive pressures from proposed systems, such as Odyssey and Globalstar, will also drive down the user cost, not to mention lower land-to-mobile service competition from cellular, PCS, SMR, and others. The annual cost per voice channel does seem, however, to have a high-profit margin potential. This presumes that there will be full utilization of all channels and high demand.

**3.3.5.5  Industry risk versus return matrix.**   The matrix shown in Fig. 3.10 shows the relative position of risk and return. VITA Sat shows low risk with low return, whereas LEOSAT has a lower risk with medium return. ORBCOMM, Odyssey, and Globalstar have medium risk with medium return. AMSC shows lower risk with higher return. And Celsat and Iridium have much higher risk, but will return about the same as AMSC, Globalstar, and Odyssey. This chart allows an investor to analyze the relative positions of the speculative PCSS operators. Some players will win, others will fade away.

## 3.4  Legal and Regulatory Issues

The legal and regulatory climate is complex, dynamic, and ever changing. Even though there is a global trend to have universal laws and regulation, the simple fact that there are two world-standard-based PCSS networks dictates that there will be an ANSI standard and a CCITT/ITU standard. This is further complicated by the political arena—regional, continental, and country by country.

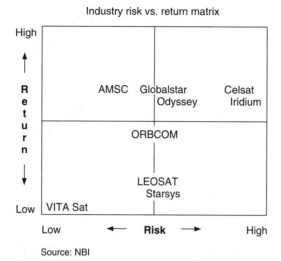

Figure 3.10 Industry risk versus return on investment matrix.

Each country's PTT will want regulation so it can realize their revenue for usage that originates, terminates, or is switched within its sovereign jurisdiction. All countries are members of the ITU, which has jurisdiction over the PSTN and PLMN. However, something as elusive as a satellite that may bypass sovereign territories and regulations will cause disruption in global deployment. Many countries are forming consortia, such as INMARSAT, and have representative signatories that vote the national, political, and economic interests of their respective countries. Voting blocks form to prevent motions that may not be in their interest, causing disruption, slowdowns, and blockage of implementation. The following sections describe the major legal and regulatory issues that will affect MSS and PCSS.

### 3.4.1 U.S. regulatory environment

The United States and Canada account for over half of the telecommunications service in the world, and have always gone their own way. Even though the United States and Canada are ANSI standard, there has been a concerted effort to work with the ITU for some type of global standardization, frequency and spectrum coordination, and universal technology going into the future. Regarding satellite telecommunications, the FCC and the DOC (Department of Communications in Canada) have given very broad guidelines. They are leaving all the details for private MSS and PCSS developers to work out.

**3.4.1.1   Status.**   In 1962, Comsat (Communications Satellite Corporation) was formed to be a nongoverning operating company under the FCC and State Department to promote satellite technology internationally. Subsequently, INTELSAT (International Telecommunications Satellite Organization) became a multinational effort and launched its first GEO over the Atlantic in the early 1960s. The United States has tried to give INTELSAT a monopoly on all international communications, but the European countries would be the losers, thus the PTT resisted this effort. The United States reversed its position in the 1980s and has encouraged the private sector to develop and deploy separate systems.

AMSC was the only FCC licensee for commercial service. They were granted exclusive rights in the United States; however, it was overturned through legal challenge. Experimental licenses were granted in September 1992 to Motorola for 11 satellites, Ellipsat for 4 satellites, Constellation for 2 satellites, and TRW for 12 satellites. Subsequent licensing began in 2Q94. There are now over 20 MSS/PCSS applicants today. In late 1994, the FCC awarded three licenses on the 1.6-GHz band to Iridium (backed by Motorola), Globalstar (a joint venture of Loral and Qualcomm), and Odyssey (a joint venture of TRW and Tele-Globe). The FCC allocated 16.5 MHz of spectrum to these systems.

**3.4.1.2   Issues.**   Frequency band allocation remains a key issue. Another issue questions whether the government should grant whole or partitioned spectrum or keep spectrum in reserve. The FCC has chosen not to standardize or coordinate the modulation scheme; therefore, PCSS carriers are planning to use different modulation schemes, including FDMA, TDMA, CDMA, and hybrid combinations. This lack of standardization ensures that users will not be able to use multiple carriers, and subscribers will be held captive to their carrier unless they switch from MS to another PCSS operator.

**3.4.2   International regulatory environment**

The ITU includes virtually every nation for standardization and coordination including both CCITT and ANSI telecommunications-based countries. Because MSS and PCSS cover such large areas, the ITU has divided the globe into three regions: Europe and Russia, Africa and the Middle East, Asia and most of the Pacific. Even though the United States and Canada, as well as Latin America, are members of the ITU, it allows the Western Hemisphere to follow the lead of the United States. The Western Hemisphere usually will stick with U.S. standards because of economic ties to the United States. However, the rest of the world relies on the ITU to coordinate PCSS and MSS standards because political diversity and sovereign interests interfere with self-imposed coordination.

A committee called WARC (World Administrative Radio Conference), now called WRC (World Radio Conference), holds regional and national conferences to coordinate and regulate frequency allocation for different types of communications. The ITR-R (ITU radio communications sector) has used permanent and ongoing programs to coordinate the use of frequencies. These programs include the Radio Regulations Board, formerly the IFRB (International Frequency Registration Board), which serves as a neutral registrar of frequencies and orbital locations by governments. Commercial communications satellites were allocated the frequencies shown in Table 3.6 for mobile satellite technology.

**3.4.2.1 WARC 1992.** The WARC 1992 world conference defined frequencies and standards for new MSS and PCSS services. They defined 1 GHz for LEOs and <1 GHz for little LEOs, and that MSS and PCSS could not restrict land-service development. They also regulated that MSS and PCSS could not bypass the local operator or the PSTN, and that local operators and the PSTN could interfere with fixed and mobile services. The defined intent of satellite telephony is to respect sovereignty and share revenues with developing countries.

**TABLE 3.6  MSS and PCSS Frequency Allocation**

| Band | Frequency range (+ or −) |
| --- | --- |
| VHF | 100 to 300 MHz |
| UHF | 300 to 1500 MHz |
| L-band | 1.5 GHz |
| S-band | 2.0 to 2.5 GHz |
| C-band | 4 and 6 GHz |
| Ku-band | 12 and 14 GHz |
| Ka-band | 20 and 30 GHz |

NOTE: Uplinks are typically higher frequencies than their associated downlinks because of the limitations of a satellite to generate electricity (this is opposite of terrestrial cellular where the base station is not limited by power). Frequencies are paired; for example, if a satellite is allocated C-band, the uplink will be 6 GHz, and the downlink will be 4 GHz.

## 3.5  Technology

It's kind of interesting that most of the PCSS players come from the defense industry. Since the end of the cold war, defense contractors with dried-up defense dollars are seeking commercial ventures so that they may stay in business. They are capitalizing on their knowledge of pioneering multistage rockets, satellites, satellite RF, launching, and defense communications. Although they are experts in all these areas, they were unaware that switching, networking, and terrestrial wireless technologies are required for a PCSS to work. Those who became aware of terrestrial integration suddenly were overwhelmed with PCSS' reliance, complexity, and cost of terrestrial telephony. PCSS players had to rethink what they were doing, redesign their concepts, and look for recapitalization.

PCSS combines and will employ several distinct technologies to make a working system: terrestrial telephony switching and networking, terrestrial wireless, wireless handset, multistage rocket and launch, and several classes of satellites. This section is an overview of integrated technology that constitutes the four PCSS types by class of satellite—GEO, LEO, MEO, or HEO.

### 3.5.1  Satellite technology

Satellites have evolved from point-to-point, high-capacity trunk communications between large, costly, ground stations to multipoint-to-multipoint communications among much smaller, lower-cost ground stations. There are many newly developed multiple access methods whereby each ground station can receive communications directed to them by assigning TDMA time slots to a particular ground station.

TDMA allows the satellite to have more efficient use of the onboard power supply; similarly to terrestrial cellular service, frequency reuse schemes allow satellites to communicate with a number of ground stations. In cellular, frequency reuse is achieved by arranging the cellular system so that the same frequency is used in a different cell but separated by a certain distance and number of cells defined by the selected frequency reuse plan. This reuse plan is implemented to minimize cochannel interference.

With satellite technology, frequency can be directed by transmitting narrow beams toward a particular ground station. The distances would be so great between ground stations that there won't be any overlap of radiation, which would cause cochannel interference. Therefore, there is no need to implement a frequency reuse plan. Beam widths can be adjusted to cover areas as large as a continent or as small as 20,000 square miles, the size of a small state. Although a narrow beam's coverage is relatively small compared to the satellite's much larger footprint, it is still many times larger than any terrestrial coverage from any base station. The new class of PCSS satellites is being designed to accommodate over 40 spot beams within the same satellite.

### 3.5.2  The satellite environment

Once a satellite is launched and achieves its selected orbit there are many hazards that face it. All these factors are considered in the class of satellite that will provide service. Orbit selection is crucial to both the quality of communications, the satellite's survivability, and the economy of providing service. The farther out the satellite the more delay there is in the signal and the higher the cost to launch it; however, the satellite's life is about 15 years longer. The closer the satellite, the better the quality and the less delay and cost to deploy, but the satellite's life is about 3–5 years shorter. This means more satellites will need to be launched due to short life. The orbit selection criteria is discussed in more detail later in this chapter. There are other environmental factors that also need to be considered, for example, gravitational pull and radiation and meteor belts.

The gravitational pull of the earth and the earth's centripetal force cause the satellite to decay and burn up in the earth's atmosphere. If a satellite were too far out, it would spin out into space because of the earth's centrifugal force. There is a narrow zone 22,300 miles above the earth's surface where the gravitational pull and centripetal force of the earth are neutralized by the earth's centrifugal force. This zone is known as the *geosynchronous orbit*. At this altitude, a satellite positioned over an apparent point on earth would complete one orbit, that is, at the same time its corresponding point on earth would complete one orbit every 24 hours. In the time it takes the earth to rotate once, a satellite moving in the same direction will remain in an apparent fixed position on earth. This allows uninterrupted contact between the ground station and the satellite. AMSC positioned a satellite over Lubbock, Texas, in April 1995 to provide mobile satellite telephony. Each subscriber must aim his or her antenna in the direction of Lubbock in order to have a line of sight to the satellite.

The earth has zones in space called the Van Allen radiation belts that consist of electrons and protons trapped in a doughnut-shaped region around the magnetic equator. The belts trap cosmic rays from the sun, making them radioactive. A satellite could not send or receive any communications because there is so much interference from the radiation. Therefore, a satellite would have to be positioned on either side of or between the radiation belts to work. However, locating between the belts is not a viable solution because they wander and change due to both cosmic events and fluctuations in the earth's magnetic field.

There are other hazards in space such as asteroids and meteors. Within the earth's gravitational influence there are millions of meteors. Meteors are fragments from asteroid collisions that are trapped between the geosynchronous orbit zone and the earth. Eventually, they will be pulled to earth, but they are constantly replenished by the large asteroid belt between Mars and Jupiter. Here, there are over 1600 planetoids and asteroids that collide, spewing fragments that eventually work their way into the influence of the earth's gravitational pull. There is a concentration of meteors orbiting the earth between the geosynchronous orbiting plane and the Van Allen radiation belt. This is referred to as the earth's asteroid or meteor belt. A satellite would have to be positioned outside this belt in order to survive. The meteors would pelt the satellite destroying solar batteries, components, and the protective shell itself at speeds three times faster than a rifle bullet. Although this is the most hazardous zone, a meteor could destroy a satellite anywhere, but more likely the satellite will be pelted by small pebble-sized debris until the satellite is rendered useless. The planned life span for a satellite in the safest orbit is 15 years.

### 3.5.3    PCSS system segments

There are five main segments to a PCSS system as shown in Fig. 3.11: the satellite, its launch, the antenna and tracking system, the ground segment, and the mobile station. The launch, antenna and tracking system, the earth station, and the satellite are separate technologies, and are briefly discussed in this section. Satellite technology will be discussed only as it pertains to the feasibility and the type of PCSS deployment.

The launch is an important consideration of PCSS because of the cost and risk to put a satellite into orbit. The farther out the orbit, the higher the cost to deploy. Different classes of staged rockets are needed to launch and position a satellite in different orbits. Even the use of a space shuttle is expensive to position a rocket. Risk is always high. There are no guarantees that failure or disaster won't occur. For example, when the Challenger space shuttle exploded, the craft, people, and the payload were lost; and in the summer of 1994, when NASA launched two communications satellites for AT&T, one remained in orbit and the other was lost.

The antenna and tracking system track the satellite and siphon off the communications routed to that position. The new class of tracking systems will not only track moving satellites, but will monitor satellite-to-satellite handoffs.

The ground segment consists of the earth station and the gateway. A new class of earth stations will coordinate both satellite handoffs and

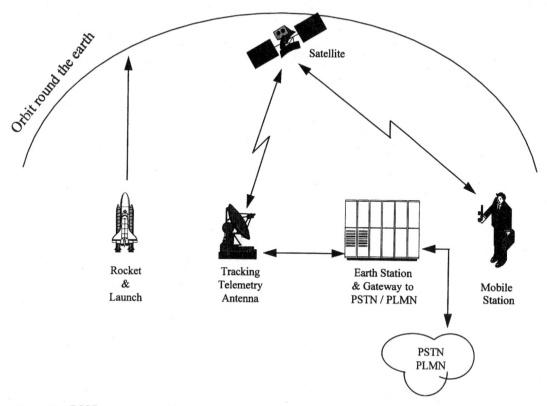

**Figure 3.11** PCSS system segments.

beam-to-beam handoffs on the same satellite. The earth station will have a mobile switch that will interface the PSTN and PLMN. The gateway is a remote switch from the earth station that will route calls to various countries, thus the gateway to the PSTN of a region or sovereign state. The gateway will also interface the PLMN to access HLR and store VLR information of a PCSS roamer.

The mobile station is an important segment of PCSS. Current MSS terminals are large and bulky, and the cost ranges to well over $20,000 per unit. The mobile unit is being reduced in size, will have more features, and will be lower priced. The mobile station will be discussed later in this chapter.

### 3.5.4 GEOs—"Old Faithful"

Geosynchronous earth orbit satellites (GEOs) have been around, and have been very successful, for several decades. They provide voice, data, and video communications. The speed of a GEO is such that it remains in the same relative position to its fixed point on the surface of the earth (shown in Fig. 3.12) so that the broadcasting station (fixed or mobile) will never lose contact with its receiving satellite. GEOs appear to be fixed over a position on earth at the height of 22,300 miles. They appear fixed because the orbital period is equal to the orbital period of a point on earth.

A satellite receives a microwave signal from a ground station on the earth (the *uplink*), then amplifies and retransmits the signal back to a receiving station or stations on earth at a different frequency (the *downlink*). These systems provide great signal clarity and can handle large amounts of data. Transmission delays of 700 milliseconds have been aided by digital-echo canceller technology. This makes the delay less noticeable to the human ear; however, any greater delay, such as a second satellite hop, transcoding delays, or switching delays, cannot be further aided with today's technology.

Today, there are several companies with GEOs. INMARSAT is considered to be the most successful, and there are a number of other satellite carriers, including Celsat, Eutelsat, the Canadian company Telesat Mobile Incorporated (TMI), and Volunteers in Technical Assistance (VITA). Terminals are typically large and fixed on land. On ships they are large with high-gain directional, high-grain steering anten-

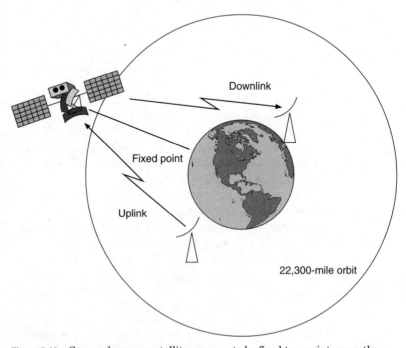

**Figure 3.12**  Geosynchronous satellite appears to be fixed to a point on earth.

nae. And portables can fit into a suitcase with retractable high-gain directional, high-gain steering antennae. Today, there is very little application for handheld terminals in maritime application; therefore, there has been little incentive to develop a handheld terminal. With the announcement of AMSC's launch in April 1995, many companies are beginning to design smaller, low-power, low-cost, handheld terminals. There are concerns, however, that quality, limited service, and coverage may be jeopardized without high-gain, directional antennae. More satellites would have to be launched to improve coverage and quality, but the cost to provide full service will considerably increase capital investment requirement, and this leads to concerns about their business case.

The advantages of a GEO are (1) it has proven and reliable service, (2) current space technology and infrastructure is in place to launch the satellites, and (3) the earth stations and network control systems are already developed. A single satellite is sufficient to provide semiglobal coverage 24 hours a day. However, there is limited visibility of the GEOs in the northern hemispheres with low angles to the horizon, so it would take three satellites to provide total global coverage. Again, this would drive up the cost of providing service.

Each satellite can accommodate 55,000 circuits, with 100 directional beams. The cost to launch into geosynchronous orbit is very high. A Titan-class multistage rocket is required to bring a GEO to its 22,300-mile orbit. Because of it distance from earth, a 700-ms delay is incurred, but can be made tolerable to the human ear with digital-echo technology. The geosynchronous orbit is relatively free from space debris and radiation; thus, a GEO's life span is 15 years. This is the longest life expectancy of the four classes of satellites, thus making the overall cost to provide service more economical.

Minimum intelligence is required in the satellite; however, moderate intelligence is needed at the earth station and in the mobile terminals. Direct connections are made from the earth stations to the PSTN or PLMN where all the switching and networking resides. Thus, GEO technology is for the most part a passive system.

### 3.5.5  The HEOs

Both high earth orbit (HEO) and the highly elliptical orbit (also called HEO) satellites fall into this category. However, there are many differences between the two, other than just orbital height. The elliptical HEO was developed extensively by the former Soviet Union. The characteristics of the elliptical HEO is that its orbit is highly elliptical, with the outer ellipse's, or apogee, orbital height at 24,180 miles above the earth. Instead of appearing stationary, as a GEO, the elliptical HEO's location appears to be rotating. Therefore, technology of satellite hand-off will need to be developed, because as the satellite passes over the far horizon, a call or transmission will be dropped unless handed off to another satellite as it is approaching over the new horizon as shown in Fig. 3.13.

Figure 3.13 is the plan for the United Kingdom's communication engineering research satellite (CERS) project. With the high perigee, satellite service is accessible to users in densely populated urban areas because the subscriber will be in the eye of the satellite straight above, unobstructed by high buildings that would block communications between the satellite and the subscriber. However, there are a host of problems associated with HEOs. The HEO must pass twice through the Van Allen radiation belts and the asteroid field around the earth. The Van Allen radiation would cause fade-outs and/or bad quality while the HEO was passing through. Long-term radiation would cause overheating and damage to the circuitry. In addition, while going through the asteroid belt, the particles would pelt the satellite, eventually rendering it useless. The satellite would be influenced by the earth's gravitational pull, and would eventually decay into the earth's atmosphere. Expected life is about five years. Therefore, replacement satellites using a Titan-class multistage rocket would have to be launched to maintain service. Also, with each launch, there is the risk of failure.

Another disadvantage is that the apogee is so far from the earth that the link budgets and delay problems with the satellite in relative

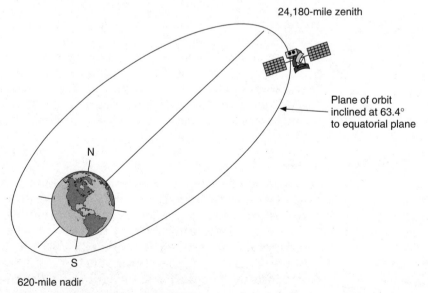

**Figure 3.13**   Highly elliptical orbit (HEO).

motion to the earth would cause Doppler shift problems as the satellite approaches and travels away from the earth. The window of coverage is about eight hours; therefore, it would take three satellites to provide 24-hour service.

The CERS program in the United Kingdom has attracted continued support and has subsequently attracted the technology satellite (T-SAT) Project 1 plus the European space agency's ARCHIMEDES Program 2.

The high-earth orbit HEO is a variation of an MEO, in that it has a spherical orbit about 10,000 miles above the earth, located between the Van Allen radiation belt and the asteroid belt. This HEO would have a large footprint on the earth's surface and similar characteristics and lifespan to a MEO. It would have slightly longer delays, cost more to launch, and would have to be equipped with additional plating for radiation and pelting from space debris.

### 3.5.6   MEOs—a new class of satellite

The higher the satellite, the larger its footprint for terrestrial coverage, but it has longer delay. The lower the satellite, the smaller the footprint, but the delay is shorter. The delay comes from the distance the signal has to travel to and from the earth at the speed of light. However, one can't deploy a satellite anywhere just because it is closer to the earth; there are the problems of gravitational pull, meteor belts, and radiation belts. The MEO has taken into account all these hostile conditions and found a friendly environment at about 6000 miles above the earth's surface. This altitude gives the MEO a large footprint location between the Van Allen radiation belt and the residual atmosphere that would burn up a satellite.

MEOs have a spherical orbit as shown in Fig. 3.14. It takes six hours to rotate around the earth. Each MEO would have 3000 circuits with up to 40 beams that could be focused on high-density areas. Delay is minimal at 80 milliseconds compared to GEOs. Six satellites would cover 95 percent of the world's population, and it would take 10 satellites for full global coverage. Eventually, the earth's gravitational pull would place the satellite within the earth's decay. Considering decay, space debris pelting, and radiation, the expected life of a MEO is 15 years.

The intelligence for MEO satellites is relatively low; however, the intelligence would reside in the earth station where it would control the beam focus, beam-to-beam handoff, satellite-to-satellite handoff, and connection to the PSTN and PLMN for switching and networking.

### 3.5.7   LEOs—the big gamble

LEOs are those satellites that orbit less than 1000 miles above the earth's surface, between the atmosphere and the Van Allen radiation

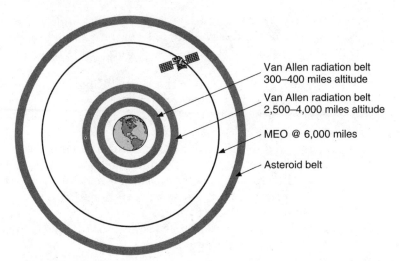

**Figure 3.14**   The relative position of a MEO between the Van Allen radiation belts and the asteroid belt.

belt. There are two types of LEOs: (1) big LEOs, which operate at >1 GHz for voice and data and (2) little LEOs, which operate at <1 GHz for data communications only. LEOs are more technically complex than GEOs, HEOs, and MEOs. Their orbit is so close to the atmosphere that at the speed they travel there is enough friction to overheat electrical components and burn up the satellite. Also, LEOs are so close to the earth that the gravitational pull will cause decay in just a few years. Expected life of a LEO is five years.

Even though the life expectancy of a LEO is short, there are great cost savings because launching is relatively simple and inexpensive when compared to that of a risky multistage rocket. A LEO can be launched from an L-1011 jumbo jet aircraft over the Pacific Ocean, thus lower cost and lower risk.

The new, big LEO, which everyone is talking about, has unproven technology and exists only on paper. Each satellite is extremely intelligent, i.e., contains a complete switching program, tracks the user, provides user features, provides networking, and performs satellite-to-ground, satellite-to-satellite, and ground-to-satellite handoff. Each big LEO will provide 7600 circuits, with a 1.5-h rotating orbit. Each LEO footprint on the earth's surface is large (about 1000 miles across traveling at over 1000 mph) compared to GEO (semiglobal), but still much smaller than conventional terrestrial cell sites; thus, it will require about 70 LEO satellites for full global coverage. A fully operational LEO sky network would have to be launched gradually, perhaps one per month, until the whole sky network is deployed. Then because of heat damage from friction caused by passing through residual air molecules at an extremely high rate of speed, and because of decay into the earth's atmosphere, new satellites would have to be launched one per month just to keep a constant number of satellites in orbit. Thus, the cost of maintenance is very high and there is a much higher cost per satellite because each one contains an intelligent switch that needs special heat protection. Table 3.7 shows the technological differences in the major classes of satellites.

Big LEO will only be referred to as a LEO since a big LEO is the only satellite of this class that will provide PCSS service.

The only similarity between a MEO and a LEO is that they are both rotating and are not synchronous to a fixed point on earth. However, there are great differences between LEO and MEO technology. A MEO completes an orbit every six hours and passes an apogee every 90 minutes. A LEO has a 90-minute orbit and passes an apogee every 10 minutes. This means that a LEO will perform a satellite-to-satellite handoff every 10 minutes, whereas with a MEO it is very likely that one would remain on the same satellite for an entire call duration since there is satellite coverage overlap. A MEO will have about 40 directional beams that can be dynamically focused on high-traffic areas with a very large footprint. A LEO can have about 15 beams, but, because the satellite is moving so fast, coverage will appear to have a push-broom pattern where every beam will sweep across its footprint performing a beam-to-beam handoff, then a satellite-to-satellite handoff.

The maintenance cost of a MEO is low because it has a safe orbit, providing a lifespan of 15 years. A LEO will encounter high heat and gravitational pull, so it has an expected life of five years. Whereas the cost of a MEO is lower because it is a passive device, each LEO will

**TABLE 3.7  Satellite Technology**

|                      | LEO            | MEO                   | GEO        |
| -------------------- | -------------- | --------------------- | ---------- |
| Orbit (miles)        | <1,000         | 10,000                | >22,000    |
| Type                 | Rotating       | Rotating              | Stationary |
| Footprint (sq. miles)| Large (±3,000) | Very large (±10,000)  | Semiglobal |
| Quantity*            | 70             | 10                    | 3          |
| Cycle (hours)        | 1.5            | 6                     | 0          |
| Intelligence†        | Very smart     | Smart                 | Some       |
| Circuits             | 7,600          | 3,000                 | 700        |
| Beams                | TBD            | 40                    | 100        |
| Delay (ms)           | 10             | 80                    | 700        |
| Life cycle (years)   | 5              | 15                    | 15         |
| Cost                 | $1.8 b         | $.9 b                 | $1.0 b     |

\* Quantity of satellites is for total global coverage.
† Smart with embedded intelligence as well as in board processor.

have embedded switching and cost much more. Also, more satellites will be required for coverage, but this will be offset by the relatively low cost of a launch. Even though there are three classes of satellites, there are just two main categories: LEOs, which include all rotating classes including all LEOs, MEOs, and HEOs; and GEOs, which are stationary and considered a separate category. For comparison purposes, this section will show the differences between the MEO and LEO technology, and then discuss the differences between the LEO/MEO category and the GEO category. Table 3.8 shows the comparison between a MEO and a LEO.

In order to determine GEO technology or LEO/MEO technology for PCSS, there are several criteria to weigh. Whereas a LEO/MEO will provide better voice quality without delay, a GEO cost is much lower to deploy and maintain. Table 3.9 shows the price/quality trade-offs between GEO and LEO/MEO technology.

### 3.5.8  Operating environment

The satellite communications operating environment for PCSS for the terminal or mobile station (MS) uplink is L-band (28 MHz, each link at

**TABLE 3.8    MEO versus LEO Technology**

|                   | MEO                   | LEO                |
|-------------------|-----------------------|--------------------|
| Orbit             | 6 hrs                 | 90 min             |
| Pass              | 90 min                | 10 min             |
| Beams             | 37                    | ±15                |
| Beam spread       | 1,000 miles           | <100 miles         |
| Pattern           | Fixed-point steering  | Push broom         |
| Satellite handoff | None                  | Required           |
| Cost              | Low                   | High               |
| Altitude          | 6,500 miles           | 1,000 miles        |
| Intelligence      | Passive               | Embedded switching |
| Life              | >15 years             | 3–5 years          |

**TABLE 3.9    Orbit Selection Criteria**

| GEO                                        | LEO/MEO                                       |
|--------------------------------------------|-----------------------------------------------|
| Limited voice quality                      | Better for voice service                      |
| Service prices as low as $.25/minute       | Service priced up to $3.00/minute             |
| Continuous operation with 1 satellite      | Continuous operation with 6–77 satellites     |
| Cost about $220,000 each                   | Cost $450,000 to $2.1 million each            |
| Highest-power (1 watt) mobile station      | Low-power (0.05 Watt) mobile station          |
| Earliest possible deployment               | Unproven technology, not authorized by FCC    |

the 1.61–1.62 GHz frequency range). The MS downlink is S-band (frequency range 2.48 to 2.50 GHz). The ground station up-downlink is in the Ku-band (100 MHz, each link at the 13-GHz frequency). Data services from little LEO will operate below 1 GHz. Figure 3.15 shows a typical PCSS configuration. The frequency bands of each operator are coordinated by the ITU and regulatory bodies of each country; therefore, the frequencies will vary within these frequency bands.

The spectrum utilization in North America is shown in Table 3.10 and the frequency allocated is shown in Table 3.11.

### 3.5.9  Mobile stations

There is an assortment of mobile stations. An MS is the transceiver terminal used by each MSS subscriber. Early MSs were large, and were permanently mounted on ships or earth stations. As the market demand grew, the MS equipment became smaller, as did the antennae. The antennae were fixed with an eye toward the GEO. Mobility required the antenna to move so that its eye could focus toward the direction of a GEO. Today, the MS can be carried in a suitcase or mounted in the trunk of a car. It has either a 2-inch diameter, moveable antenna aimed at a GEO or a new-generation electronic, omnidirectional antenna that automatically seeks the direction of a GEO.

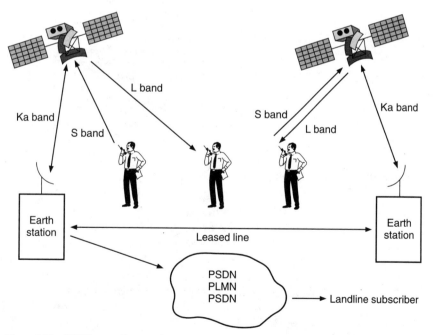

**Figure 3.15**  PCSS operating environment.

TABLE 3.10  Spectrum Utilization in North America

| Terrestrial bands | Music | VLF | 1 kHz |
|---|---|---|---|
| | | LF | 100 kHz |
| | AM/FM | MF | 100 kHz |
| | | HF | 1 M |
| | Television | VHF | 10 MHz |
| | | UHF | 100 MHz |
| Satellite bands | | L | 1 GHz |
| | | S | — |
| | | C | 6/4 GHz |
| | | X | — |
| | Microwave | Ku | 14/12 GHz |
| | | Ka | 30/20 GHz |
| | | — | 100 GHz |

TABLE 3.11  Frequency Usage

| | |
|---|---|
| 88–108 MHz | Standard FM broadcast |
| 535–1605 kHz | Standard AM broadcast |
| 470–512 MHz | Land mobile service, TV channels 14–20 |
| 608–614 MHz | Shared government/nongovernment, radio astronomy, no broadcasting allowed |
| 826–845 MHz | Cellular telephone |
| 870–890 MHz | Cellular telephone |
| 902–928 MHz | Government radio location, ISM and amateur radio on a secondary basis |
| 960–1215 | Allocated worldwide to aeronautical navigation |
| 1.2 GHz | L-band satellite |
| 1850–1990 MHz | Nongovernment fixed service— >8100 assignments in power, petroleum, railroad, and local government |
| 1850–1900 MHz | Bands allocated to PCS |
| 6/4 GHz | C-band satellite |
| 14/12 GHz | Ku-band satellite |
| 30/20 GHz | Ka-band satellite |

### 3.5.10  Terrestrial technologies

There is little doubt that all PCSS protocols will be digital. The question, then, is which digital technology will be used? The terrestrial access protocols, or air protocols, will employ frequency-division modulation access (FDMA), time-division modulation access (TDMA), code-division modulation access (CDMA), variations of the protocols, and/or hybrids. PCSS service will probably be required to be provided over a dual-mode MS. However, dual-mode is not a technical requirement; PCSS can be maintained as a stand-alone system fully independent from the PSTN or PLMN.

TDMA access protocol can be GSM, DECT, or other variations, for example, E-TDMA, a variation of TDMA. CDMA can have just as many variations, such as wideband, broadband, narrowband, enhanced-CDMA (the ITU's version of CDMA for CCITT- and CEPT-based countries as defined by the RACE Committee), and of course, proprietary variations.

PCSS switching and service platforms are most likely to be either ANSI-based or GSM-based, since these will be the most dominant global standards for switching platforms.

A unique network possibility is that PCSS can not only connect to either the ANSI or the CCITT network, but it can entirely bypass the PLMN and PSTN with all networking, switching, and routing from origination to destination entirely in the sky. The bypass issue could be a major roadblock for global implementation, as discussed earlier in the legal and regulatory issues section. The Iridium architecture is an example of a PCSS system that could be an entire network of switches without a call ever having to be on the PSTN or PLMN. When an Iridium MS communicates to another Iridium MS, the call switches and passes from one satellite to the next until the call reaches its destination. In this scenario, it would benefit the Iridium operator (Motorola) because (1) they would not have to provide revenue sharing (separations) with the PTTs, (2) network and protocol conversions would not have to occur, saving that cost, and (3) there would be cost savings associated with multiple network conversions terrestrial equipment, i.e., ANSI, CCITT, TUP+, ISUP+, R2+ (+ = country variant), as well as different HLRs and VLRs.

## 3.6  Summary

The race is on. On April 4, 1995, ORBCOMM (Orbital Sciences Corporation) launched the first two of 36 LEOs that orbit at 455 miles above the earth for global coverage. Three days later on April 7, 1995, AMSC launched a $120 million satellite to provide mobile telephone, voice, fax and data service for air, maritime, and fixed land-based customers throughout the United States, its territories, and 200 miles of coastal waterways.

So, will PCSS fly? Even though there is skepticism in the high-tech investment market, especially about MSS and PCSS; even though there are technical obstacles; even though there are regulatory and legal issues; even though there are major concerns on frequency and spectrum; and even though there is a great deal of competition, there will be PCSS. In fact, by the year 2000, it is estimated that 14 percent of all mobile-data users will use PCSS. The U.S. Department of Commerce, in "U.S. Industrial Outlook—1994," projects that mobile satellite revenue will reach $473 million by 1996.

Indeed, PCSS is in its infancy and there is a niche market for seamless, ubiquitous, universal service; a satellite can provide coverage for fixed applications and remote usage more easily, over wider areas, and at a lower cost than land-based wireless telecommunications. PCSS can complement existing land-based telecommunications services.

The night sky, as we know it now, will be totally different from the night sky we'll know in the year 2000. Where stars seem to be static today, in the future sky, dozens of apparent stars will rapidly traverse the sky followed by another, and another, and another—all in 10-minute intervals. Each of these new stars will be a mesh of technologies handling thousands of communications spanning the entire globe.

# Paging Systems

## 4.1 Paging System Overview

Paging has been looked at from two points of view: (1) traditional telephony and (2) the emerging wireless telecommunications market. However, paging is indeed an integral part of both wireless and personal communication services. While all the hoopla was being focused on cellular, SMR, and PCS, the paging revolution was continuing on a path of its own. Paging has come of age and is one of the fastest-growing market segments within wireless telecommunications.

Traditional paging is a one-way, wireless communications device that allows the user to have continuous accessibility away from the wired communications network. Essentially, a person carries a palm-sized device that alerts the user to call for a message. There are several methods of message notification that will be described in this chapter. Today, a person carrying a paging receiver can be alerted anytime and anywhere within the coverage area. Enhanced features of the paging device, such as different tones, pulsating lights, and vibrations, notify the user that there is a message. Also, alphanumeric pagers transmit a number to call, a short message, important information, or data to the paging receiver's display.

Paging systems make efficient use of radio spectrum, which keeps the cost of a paging system and the cost per subscriber low. In recent years, the paging industry has exploded, due to low cost, enhanced features, and better coverage. Two-way paging will be implemented with narrowband PCS technology, which will allow advanced applications such as e-mail, digital voice mail, fax, connection to on-line services, and data transmission. The paging revolution is not limited to North America. It is consumer pent-up demand in the Asia and Pacific region and is expanding at incredible rates. By the end of 1993, it was reported that North America and Asia and Pacific accounted for over 90 percent of all pagers worldwide, and by the end of 1999, Asia and Pacific, alone, will account for 55 percent of all pagers worldwide.

This chapter will describe the history of paging and the current market, primarily in the United States since it the largest market in the world. Industry needs will be assessed along with the advantages and obstacles of paging. Paging, as a form of PCS, will be compared to both cellular and PCS wireless communications. The legal and regulatory environment will be discussed. Economic considerations will be reviewed along with both the equipment manufacturers and paging operators. And finally, the service offerings and technology will be discussed.

## 4.2   History

Throughout the history of humankind there has been the need for paging. The ancients sent runners to summon someone for a meeting or to deliver a short message, such as "the battle was won." In the Middle Ages, paging became so important that monarchs had several pages as part of their court whose sole function was to summon someone or to deliver a short message. Only the privileged, the aristocrats, and the military had the need for paging. As commerce flourished, the need for paging became a necessity for merchants who wanted to check on the status of a caravan of goods, to arrange meetings with their suppliers, and to check up on prices.

Still, throughout the ages, paging was for the privileged. It became so important that innovators tried to apply technology to paging communications. Courier pigeons with bands on their legs accelerated paging and short-message service. The time it took to receive a message decreased from months to days. Telegraph and ticker tape allowed more people faster information. Still, until the end of World War II, it was common for a page or bellhop to walk through a hotel yelling, "Paging Mr. Jones." In the 1950s and 1960s, department stores used a series of gongs to summon a manager or to notify security people. Intercom and speaker systems were also introduced about this same time, but using them for paging and messages was disruptive.

Wireless paging was pioneered by the Detroit Police Department in 1921 with the concept of one-way information broadcasting. The use of radio paging with powerful transmitters became widespread in the 1930s by police and fire departments, government agencies, and the armed forces. Voice broadcast paging evolved into digital addressing, where messages were addressed to specific pagers. The first pagers in the early 1970s had a simple beep that alerted the user of the paging device to call in for a message. As tone, numeric, and alphanumeric pagers were evolving, standardization became necessary, and in 1976 the POCSAG (Post Office Standardization Advisory Group) standard radio paging code, developed by engineers from several countries, was accepted internationally. There are some leftovers of GSC (Golay Standard Code), 5/6-tone codes, CEPT, and RDS, which were implemented before POCSAG, still being used in some markets. Display pagers were introduced in the 1980s.

The need for paging and message service became integral with the popularity and deployment of the telephone and pay phone. The whole industry of message service still thrives today. Professionals and executives subscribed to message services, but always had to check in to retrieve their messages. Those who wore pagers were typically those who needed to provide a quick response or immediate care, such as doctors, nurses, plumbers, electricians, and computer repair technicians. Missed messages with this system created turmoil. Medical care was impacted if physicians could not respond to their messages fast enough. Business deals were ruined. Quality of service depended on promptness, but what is considered a slow response today was the accepted way of life.

Physicians, professionals, and executives had personal lives. They weren't always available to call in to their message services to retrieve messages. It was not uncommon to hear an emergency call for a doctor at a basketball game, theater, restaurant, or other event. In the 1950s the first wireless pagers were introduced. They were bulky, had low battery life, and were very expensive. Most were limited to a specific area such as a hospital building. The paged party heard a beep and knew that he or she needed to find a telephone to call the message service. This procedure to retrieve a message was time-consuming and cumbersome.

As technology continued to increase, paging devices became smarter, smaller, and had longer battery life and range. Tones were replaced by one-way, short voice messages, by combination tone/voice messaging, by a display showing a number to call, or by alphanumeric display where a short message can be shown. Paging has evolved from small, local areas to whole metropolitan areas, whole regional areas, national coverage, and now international coverage. So now what? The world is about to embark on the new generation of paging.

Paging has been growing throughout the world at an average of 20 percent per year, and people are wearing pagers in airports, businesses, shopping malls, banks, at sporting events, and places of entertainment. Even the younger generation are wearing pagers as both a status symbol and as a means of keeping in touch with their friends. Social and business trends are making the population even more mobile. Telecommuting and the desire to just be accessible are creating tremendous demand for paging services. A tone or even a display with a number to call is no longer adequate. Consumers want more information, more data, and the ability to respond; the calling party wants to know that the paged party actually received the page (i.e., paging confirmation). More information and more users mean that more airtime is required. In order to keep up with the increasing demand, the paging operators are increasing the data rate with baud rates exceeding 6400 kbps, and in 1993 the FCC granted a Pioneers Preference license to M-Tel for enhanced paging. Prior to this, the paging industry was left relatively unregulated.

M-Tel's new enhanced-paging system was characterized by higher speeds, frequency of reuse as determined by the subscriber's location, and acknowledgment or confirmation that the page was received (referred to as two-way paging). The success of M-Tel led to the narrowband auctions in 1994 where would-be operators bid and paid handsomely for the right to provide enhanced paging. Several large companies spent millions of dollars to acquire these licenses. Since then, the narrowband operators have been testing their two-way paging service offerings. Most notable is PageNet with VoiceNow two-way paging service announced in September of 1995. VoiceNow allows the paging subscriber to hear and repeat the message from the caller over a pager that resembles a miniature, wireless, portable answering machine. The subscriber can select from up to 16 customized, predetermined responses that can be directed to a person or a group of people. The new-generation paging will go beyond simple two-way communications to wireless electronic mail, wireless electronic document, and wireless data transfer anywhere, anytime. Advanced computerized hardware, software, and switching systems used in radio paging have evolved from simple operator-assisted systems to integrated paging systems that are fully automated with store-and-forward voice and data messaging, voice mailbox, and other enhanced features.

Paging, in itself, does not compete with PCS. Paging is the text and data form of personal communications service expanding the need for simple communications, which has been needed throughout history. The cost of paging will be much less than cellular and PCS because these forms of wireless personal communications integrate text and data communications with their main function of very intensive voice communications. Today, paging is a supplement to the landline and cellular telephones. Paging will continue to supplement all forms of telephone service where only simpler notification is needed. Paging will finally bring inexpensive PCS to the common person rather than the privileged few. It will also make all wireless networks more efficient by offloading the wireless airways with short messages, by qualifying the calling party before returning the call, and by electronic mail and document transfer over noncellular and PCS channels.

### 4.2.1  Types of paging systems

There are two types of paging systems, *radio common carrier* (RCC) and *private* systems. An RCC system is also called a *subscriber* system, which is a licensed, public paging company providing paging services to the public. RCCs provide service for local, statewide, nationwide, and international paging transmission. The major RCCs are PageNet, Sky-Tel, MobileComm, AirTouch, MetroMedia and US West. The RCCs have a coverage area and network infrastructure. Typically, a subscriber pays for a one-time activation fee, a monthly fee, and an optional rental fee for a pager. Service charges vary on coverage requirements and value-added services such as alphanumeric messaging, voice mail, and text messaging.

A private paging system involves customer-owned equipment for private-use paging. Many firms implement their own private systems because the cost of an RCC is too high or coverage is not adequate. A private paging system allows an organization to control paging communication and ensure reliable communications for mobile personnel within a building, campus, or geographic area. Large systems must be licensed by the FCC according to Part 90—Private Land Mobile Radio Services, Section 8. The private system must also coordinate frequency and must be filed with all appropriate agencies. Smaller, low-powered ($\leq 2$ watts) systems used for in-building or on-site applications need not file with the FCC if the paging equipment supplier has a National Shared License Agreement from the FCC.

### 4.2.2  Paging systems' evolution

Paging systems have evolved from the nonselective operator-assisted voice paging system where centralized control operators received and taped incoming voice messages. These taped messages were intermittently broadcast to all subscribers as shown in Fig. 4.1. Subscribers had to listen to all broadcast voice messages at predetermined times. Although revolutionary, this was inconvenient and messages had no privacy. Addressing led to selective operator-assisted paging where the message was directed to a specific subscriber as shown in Fig. 4.2. With an alert tone, the subscriber was notified to call in to the operator for an urgent message.

With advancements in paging technology and the use of computers, automatic paging was developed. Each pager was given its own telephone number and the calling party could call a specific person. The pager would automatically page the called party with the address code

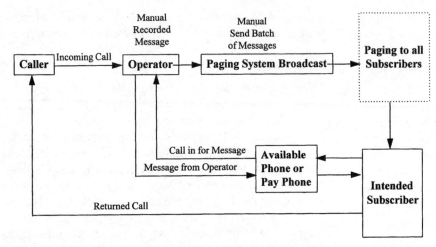

**Figure 4.1**   Operator-assisted voice paging call flow.

and transmit a voice message. Automatic paging eliminated the need for operator assistance but consumed a lot of airtime. As the paging market grew, frequencies became congested, causing lengthy delays of messages.

Tone pagers of the mid-1970s alerted the subscriber with an alert tone to call a predetermined telephone number for a message, and by the mid-1980s numeric pagers displayed the calling number on the pager. Numeric pagers alert the user with a tone or vibration, signaling receipt of a telephone number or message. Numeric paging is fast and convenient, eliminates frequency congestion of the airways, leads to much lower subscriber costs, and is the most popular form of paging today.

Alphanumeric pagers alert the subscriber with a tone or vibration signaling receipt of an alphanumeric text message or news item with more detailed information. Ideographic pagers use firmware within

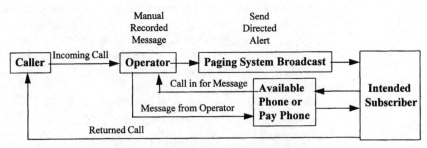

**Figure 4.2**   Selective operator-assisted voice paging call flow.

the pager to convert a text message to ideographic languages, such as Chinese or Japanese. The new generation of paging will be narrowband PCS, which will enable two-way messaging and cellularlike frequency reuse at speeds of up to 112,000 bps per channel.

The paging systems that transmit information and data offer the greatest potential for market expansion in the paging industry. Equipment manufacturers are developing new transmission techniques with greater speeds and information rates and increased coverage and reliability in order to keep up with increased competition and the demand for quality. The trend is to alphanumeric paging for more data and message services. The trend will be toward two-way paging and combining paging with other applications and technologies such as cellular and PCN. Competition will drive down the price, and as the price goes down more users will subscribe to advanced paging services. Operators will continue to try to attract new subscribers by adding more features and reducing prices. This trend of lower prices, more features, and more subscribers will continue to spiral well into the next century.

## 4.3  The Paging Market

Paging service is an easy-to-use, affordable messaging solution. A pager is easy to use and maintain. Prices range from $6 to $100 depending on services and coverage selected by the subscriber. Because of ease of use, low cost, mobility, and the need to keep in touch, paging is now used throughout the business and residential community. It is reported that about 40 percent of all pagers are used by service industries in which quick response or immediate care is important. Hospitals and universities account for about 25 percent of all pagers in operation; sales, health, and construction account for about 15 percent; and about 60 percent are credited to real estate, transportation, and personal use. Of all the new pagers going into service about 50 percent are purchased for personal use (e.g., a mother contacting her children or students coordinating their social activities).

The paging operators are taking a wait-and-see attitude as to which way the market will swing. Industry experts maintain that the current low price in the one-way paging market will continue for many more years, but many subscribers will change over to two-way pagers because of enhanced features and immediate response. The most profound growth for two-way paging will be for machine-to-machine communications, which currently account for less than 1 percent of messaging communications.

Market reports and industry analysts indicate that the paging industry grew 38 percent in 1994, adding 7.5 million users for a total of 27.3 million subscribers in the United States. Driving the growth was the lower cost of paging service and the decrease in rental cost from an average of $14.20 to $13.10 per month. Digital display pagers accounted for 87 percent, with alphanumeric remaining stable at 7 percent, but industry analysts project that alphanumeric will report an increase of 39 percent by 1996. Tone accounted for 3 percent, and tone/voice for 2 percent. Analysts also project that the paging market will continue having market growth due to narrowband PCS, advanced services such as two-way messaging, message acknowledgment, and data applications.

There were about 75 million paging subscribers in the world in 1995, and that number is expected to rise to over 130 million in 1999 as shown in Table 4.1.

The distribution of paging growth is shifting from the United States which had 44 percent of the world market in 1993 to 31 percent in 1999 as shown in Table 4.2.

At the end of 1993, 81 percent of all pagers were numeric, with tone accounting for 13 percent, alphanumeric accounting for 5 percent, and voice accounting for 1 percent. Sixty-five percent of the pagers operated in the VHF (150 MHz) frequency range, 26 percent in the 900-MHz frequency range, and 9 percent in the UHF frequency range.

There are about 90 million paging subscribers in the world, and the market is growing rapidly. The Asia and Pacific region has huge populations, and the economy is robust and expanding. The majority of

**TABLE 4.1    Number of Subscribers by Region**

| Region | 1995 | 1996 | 1997 | 1998 | 1999 |
|--------|------|------|------|------|------|
| North America | 29,775,000 | 33,325,000 | 36,700,000 | 39,700,000 | 42,350,000 |
| Latin America | 1,250,000 | 1,650,000 | 2,100,000 | 2,650,000 | 3,250,000 |
| Western Europe | 4,425,000 | 5,400,000 | 6,600,000 | 8,000,000 | 9,700,000 |
| Asia and Pacific | 37,500,000 | 46,000,000 | 54,300,000 | 62,300,000 | 70,750,000 |
| Rest of world | 1,300,000 | 1,950,000 | 2,760,000 | 3,600,000 | 4,800,000 |
| Total world | 74,250,000 | 88,325,000 | 102,460,000 | 116,250,000 | 130,850,000 |

**TABLE 4.2    Distribution of Paging by Country**

| Country | 1993 | 1999 |
|---------|------|------|
| United States | 44% | 31% |
| Japan | 16% | 26% |
| China | 13% | 10% |
| South Korea | 6% | 7% |
| Taiwan | 3% | 3% |
| Rest of world | 18% | 23% |

growth will occur in the Asia and Pacific region meeting pent-up demand for messaging service.

### 4.3.1  Market trends

Pagers will be designed to be more visually appealing and user-friendly to the nonbusiness personal-use consumer, such as families and students. One-way pagers will be designed to be integrated into watches, pens, credit cards, calculators, and jewelry as a fashion statement. Pagers will be smaller, lighter weight, and have much longer battery life.

Pagers will be integrated with other technologies such as cellular, PCS, and CT2 wireless telephones. The integrated paging capability will allow a cellular or PCS subscriber to screen calls via the digital display on the cellular phone. This will reduce airtime charges by returning or answering only the calls that require immediate attention. In CT2 phones, the user can be alerted by page to make a call, since CT2 phones can make only outgoing calls. Paging capability is being integrated into pen-based and handheld computers with the introduction of small radio modems.

Pagers will evolve into multiple-frequency devices that will have automatic scanning receivers. These sophisticated pagers will be used by a roaming subscriber who passes through multiple systems with different protocols and speeds of transmission.

The traditional distribution channel throughout the world has been the paging operator. In recent years, subscribers can purchase their own paging receivers from retail stores and outlets for as low as $69. By owning one's own pager the monthly rental and insurance cost is eliminated. The subscriber simply pays an activation fee and a monthly fee depending on features and coverage. This mass-marketing distribution has spread from North America to the rest of the world. This marketing approach promotes higher awareness of paging and increases the perceived need for a pager.

Many new features will be developed, such as mobile data and two-way paging, in order to expand the market. Other vertical features will include both high- and low-resolution graphics, video, e-mail, fax, digitized voice, message acknowledgment, and encryption. As more messaging increases airway congestion, faster speeds and better coverage will necessitate new infrastructure paging equipment technology.

CallMax, an Eindhoven, Holland–based company, has combined traditional paging with high-tech messaging to produce Maxing, an advanced wireless service that will be marketed commercially in 1996. With Maxing, subscribers will be able to store or forward faxes and access e-mail and the internet. The subscriber device is called the

Maxer, which will provide a tone, numeric, or alphanumeric display with a message box. CallMax will be able to bill the person calling the pager through an advanced network interface.

Many American companies have developed software needed to provide information services over paging systems, such as financial news, stock quotes, headlines, sports scores, news, weather advisories, and traffic. These companies connect to information sources such as Reuters, The National Weather Service, and sports networks. Their software incorporates information into a program that can be transmitted over a paging network. The paging operators bundle these services into their paging product offerings.

### 4.3.2 Traditional paging versus narrowband PCS

1995 was a year of change and rapid growth for the paging industry. Public paging operators crossed market borders, and there were a lot of mergers and new ventures. The growth rates were 29 percent at the end of 1993 and a 38 percent growth rate by the end of 1994, a 9 percent increase over the previous year. Double-digit growth rates are expected through the end of this decade.

The paging industry is shifting its traditional targeted market from the businessperson who has a need for urgent messages to the consumer market. Businesspeople have used paging as a form of personal communications for several decades. The focus has shifted to family members: to keep in touch while being mobile; to coordinate student and social activities; to receive news headlines, stock reports, and sports scores; and so forth. Low-priced pagers are available in retail stores.

The paging industry must be properly positioned with the infrastructure in place to offer advanced messaging services such as narrowband PCS. In order for the narrowband PCS to survive broadband PCS, the paging companies must be large enough to take advantage of economies of scale in order to compete with the broadband PCS carriers. Many companies are merging and consolidating as they position themselves for competing technologies and market share.

Mergers and acquisitions are being driven by the need for capital. The larger the subscriber base, the greater the incoming cash flow. The greater the capital, the greater the economies of scale, and the greater the value to shareholders. The greater the value, the more attractive the company is to speculators and investors. The strategy is to grow, buy, and expand.

Acquiring a company is desirable because the infrastructure is already in place for accounting, billing, engineering, and marketing.

This gives a company an embedded base to grow on. Growth occurs when the company expands to adjacent markets. Adjacent markets, about 30 miles away from the established market, are desirable because new infrastructure isn't required. Also, the existing sales force and engineering technicians can be used.

Merging is desirable for small providers because they do not have the capital to implement the new paging technologies. The small operators also realize that they cannot compete with the newer, advanced messaging that is inevitable. Therefore, merging is a win-win situation where all parties will be stronger and better positioned to compete with broadband PCS.

There are two positioning strategies for paging operators. One is to concentrate on traditional one-way paging and allow narrowband PCS to be the vehicle for new and enhanced messaging and two-way paging. The other positioning strategy is to be fully committed to narrowband PCS. Companies have paid over $1 billion to the government for nationwide licensing as shown in Tables 4.3 and 4.4. Nationwide narrowband PCS licenses awarded by the FCC are shown in Table 4.3.

The list of regional narrowband PCS licenses awarded by the FCC is shown in Table 4.4.

Two-way paging is expected to cost about 50 percent more than traditional one-way paging. Several operators are experimenting with Motorola's ReFLEX products from both Motorola and Glenayre. Glenayre is licensed by Motorola to manufacture paging systems employing the FLEX and ReFLEX technology. AT&T is developing P-act, AT&T's own open standard for paging. AT&T is competing with Motorola for the de facto standard.

**TABLE 4.3    Nationwide Narrowband PCS Licenses**

| License | Type | Winning bidder | Bid ($M) |
|---------|------|----------------|----------|
| 50/50 kHz | Paired | Paging Network of Virginia | 80 |
| 50/50 kHz | Paired | Paging Network of Virginia | 80 |
| 50/50 kHz | Paired | KDM Messaging Company | 80 |
| 50/50 kHz | Paired | KDM Messaging Company | 80 |
| 50/50 kHz | Paired | M-Tel | 80 |
| 50/12.5 kHz | Paired | AirTouch Paging | 47 |
| 50.12.5 kHz | Paired | BellSouth Wireless | 47.5 |
| 50/12.5 kHz | Paired | M-Tel | 47.5 |
| 50 kHz | Unpaired | Paging Network of Virginia | 38 |
| 50 kHz | Unpaired | PageMart II, Inc. | 38 |
| 50 kHz | Unpaired | M-Tel (Pioneer's Preference) | 33.3 |
| Total | | | 650.3 |

TABLE 4.4    Regional Narrowband PCS Licenses

| License | Type | Winning bidder | Bid ($M) |
|---|---|---|---|
| **Region 1** | | | |
| 50/50 kHz | Paired | PageMart II, Inc. | 17.5 |
| 50/50 kHz | Paired | PCS Development Corp. | 24.7 |
| 50/12.5 kHz | Paired | MobileMedia PCS, Inc. | 9.5 |
| 50/12.5 kHz | Paired | Advanced Wireless Messaging | 8.9 |
| 50/12.5 kHz | Paired | AirTouch Paging | 8.7 |
| 50/12.5 kHz | Paired | Lisa-Gaye Shearing | 17 |
| **Region 2** | | | |
| 50/50 kHz | Paired | PageMart II, Inc. | 18.4 |
| 50/50 kHz | Paired | PCS Development Corp. | 31.3 |
| 50/12.5 kHz | Paired | MobileMedia PCS, Inc. | 11.8 |
| 50/12.5 kHz | Paired | Advanced Wireless Messaging | 11.5 |
| 50/12.5 kHz | Paired | Insta-Check Systems, Inc. | 8 |
| 50/12.5 kHz | Paired | Lisa-Gaye Shearing | 18.8 |
| **Region 3** | | | |
| 50/50 kHz | Paired | PageMart II, Inc. | 16.8 |
| 50/50 kHz | Paired | PCS Development Corp. | 28.9 |
| 50/12.5 kHz | Paired | MobileMedia PCS, Inc. | 9.3 |
| 50/12.5 kHz | Paired | Advanced Wireless Messaging | 10 |
| 50/12.5 kHz | Paired | AirTouch Paging | 9.5 |
| 50/12.5 kHz | Paired | Lisa-Gaye Shearing | 18.8 |
| **Region 4** | | | |
| 50/50 kHz | Paired | PageMart II, Inc. | 16.8 |
| 50/50 kHz | Paired | PCS Development Corp. | 28.6 |
| 50/12.5 kHz | Paired | MobileMedia PCS, Inc. | 8.3 |
| 50/12.5 kHz | Paired | Advanced Wireless Messaging | 8.8 |
| 50/12.5 kHz | Paired | AirTouch Paging | 8.3 |
| 50/12.5 kHz | Paired | Benbow P.C.S. Ventures, Inc. | 17.5 |
| **Region 5** | | | |
| 50/50 kHz | Paired | PageMart II, Inc. | 22.5 |
| 50/50 kHz | Paired | PCS Development Corp. | 38 |
| 50/12.5 kHz | Paired | MobileMedia PCS, Inc. | 14.9 |
| 50/12.5 kHz | Paired | Advanced Wireless Messaging | 14.3 |
| 50/12.5 kHz | Paired | AirTouch Paging | 14.3 |
| 50/12.5 kHz | Paired | Benbow P.C.S. Ventures, Inc. | 18.2 |
| Total | | | 488.8 |

NOTE: *Paired* is two frequencies, one for the transmit frequency and the other for the receive frequency. Paired frequencies are on the opposite sides of the allotted bandwidth so as not to interfere with one another or cause cochannel interference.

### 4.3.3  Paging system operators

There are paging operators in virtually every country of the world. Table 4.5 is a list of the top-20 largest international public markets as reported by the U.S. Department of Commerce for 1995. It should be noted that even though the Asia and Pacific region started paging almost a quarter of a century after the United States, Asia and Pacific ranks near the top in number of subscribers. Singapore has almost double the penetration rate of the United States. The explosive paging growth in the Asia and Pacific region is due to pent-up demand for communications and a dynamic economy.

Some paging operators have announced their plans for implementation of narrowband PCS. Others are biding time to determine market potential and how to position themselves for the market. Their rollout strategies are kept under wraps and flexible to accommodate market fluctuations and swings.

## 4.4   Regulatory Issues

Global paging and narrowband PCS are coordinated by the WRC (World Radio Committee), part of the ITU (International Telecommunications Union). International countries regulate paging through their PTTs; however, the United States regulates through the FCC. Public and large, private paging systems must be licensed by the FCC, under Part 90—Private Land Mobile Radio Services, Section 8. The private system must also coordinate frequency and file with all appropri-

**TABLE 4.5   Top-20 International Paging Markets**

| Rank | Country | Subscribers | Penetration | Start | Pager types | Signaling |
|---|---|---|---|---|---|---|
| 1 | United States | 34.1 M | 12.9% | 1950s | Tone, tone and voice, digital, alpha | POCSAG, Golay, FLEX, ReFLEX |
| 2 | China | 11.2 M | 0.9% | 1984 | Tone | POCSAG, FLEX |
| 3 | Japan | 8.8 M | 7% | 1968 | Tone, alpha | POCSAG |
| 4 | South Korea | 3.5 M | 7.7% | 1984 | Tone, digital | POCSAG |
| 5 | Taiwan | 1.6 M | 7.2% | 1975 | Digital, ideograph | POCSAG |
| 6 | Thailand | 1.4 M | 2.1% | 1986 | N/A | POCSAG, FLEX |
| 7 | Hong Kong | 1.4 M | 22.6% | 1971 | Digital, alpha | POCSAG, Golay |
| 8 | United Kingdom | 1.3 M | 2.2% | 1972 | Alpha, tone, digital | POCSAG |
| 9 | Singapore | 815,400 | 28.5% | 1973 | Alpha, tone, digital | POCSAG, FLEX |
| 10 | Canada | 700,000 | 2.5% | 1960s | Alpha, tone, digital, tone and voice | POCSAG, Golay |
| 11 | Germany | 465,000 | 0.6% | 1974 | Alpha, tone, digital | CEPT, RDS, POCSAG, ERMES |
| 12 | Netherlands | 421,000 | 2.7% | 1963 | Alpha, tone, digital, tone and voice | POCSAG, ERMES |
| 13 | Australia | 340,000 | 1.9% | 1978 | Alpha, tone, digital | POCSAG, Golay |
| 14 | France | 307,070 | 0.5% | 1975 | Tone, alpha | CEPT, RDS, POCSAG, ERMES |
| 15 | Brazil | 243,300 | 0.2% | 1960s | N/A | POCSAG |
| 16 | Belgium | 225,500 | 2.2% | 1980 | Alpha, tone, digital | POCSAG, ERMES |
| 17 | Sweden | 222,000 | 2.5% | 1978 | Tone, alpha | POCSAG, ERMES, RDS |
| 18 | Italy | 217,000 | 0.2% | 1980 | Tone | POCSAG, Golay, ERMES |
| 19 | Colombia | 195,000 | 0.5% | 1972 | Alpha, tone, digital | POCSAG |
| 20 | Mexico | 172,000 | 0.2% | 1970s | Alpha, tone, digital | POCSAG, Golay |

ate agencies. Smaller, low-powered (≤2 watts) systems used for in-building or on-site applications need not file with the FCC if the paging equipment supplier has a National Shared License Agreement from the FCC. Narrowband PCS is regulated by Part 24, Subpart D.

Part 24, Subpart D specifies that any one ownership must have interest in no more than three of the 26 channels. There is ownership restriction in that, "Narrowband PCS licenses shall not have an ownership interest in more than three of the 26 channels listed in §99.129 in any geographic area. For the purpose of this restriction, a narrowband PCS license is any person or entity with an ownership interest of five or more percent in any entity holding a narrowband PCS license." The purpose of ownership restriction is to even the playing field, to foster fair competition, and to prevent domination by any one entity. The FCC believes that this kind of even competition will prevent price-fixing and keep the cost of service low for the consumer while stimulating aggressive network implementation and increasing the demand for new technology and features.

Service areas are defined the same as the PCS service areas: "Narrowband PCS areas are nationwide, regional, major trading areas (MTAs) and basic trading areas (BTAs)." MTAs and BTAs are based on the Rand McNally 1992 *Commercial Atlas & Marketing Guide.* The FCC's construction requirements for narrowband PCS are as follows:

> Nationwide narrowband PCS licensees shall construct base stations that provide coverage to a composite area of 750,000 square kilometers or serve 37.5 percent of the U.S. population within five years of the initial license grant date; and shall construct base stations that provide coverage to a composite area of 1,500,000 square kilometers or serve 75 percent of the U.S. population within ten years of original license grant date.
>
> Regional narrowband PCS licensees shall construct base stations that provide coverage to a composite area of 150,000 square kilometers or serve 37.5 percent of the population of the service area within five years of initial license grant date; and shall construct base stations that provide coverage to a composite area of 300,000 square kilometers or serve 75 percent of the service area within ten years of initial license grant date.

> MTA narrowband PCS licensees shall construct base stations that provide coverage to a composite area of 75,000 square kilometers or 25 percent of the geographic area, or serve 37.5 percent of the population of the service area within five years of initial license grant date; and shall construct base stations that provide coverage to a composite area of 150,000 square kilometers or 50 percent of the geographic area, or serve 75 percent of the population of the service area within ten years of initial license grant date.
>
> BTA narrowband PCS licensees shall construct at least one base station and begin providing service in its BTA within one year of initial license grant date.

The frequencies allocated to narrowband paging are consistent with WRC 95 frequencies for global coordination. Operations in markets or portions of markets that border other countries, such as Canada and Mexico, will be subject to ongoing coordination arrangements with neighboring countries. There are 26 paired frequencies with band separations of 50 and 12.5 kHz symmetric and asymmetric pairing. There are also 13 unpaired channels at 12.5 kHz for BTA and at 50 kHz for nationwide and MTA usage. All frequencies are between 901 MHz and less than 941 MHz.

The FCC is now proposing that there be no differences between common carriers and private carriers (Part 22 and Part 90 paging entities). The industry debate will focus on the following issues:

- Should geographic licensing replace site-by-site licensing?
- What type of process and procedure should be used to resolve mutually exclusive licenses?
- How should licensing be handled during the transition?

There is a trend for the FCC to step away from its past methods of regulating the industry by streamlining the regulatory process and bringing the paging regulation in sync with other CMRS (Commercial Mobile Radio Service) regulations. The FCC is shifting from tight control and intrusive regulation of the market to monitoring the market and stepping in only when necessary.

## 4.5   Technology

This section will describe the basic paging system and its elements: the input source, the wireless telephone network, the encoding and transmitting device, and the paging receiver. This will be followed by the engineering elements of system coverage: the base station, paging frequencies, technology evolution, and narrowband technology.

### 4.5.1   Basic paging system

There are four segments of a paging system: the input source, the PSTN, the encoding and transmitter control equipment, and the pager. Figure 4.3 shows a basic paging system.

**4.5.1.1   Input source.**   The input source can be entered from any telephone device, a computer with a modem, a page-entry device, or an operator. The page goes through the PSTN to the paging terminal for encoding and broadcast transmission over the paging system.

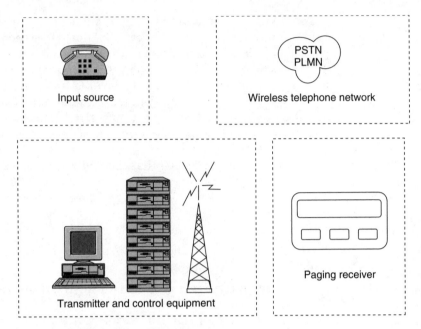

**Figure 4.3** Basic paging system.

**4.5.1.2 The wire-line telephone network.** The PSTN (Public Switched Telephone Network) is a key element of a paging system. The PSTN provides access for the calling party into a manual paging system as well as direct inward dialed (DID) numbers for automatic paging systems. A DID is a block of assimilated-facilities group numbers that identify each calling station or paging receiver. Paging systems work similar to PBX. DID numbers are commonly used in PBX systems to call in to a specific party even though there is not a dedicated trunk to that telephone extension. In a paging system, there are a few trunks that will handle all traffic coming in from the PSTN; however, a dedicated seven-digit DID telephone number is assigned to each pager. The DID telephone interfaces the encoding and transmitting device bringing the incoming call from the PSTN. Figure 4.4 shows how a DID call is routed to the paging receiver.

**4.5.1.3 Encoding and transmitting device.** The encoder is a device that converts the paged number into codes that can be transmitted either manually or automatically. A paging operator enters the number via a message keypad in a manual system. In an automatic encoding system, a caller dials into an automatic paging terminal and uses the telephone's Touch-Tone pad to enter the paging number.

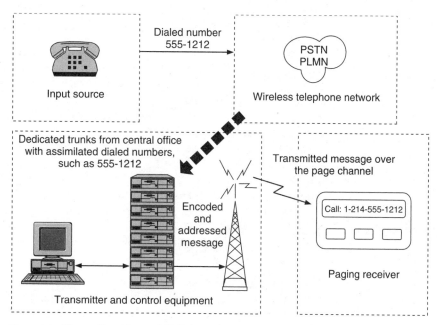

**Figure 4.4**  How a direct inward dialed number is routed to the paging receiver.

The encoder accepts the incoming message, checks whether the paging number is a valid subscriber, and looks up in the database the *cap code* or pager address to locate the pager. The encoder then converts the message to the appropriate page-signaling protocol. The encoded message is sent to the transmitter base station and broadcast across the coverage area on the specified frequency.

The base-station transmitter is a radio system that transmits page codes on an assigned radio frequency. It consists of the radio transmitter and antenna subsystem.

There are two types of paging systems. In a *manual paging system,* an operator receives the call and transmits the telephone number for the paged party to call over a 12-digit numeric pad or a text message via a computer keyboard. The encoder converts the message into the paging format and transmits the code via RF. An operator will monitor the channel and manually send the page when the channel is free. All pagers within the coverage area will receive the RF signal; however, only the paging receiver with the specific cap code or address will activate an alert.

*Automatic paging systems* answer a telephone line and allow the caller to enter voice, numeric, or alphanumeric text message from any Touch-Tone telephone or text entry device. The encoder will convert the message into the paging format and transmit the page without operator intervention.

After receiving the incoming message and validating it, the encoder will apply the address and convert the message into the page-signaling protocol. The signaling protocols can be manufacturer-specific or open standard protocols. All signaling protocols fall into two formats whether they are analog or digital: tone coding format or binary coded format. In tone coded format a sequence of tones is used to identify each pager, but this can be slow and supports fewer subscribers. In binary coded format the address code and information are coded in binary form with a string of 1s and 0s. They are modulated by frequency shift key (FSK) and then demodulated and decoded by the paging receiver. Binary coded formats offer a larger number of codes with multiple addressing capability, multiple function capability, faster signaling speeds, and error correction.

**4.5.1.4  Paging receivers.**  The pager is an FM receiver tuned to the same frequency as the base-station radio transmitter. Each pager has a built-in decoder that will recognize its unique cap code or paging address. All other codes are rejected. A paging receiver can be programmed to receive its own address, a group address, or up to four addresses. The basic function of a pager is to alert the subscriber, by an audible tone, flashing light, or vibration, that there is an incoming message. The message can be voice, numeric, or alphanumeric over a digital display.

A paging receiver consists of a receiver that receives and demodulates the paging signal, a decoder that decodes binary information, a display that displays the message or other information, a controller that allows the user to set parameters and access information, and a battery that is the power source.

One-way pagers can consist of just a simple beep or include a plethora of features available on today's modern pagers. Motorola is one manufacturer who is a leader in the technology of the pager. Some of the features on today's pagers are as follows:

- *Battery saver.*  To conserve battery power, battery-saving techniques are incorporated into the paging format and the pager. These techniques switch the pager into low-power mode for short intervals and therefore reduce the frequency of battery replacement.

- *Group call.*  The pager is equipped with two codes. One code is for individual calls and the other for group calls. Group call allows a selected group of individuals within the system (e.g., an emergency fire and rescue team) to be paged simultaneously.

- *Error detection/correction.* Most digital protocols are equipped with error detection and correction codes to ensure reception of the transmitted data. This involves sending additional information (along with the data), which enables the receiving station to check for the presence of errors in transmission and perform any corrective actions necessary.

- *Silent alerting.* A pager uses vibration to silently alert its user, which is useful in situations or places where tone alerts are not acceptable (e.g., meetings, church services).

- *Duplicate message detection.* If a new message is identical to a stored message it will not be entered into memory. There will be a "duplicate" prompt to signify that the message has been received more than once. This prevents depletion of memory capacity by duplicated messages.

- *Memory retention (nonvolatile memory).* Messages and pager settings will be retained in memory even when the pager is turned off or when a battery is being changed.

- *Unread message counter.* Each time a message is received, the "standby" display is updated to reflect the number of unread messages in memory.

- *Message protection.* A certain number of messages can be locked into memory. This prevents selected messages from being overwritten by incoming messages or erased by mistake.

- *Message freeze.* The user can freeze a message on the screen by holding down the Read button. This allows a user to keep a message on the screen while making a phone call or writing it down.

- *Accelerated life test (ALT).* A process developed by Motorola to simulate five years of field stress in a few weeks. The ALT is conducted to assure product quality and reliability.

- *Backlit display.* A switch can be depressed to illuminate the display. This permits reading in low-light conditions.

- *Chirp in MEM-O-LERT.* When the pager is switched to the silent position, a short beep (or chirp) will be emitted on reception of a page. This is in place of a full alert, which might be disturbing.

- *Overflow indicator.* When the number of unread messages is greater than the number of available memory slots, "overflow" replaces the unread message indicator. This tells the user that the oldest unread message has been pushed out of memory by a more recent message.

- *Memory-full indication.*   When the number of messages or characters stored reaches capacity, the pager will display a "memory full" screen. This gives the user the opportunity to lock in any new pages that they want to save, as existing messages will now be pushed out of memory by a new incoming page.

- *Over-the-air (OTA) programming.*   The pager's EPROM is reprogrammable via a PC interface and/or Motorola's over-the-air programming protocol. These methods can be used to reconfigure the pager's code and options via RF transmission without subscriber inconvenience. The pager no longer has to be brought in for service to be reconfigured.

- *Source indicator.*   When selected, a source indicator will appear at the end of each data message. The source indicates which phone number was called to send the page and provides additional information to the user.

- *Automatic alert reset.*   Pager alerts automatically reset after a designated period of time. This minimizes battery drain and the inconvenience of having to manually reset the alerts.

The major manufacturers of paging receivers are Motorola, Glenayre, NEC, Ericsson, and Uniden. All these manufacturers have been licensed by Motorola to use the FLEX family of open protocols for both one-way and two-way messaging. Ericsson will also implement AT&T's P-act protocol, which will have symmetrical two-way messaging.

## 4.5.2   Paging system coverage

The paging system coverage is the geographical area in which the paging receiver can reliably receive the transmitted signal from the base station. Private paging systems can range from in-building paging systems to those covering large geographic areas. RCC paging systems can also vary, covering small towns or large metropolitan areas that require multiple transmitters for reliable coverage. When multiple transmitters broadcast a page, it is called a *simulcast* or simultaneous broadcast. A simulcast allows a subscriber to roam from his or her home system to anywhere the paging systems are networked, either nationally or internationally, via PCSS (Personal Communications Satellite Systems) technology.

The factors that affect pager coverage and reliability are the transmitter power, antenna gain, antenna height, frequency, path loss, fading, and receiver sensitivity, as discussed in the following section.

### 4.5.2.1   The base station.   The base station consists of the encoder element, radio transmitter element, and the antenna element. The output

from the encoder goes to the radio transmitter. Coverage can be increased by increasing the RF output power, but doubling the RF output increases power by only 1.4 times field strength. A more efficient way to increase coverage is to increase the antenna height, which almost doubles the increase in field strength; for example, doubling the height of the antenna doubles the effective coverage area. Therefore, paging antennae are typically positioned on a tall building, a tower, or a hill.

Another way to increase coverage is to engineer directionalized antennae in the system. An omnidirectional antenna radiates in all directions, and therefore the signal is diluted. The broadcasted signal can realize effective radiation gain by focusing the RF beam in the desired direction. The RF beam is concentrated, and therefore the lobe will cover a greater distance from the antenna.

System engineers study propagation coverage and contour maps to determine the most effective positioning for the antennae, as well as antennae gain techniques, antennae height, and simulcast. Coverage will be extended to other paging topologies through network paging systems over the PSTN.

**4.5.2.2  The airwave.** The frequency and RF transmitted power is referred to as the *airwave*. Coverage can be influenced by the frequency of transmission. Higher frequency penetrates buildings and will attenuate or propagate from the radio transmitter to the paging receiver through trees and foliage, whereas lower frequencies tend to be absorbed or blocked.

As the transmitted power travels away from the antenna, the signal becomes diffused or spread out. The signal also passes over buildings, hills, trees, and other obstructions causing diffraction. This loss of the broadcasted signal is called *path loss*. Atmospheric conditions can bend the broadcasted signal up or down, which can vary the strength of the broadcasted signal. This phenomenon is called *multipath propagation*. RF engineers will perform field-strength measurements throughout the coverage area to fine-tune the antennae's positioning, height, power transmission, and characteristics of gain.

### 4.5.3  Technology evolution

Narrowband paging is a natural extension of one-way paging. There are two camps that have emerged among advanced messaging providers. Some are going all out in development and deployment of an advanced paging network such as Paging Networks, Incorporated, and Mobile Telecommunications Technologies Corporation. Others are taking a wait-and-see attitude. Those who are going all out are going beyond the beeper to new paging devices that feature voice and two-way data capabilities, customized response function, high-speed transmission, frequency reuse, and connections to on-line information services and e-mail messaging. The traditional definition of paging has changed from a one-way communications device to a two-way device that sends and receives data.

### 4.5.4  Narrowband technology

Motorola is the leader in narrowband paging technology with its FLEX family of paging products. Motorola has taken the lead to open its protocol in a race to become the de facto standard protocol for the integration of data and computers; however, it is competing with AT&T for the paging standard protocol. Each manufacturer has a vested interest in its protocol becoming the standard because the winner would save development cost, shorten product-to-market time, and have industry prestige. The loser will have to develop software, and possibly hardware, to the accepted paging standard protocol.

Currently, each manufacturer has its own proprietary protocol, such that an operator will have a *closed system*. A closed system means that an operator is locked into one manufacturer to supply the switch system, the paging system, and the pagers. The most dominant standard is POCSAG, which is a one-way page transmitting at 2400 bps. FLEX is Motorola's newer version of one-way paging that transmits at 6400 bps and has been widely adopted by the Asia and Pacific region. ERMES is the European standard transmitting at 6250 bps; however, Motorola's FLEX products are being more widely deployed throughout Europe.

Motorola is expanding the FLEX platform for paging-enhanced services of narrowband paging and messaging. FLEX is the platform for the ReFLEX 25, ReFLEX 50, and InFLEXion products. The ReFLEX 25 transmits outbound at 6400 bps on a 50-kHz channel, ReFLEX 50 transmits outbound at 25,600 bps on a 50-kHz channel, and InFLEXion transmits outbound at 112,000 bps on a 50-kHz channel. The ReFLEX protocols support message acknowledgment, downloading information from computers, and menu-based responses. Both the ReFLEX 50 and InFLEXion products will incorporate cellularlike frequency reuse. The topologies will be mixed with regional and local transmitters at much lower power than traditional paging systems. This will allow the operator to realize network cost efficiencies and each subscriber to have a pager with a longer-battery life.

AT&T Wireless Systems' messaging division is developing its own narrowband PCS paging system with an open, advanced messaging protocol that will support message acknowledgment and menu-based response, as well as initiating predetermined text message responses and voice paging capabilities. AT&T's P-act paging system will have an open standard protocol, which became available in the fourth quarter of 1996. Its technology is expected to be as advanced as Motorola's. Details of P-act are not readily available. The AT&T protocol will be symmetric with equal inbound and outbound data transmission. Table 4.6 shows the predominant public paging protocols.

Glenayre Technologies, Incorporated, is one of the world's leading suppliers of paging equipment. It has lined up with Motorola's open standard protocol and has agreed to standardize some key interface points in its infrastructure equipment, similar to GSM's open interfaces. This would allow a paging operator to mix and match different manufacturers' equipment within the network for custom optimization.

## 4.6 Summary

Paging has come of age because high tech has caught up with the age-old problem of sending a message; and messaging has changed from the exclusive use by businesspeople to social and recreational use within families. Two-way paging and advanced messaging enable users to store or forward e-mail, faxes, and text messages with simple palm-sized devices. From any location, people can tie in to networks or the Internet with their laptops. And, with PCSS, a subscriber can receive or send a message anywhere in the world.

Pagers have become very small and colorful. They are available in retail stores throughout the world. Pagers have become a commodity

**TABLE 4.6    Public Paging Protocols**

| Protocol | Speed out | Out channel | Speed in | In channel | Applications |
|---|---|---|---|---|---|
| ERMES | ≤6,250 bps | 25 kHz | N/A | N/A | One-way data |
| POCSAG | ≤6,400 bps | 25 kHz | N/A | N/A | One-way data |
| FLEX | 6,400 bps | 25 kHz | N/A | N/A | One-way data |
| ReFLEX 25 | 6,400 bps | 50 kHz | 9,600 bps | 12.5 kHz | Two-way data frequency reuse |
| ReFLEX 50 | 25,600 bps | 50 kHz | 9,600 bps | 12.5 kHz | Two-way data frequency reuse |
| InFLEXion | 112,000 bps | 50 kHz | 9,600 bps | 12.5 kHz | Two-way data frequency reuse |
| P-act | * | * | * | * | Two-way data frequency reuse |

* Symmetrical inbound and outbound speed, frequency undisclosed by AT&T. Product introduction expected in fourth quarter of 1996.

item in North America, and fill a pent-up demand for low-cost PCS in the Asia and Pacific region. Pagers will be put into pens, watches, or jewelry, thus extending today's pager as a fashion statement. The new marketing targets are housewives, students, hunters, and the average person. Increasing awareness of paging produces more paging subscribers, which fosters lower prices. Lower prices mean even more subscribers. And the spiral continues, with double-digit, worldwide growth.

Paging companies are merging and consolidating, positioning themselves for the rapid growth and new technologies brought about by narrowband PCS. By consolidating, they have larger subscriber-revenue streams to capitalize further expansion and the implementation of advanced messaging and new features.

Technology companies are planning for growth in the paging industry by developing new, even higher speed infrastructure, two-way paging devices, smaller pagers with very long battery life, and other new features. Retail will become an important link in the manufacturer's distribution channel. Retail competition will push the price of pagers down even further, creating a greater demand for more modern pagers.

The paging industry appears to be robust. It is enriching the lives of the average person throughout the world by fostering low-cost personal communications for anyone, anywhere, anytime. The continued rapid growth of paging appears to be a bright spot for investors and a stimulus to the economy.

# PCS Technology

# 5

# PCS Overview

## 5.1 Introduction

Personal Communications Services (PCS) means different things to different people. Consumers, vendors, operators, regulators, analysts, and others have their own views of PCS. The scenario is similar to the fabled description of an elephant by a group of blind men; each person described the elephant differently, according to the part that was touched. The FCC defines PCS as ". . . a family of mobile or portable radio communications services which provides services to individuals and business and is integrated with a variety of competing networks." Vendors and operators (noncellular) view PCS as services provided in the 1900-MHz frequency band. Many vendors debate endlessly over which architecture and technology is best suited for PCS. The battle for different technology and architecture is shown in Fig. 5.1. The subscriber does not care about the "holy war" among the vendors and operators. The subscriber cares about the features and functions—not about network access protocol, spectrum, bandwidth, RF access protocol, and network topology.

In its grandest vision, PCS will allow communications with anyone, anywhere and anytime, with the help of a single, personal number. With that single, personal number a person can be reached on a residential wire-line phone, on the office phone, on a cellular phone, or on a pager while at a business meeting. To reach a person, people will not have to memorize work, home, and car phone numbers, but a single number assigned to that person. People may be assigned personal numbers at birth, which will be theirs for life. Personal numbers can best be conceptualized as being similar to Social Security numbers or International Standard Book Numbers (ISBN), each of which is unique.

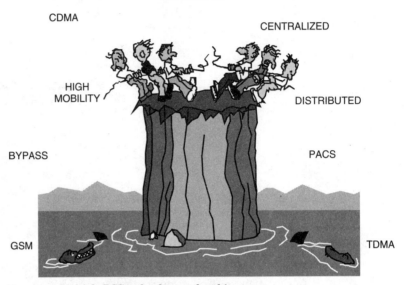

**The Battles...**

Figure 5.1    Multiple PCS technology and architecture.

This chapter provides an overview of the idealized PCS requirements. This is followed by a discussion on the industry structure and
technology of U.S. PCS, defined narrowly as a wireless telecommunications service within a specified frequency range, specifically, the 1900-
MHz range. PCS, at least for now, resembles cellular, albeit at a
different frequency range and utilizing digital technology.

## 5.2   PCS Requirements

PCS encompasses the concepts of terminal and personal mobility. *Terminal mobility* is the ability of a terminal to access telecommunications services from different locations, while in motion, and the
capability of the network to identify and locate the terminal. *Personal
mobility* is the ability of a user to access telecommunications services
at any terminal and the capability of the network to provide those services according to the user's subscription. Personal mobility implies the
ability of the user to register on any terminal for incoming and outgoing calls. The above concepts of terminal mobility and personal mobility provide the grandest vision of PCS.

It has to be understood that terminal and personal mobility are independent of each other. Terminal mobility can exist without personal
mobility and vice versa. In today's cellular network, terminal mobility
exists. A user who has a cellular phone can access the network from different locations. An example of personal mobility is the use of tele-

phone cards. Using a phone card, a user can access telecommunications services from residential, wireless, and pay phones.

### 5.2.1  Terminal mobility

One of the necessary factors for terminal mobility is the need for the link between the terminal and the network to be wireless. This factor enables the terminal to be mobile. It is also the most difficult to implement, provision, maintain, and engineer.

In a wire-line network, the number dialed by a user is for routing purposes. The dialed number uniquely identifies a port on a switch in the network. Each wire-line telephone user in the PSTN is connected via a local loop to the central office switch. The local loop is the physical wire connecting a phone jack in a home to a local, central office switch. Each telephone directory number is linked to a port on a central office switch where the local loop terminates. When a caller dials the directory number of the called party, the directory number does not identify the called party but the called party's local loop termination port, called a *network access/termination point*. The network is able to route the call from the caller's central office to the called party's central office switch because the directory number, apart from identifying the called party's network access/termination point, also identifies the terminating central office switch. Thus, in a wire-line network, the directory number is not associated with a person or a terminal, but with a network access/termination point. There is an implicit association between this network access/termination point and the called party. Thus, when a call is placed to Joe's home it is not known who will answer the call—whether it will be Joe, Joe's wife, Joe's son, or his dog. The network guarantees only that the call will terminate at Joe's home. The function of a wire-line network is that of routing and transport. These functions, however, are not sufficient to support terminal mobility. The PCS network has to perform additional functions to support that.

Terminal mobility requires that the mobile be small, lightweight, and easy to use. The phone has to be able to access the network from its current location, which can be in a home, car, office, or a multitude of different places. The phone could be mobile for the duration of a call. The requirement of network accessibility and mobility during a call make it very difficult for the wire-line network to support terminal mobility. The wire-line network can support limited forms of network accessibility, but never mobility during a call. Consider the usage of laptop computers with a modem. The laptops are lightweight and portable and the modem can be used at home, in a hotel room, or at an office by connecting it to a standard phone jack. It seems to satisfy network accessibility requirements. However, it is impossible to use a laptop connected to a phone jack and move around from room to room or even outside the premises. The factor preventing wire-line networks from supporting terminal mobility is the fixed, physical, copper-wire nature of the local loop. The necessary condition for terminal mobility is a wireless link between the mobile terminal and the network.

There are several issues involved in making the link between the terminal and the network wireless. The most important are identification of the terminal location, handoff, identification of customer profile, security, power, and service quality.

**5.2.1.1  Identification of terminal location.**  The wireless link between the mobile and network enables terminal mobility, but it introduces the problem of identifying terminal location. The wireless network needs to locate the mobile for terminating a call. This is done by a process called *paging*. The network, upon receipt of a call, will broadcast a message called a *page message* to all cells or some limited number of cells. The network locates the mobile when the mobile responds to the page signal.

**5.2.1.2  Handoff.**  Another condition for terminal mobility is that the terminal should be able to access and continue calls while it is mobile. Handoff deals with continuing preexisting calls as the mobile terminal moves from location to location. With a call in progress, a mobile terminal sends and receives traffic to and from a particular base station attached to the fixed network. As a result of the terminal moving, the radio circuit quality can be degraded as the mobile terminal reaches the cell boundary. It becomes imperative for the radio circuit to be changed in order to maintain acceptable signal quality. The new radio circuit can be another circuit connected to the same base station or to another base station. The process of changing a radio circuit connected to a base station is called a *handoff.* It is one of the most complicated and signaling-intensive procedures in the wireless network.

**5.2.1.3  Identification of customer profile.**  A customer profile is a snapshot of the currently subscribed services and features of the user. These may include subscribed features such as call-waiting, three-way calling, and call forwarding. During origination and termination of a call the serving switch determines from the customer profile if any special processing is required. An example is that of a customer who has call-waiting activated. Whenever a call is to be terminated to this subscriber, the switch has to do special processing, i.e., provide a call-waiting tone to the subscriber as an alert indication for another incoming call. In a wire-line switch there is a fixed association between each network access/termination point and the customer profile.

In wireless switches there is no association between a subscriber and the network access/termination point. Thus, the traditional wire-line approach of provisioning features based on network access/termination points will not work in wireless networks. The wireless networks resolve this by having a central database linked to the wireless switch. This database is a repository of the customer profiles of all subscribers. There is no fixed association between the network access/termination point and the customer profile. Whenever a mobile terminal originates or terminates a call, a database query is made to this central database to search and download the customer profile. Since mobile terminals can access any available network access/termination point, the mobile terminals have to provide the network with their identity. The mobile identity is used to access the customer profile.

The mobile can access potentially any network access/termination point within the network. This requires that all of the features that the user has subscribed to function identically across all switches within the network; i.e., feature transparency is essential. Similarly, any new services or features need to be introduced on a networkwide basis.

**5.2.1.4   Security and fraud.**   The wireless link between the mobile and the network leads to several security-related issues, such as eavesdropping, fraudulent mobile usage, and privacy of location.

The wireless link is a shared link, unlike the wire-line network where the local loop is dedicated to a single user. Anyone with access to the wireless link can listen to or transmit on it. The conversations are not private. An eavesdropper can capture an origination message and retransmit it later as a fraudulent call. Measures to counter eavesdropping in wireless systems concentrate mainly on making the conversation unintelligible to eavesdroppers rather than eliminating it. Elimination of eavesdropping would be difficult because of the shared nature of the wireless link. In modern digital wireless systems, powerful encryption techniques are used to encrypt the conversation over the wireless link. Moreover, the encryption codes used vary from call to call for the same user. This ensures that even if an eavesdropper is able to decipher the code for one call, he or she will not be able to monitor subsequent calls.

The second aspect of security concerning wireless systems is mobile fraud. As previously discussed, the network does not have any prior information about the identity of a mobile terminal until a call has been placed. When a call is placed, the mobile terminal provides its own identity to the network in order to enable the network to do some special processing before connecting the call, such as look up a customer profile, determine if the caller is a valid customer, etc. Since the network does not have any information about the identity of a mobile, all information necessary to identify a mobile must reside within the mobile itself.

The first kind of fraudulent mobile use occurs when valid mobile terminals are stolen and used to make calls. Since there is no association between the valid subscriber and his or her own mobile phone, the network has no way of determining if the current user is legitimate. This is not a problem in wire-line networks since the network has the subscriber's identity instead of the phone's. Thus, anyone who steals a wire-line phone has to procure a valid network access/termination point to make calls. There are instances in the wire-line network where the network does not have prior knowledge of the subscriber identity, such as in pay phone use. In such cases, the subscriber is almost always made to pay before placing a call. Measures to counter stolen mobile use include schemes to form associations between the valid user and the terminal, for example, having the user enter a PIN prior to originating or receiving a call.

The second kind of fraudulent mobile use occurs when the mobile's transmission over the wireless link is intercepted and programmed into another mobile terminal which impersonates the first. Digital wireless systems use advanced authentication algorithms to validate the mobile terminal. Almost all of these schemes make use of certain unique, shared, private information stored in the terminal and the network. This information differs from mobile to mobile and is not transmitted over the wireless link. When the network needs to validate a mobile, it requests the mobile to make use of the private information in formulating a response to the network query. The network matches the response with its own internally generated response. The mobile is validated when the network's generated response matches that of the mobile. If not, the mobile is tagged as a fraudulent mobile.

The third aspect of security concerning wireless systems is privacy of location. The mobile terminal sends its identity over the wireless link to the network, where it is capable of being intercepted by a person who is eavesdropping. If the network uses encryption and authentication, the eavesdropper cannot decipher the conversation, but because the mobile identity is sent in the clear, the user's location can be determined. This constitutes a breach of the user's privacy of location. In order to provide for this type of privacy, the network uses an alias for the mobile's actual identity. When a mobile accesses a network, it will send its alias in place of its actual identity. The network and the mobile are the only entities that are aware of the association between the alias and the terminal's real identity. Even if eavesdroppers can intercept the alias, they cannot relate the alias to an actual mobile terminal identity.

**5.2.1.5 Billing.** Due to the fixed nature of the local loop, billing in wire-line networks is a relatively simple issue. Since all calls are originated and terminated to the same network access/termination point, billing is simply done on the directory number. Billing is further simplified by the fact that in wire-line networks the caller pays for the call. There are exceptions to this, such as 800 and 411 numbers. However, primarily for all local and long-distance calls, the caller bears the cost of the call. Also, due to the fixed nature of the local loop, the approximate price for a call to that directory number is known in advance.

The situation is radically different in wireless networks. Since the network access/termination points are different, there can be no single point for billing. This issue is further complicated by the fact that a mobile terminal can roam into nonhome networks and still expect a single consolidated bill for all calls. These nonhome networks may have a pricing structure different from the subscriber's home network. Furthermore, in wireless systems the caller pays only for the wire-line portion of the call and the called party pays for airtime; i.e., time spent on the wireless link for both originating and terminating calls. In case the mobile terminal is currently visiting an operating area different from the home area, the called party bears the costs of redirection of the call from the home switch to the visiting switch. Wireless networks resolve these billing issues typically by having clearinghouses take care of roaming-related billing issues.

**5.2.1.6 Power.** Due to the existence of the copper local loop in a wire-line system, the local phone company is able to provide power to the wire-line phone set. However, due to the lack of a copper wire in a wireless system, the power source has to be part of the wireless phone. There are several issues related to having the power source attached to the wireless phone. First, the power source has to be small, compact, and lightweight. This is especially true for wireless phones that are classified as portables. Having a lightweight phone serves no purpose without a lightweight power source to go along with it. Since these kinds of power sources are expensive, they have to be reusable to be economical over time. Second, the power source should allow the wireless phone to operate for long periods of time before having to be recharged. The industry is working on this aspect from two angles. Power sources that can operate for longer periods of time are continuously being developed. Also, the wireless technology is being modified by having special features on the wireless link that will conserve battery power, resulting in longer-lasting batteries.

**5.2.1.7 Service quality.** Measuring quality of the circuit is a particularly complex task. Wire-line networks were plagued with problems during the initial stages of implementing the network, most of them related to the local loop. Problems such as cross talk, echo, and noise were very common. With the advent of digital telephony, most of these problems have been solved. This is primarily because digital signals can be regenerated. Regenerative repeaters are the most effective devices ever used to eliminate signal degradation. Regenerative repeaters are interspersed on a wire-line digital circuit, where they detect the incoming digital pulses, process the information, and then retransmit a new pulse. The closer the regenerators are on a digital circuit, the better the voice quality. Unfortunately, due to the nature of the wireless link, regenerative repeaters cannot be used in wireless systems. Thus, wireless systems are forced to use complex channel-coding schemes to make the digital signal more robust and resistant to signal degradation. These schemes make use of redundancy and require more spectrum bandwidth, which is a limited resource in a wireless system. Moreover, the mobile environment is very unpredictable and much more destructive when compared to the gradual signal degradation that occurs in wire-line systems due to distance. Radio spectrum is a limited resource because it has to be shared among several users. The less of the spectrum that is used by each subscriber for a call, the larger the number of subscribers that can be supported on the system. But according to the information theory, there is a minimum signal bandwidth that is needed to exactly reproduce a signal. Therefore, all efforts in wireless systems are concentrated on improving voice quality without increasing signal bandwidth. Improper frequency planning and frequency reuse can also degrade voice quality by increasing the total interference level in the wireless channel. All these factors add up to make the voice quality of wireless systems far inferior to that of wire-line telephony.

Another factor determining service quality is its availability. In a wire-line network, when a user picks up a phone, the network provides a dial tone. The dial tone is an indication to the user that the local central office switch has allocated resources and to process the call. Typically, the only time a call is not terminated successfully in a wire-line network is if the called party's line is busy or there is network congestion. The FCC has rules governing the blocking rate in wire-line networks, which is typically 0.5 percent; that is, 1 out of every 200 attempts may be blocked. All of this blocking will typically take place either in the switch or on the interswitch circuits. Contrast this with the blocking rate specified for wireless networks, which is typically 2 percent; i.e., 1 out of 50 attempts may be blocked. Almost none of the wireless networks achieve this rate because the blocking rate is higher during peak traffic hours. Wireless networks have a higher-blocking rate than wire-line networks because they may have blocking on the wireless link. In wire-line networks, the local loop is dedicated to each user; i.e., there is always circuit availability between the user's phone and the local central office. This is not the case in wireless networks since the wireless link is shared. The wireless links have to be traffic engineered to support the predicted traffic. Traffic engineering is based on the concept that it is possible to support a larger number of subscribers with fewer circuits if a certain blocking probability is assumed. It is based on the fact that not all the subscribers will be using the circuits at the same time; therefore, it is possible to get higher trunking efficiencies than would be possible by a straight one-to-one mapping between subscribers and circuits. Typically, traffic engineering is done on the interswitch trunks within the telecommunications networks. Traffic engineering for the wireless link is more complex and has to account for the mobile nature of the terminals, the cell size, and subscriber density of the cell.

Service accessibility is another factor affecting service quality in wireless networks. This problem is unique to wireless and is not an issue in wire-line networks. A wireless system might have very good voice quality and there might be available channels in the cell; however, the mobile user might be outside the cell boundary and thus not be able to access the network. Service accessibility is directly related to the issue of coverage. Coverage in a wireless system is a metric of how large the service area is and how good the signal quality is for that particular service area. Having good coverage in the entire service area means that all mobiles can receive a strong signal and access the network.

### 5.2.2 Personal mobility

PCS will associate communications with individuals, rather than with terminals or network addresses. This requires personal mobility. Today's cellular networks provide terminal mobility to users, while the wire-line networks do not. Currently, users are tied either to a terminal or to a network access/termination point and communications are associated with such points of access to the networks rather than with the actual users. PCS will enable communications to be user-oriented rather than terminal- or network access/termination-point-oriented.

Personal mobility will allow users to originate and terminate calls based on a personal number, across multiple networks, independent of access technology, and independent of geography. A personal number will be allocated to each user and the network will route the call based on this personal number and current location. It will also use this personal number to charge the call and retrieve the user's service profile. Personal mobility can be implemented on wireless or wire-line networks. There is a popular misconception that terminal mobility is required for personal mobility. Terminal mobility is one of several ways to implement personal mobility. There are several issues that have to be resolved in order to provide personal mobility. These issues are personal addressing, user authentication, and user location/routing.

**5.2.2.1    Personal addressing.**    Table 5.1 describes the evolution of networks with regard to addressing. In the PSTN, the caller addresses the network access/termination point. ISDN networks allow the capability to address network access/termination points as well as terminals with the help of subaddressing. Cellular networks enable terminals to be addressed by mapping the terminal's logical address to the actual physical network address. This process often involves terminal-location determination, which is accomplished by the paging procedure. In a PCS network, a person is addressed, then the PCS network maps the personal address to the called party's actual physical network address that may be a terminal if the user is registered at a wireless phone, or at a network access/termination point if the user is registered at the wire-line phone.

**5.2.2.2    Subscriber authentication.**    Personal mobility requires personal authentication, i.e., verifying the subscriber. It involves ensuring that the user is authorized to access the network. There are several ways to implement the authentication process. The subscriber can access a network, and, by way of authorization, be required to enter in the subscriber's personal number followed by a key sequence, PIN, or password. The network validates the caller based on the personal number and the security code used. Alternatively, for advance systems using voice recognition the caller may be required to speak his or her personal number followed by a key sentence in order to allow the network to verify the user. The network will store a profile of the caller's speech pattern that is unique to each individual and would validate the spoken sentence against it. One of the drawbacks of this method is that it

**TABLE 5.1    Addressing in Different Networks**

| Type of network | Addressed entity |
| --- | --- |
| Public Switched Telephone Network (PSTN) | Network access/termination point |
| ISDN | Network access/termination point or Terminal |
| Wireless Networks | Terminal |
| PCS | Person |

requires the speech quality to be very good in order for the speech recognition software to function properly. This is not often the case in wireless networks.

Another method being proposed is the use of smart cards or plug-in devices. All the relevant, subscriber information is stored in these smart cards, such as personal number, calling preferences, security code. All that the subscriber needs to do is to be authorized by the network by plugging in the smart card to the telephone device being used. The concept is somewhat analogous to that of credit cards used today where, in a supermarket, a cashier swipes the card through a card reader and the purchase is made. The drawback of this approach is that it requires all terminals be able to handle these plug-in devices.

**5.2.2.3   User location/routing.**   The PCS network has to keep track of the subscriber's current location in real time so that any incoming calls can be routed to the subscriber's current location. There are several ways in which the network can keep track of the subscriber's location. The subscriber's service profile and location information usually reside in a database that is queried whenever the profile information is needed. Automatic location updating means that subscribers can roam from one place to another without having to "sign in" to the network in order to update the database. This is usually possible if the user has a mobile terminal. The terminal automatically keeps the network updated of its current location by frequently sending in its current location. The network then sends that information to the database which would update its records to correctly show the subscriber's location. Manual location updating would apply if a subscriber wished to relocate service from one wire-line terminal to another. In this case the subscriber would have to use the personal number to access the network and inform the database of the new location. PCS networks will evolve toward an intelligent network model in order to support location management, call routing, and service-profile management.

## 5.3   PCS Industry Structure

The analysis of the PCS industry is similar to the analysis of the cellular industry presented in Chap. 2. This is expected since the services offered by PCS and cellular are identical, except that the two operate at different frequency bands. The subscriber is indifferent to the frequency band as long as the services are unaffected. The services that can be offered by PCS can also be offered by cellular carriers by upgrading the network with digital technologies. Of course, cellular has a head start of more than ten years, which to some extent is compensated by the difficulty of transitioning the cellular network from analog to digital technology and migrating the customer base from analog to digital. The historical weight of the existing analog systems will delay the existing cellular carriers' transition from analog to digital. PCS, on the contrary, will start with a clean slate, using the latest digital technology, thereby avoiding the transition and migration problems.

The structure of the PCS industry, like the cellular industry, is determined by forces that are very broad and encompass social as well as economic factors. Regulation and other external market forces lead to moves that shape the structure of the industry. These outside forces affect all firms in the industry and determine the ultimate profit potential. These forces are different for different industries. As indicated in Chap. 2, the forces are intense for industries such as tires and steel but are mild for industries such as cosmetics and toiletries. These forces explain the low rate of return on invested capital in the tire and steel industry and the high rate of return in the cosmetics and toiletries industry. Knowledge of these forces will allow the reader to better appreciate the tactical, strategic, and competitive moves of the industry participants. Competition in the industry continually works to drive down the rate of return on invested capital toward the competitive floor rate of return, or the return that would be earned by the economist's "perfectly competitive" industry. The presence of higher rates of return will attract capital, either through new entry or through additional investments by existing competitors. This will lead to a rate of return moving toward the competitive-floor rate of return.

The main forces that influence the PCS industry are essentially the same as in the cellular industry. These forces are as follows:

1. Regulators

2. Subscribers

3. Equipment vendors

4. Substitution products such as cellular and MSS

5. Operators and competition among them

These forces are shown in Fig. 5.2.

### 5.3.1 Regulatory

In the early stage of the industry, regulatory forces shape the market structure. The FCC has allocated 120 MHz to broadband PCS, which has been licensed in six bands: three bands each containing 30 MHz

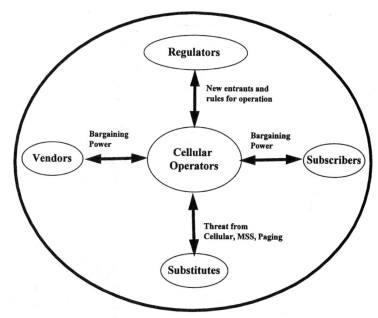

**Figure 5.2**   Forces affecting the PCS industry.

(Blocks A, B, and C) and three bands each containing 10 MHz (Blocks D, E, and F), as shown in Fig. 5.3 and summarized in Table 5.2.

The service areas for the six bands are based on major trading areas (MTAs) and basic trading areas (BTAs). MTAs and BTAs are based on the Rand McNally 1992 *Commercial Atlas and Marketing Guide,* 123d edition. Rand McNally has organized the 50 states and District of Columbia into 47 MTAs and 487 BTAs. The FCC modified the MTA

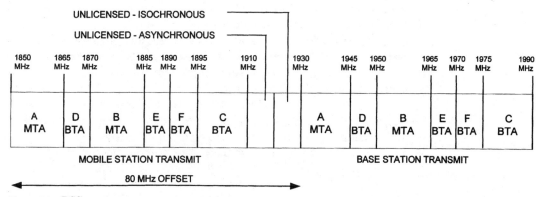

**Figure 5.3**   PCS spectrum.

**TABLE 5.2  PCS Frequency Allocation**

| License | Bandwidth (MHz) | Frequency band (MHz) | No. of licenses | Service area |
|---------|-----------------|----------------------|-----------------|--------------|
| A | 30 | 1850–1865, 1930–1945 | 51 | MTA |
| B | 30 | 1870–1885, 1950–1965 | 51 | MTA |
| C | 30 | 1895–1910, 1975–1990 | 493 | BTA |
| D | 10 | 1865–1870, 1945–1950 | 493 | BTA |
| E | 10 | 1885–1890, 1965–1970 | 493 | BTA |
| F | 10 | 1890–1895, 1970–1975 | 493 | BTA |

definition by adding four more MTAs. These additional MTAs are Alaska, which Rand McNally defines as included in the Seattle MTA, Guam and Northern Mariana, Puerto Rico and U.S. Virgin Islands, and American Samoa. The addition of these four MTAs to the original 47 MTAs defined by Rand McNally leads to a total of 51 MTAs for licensing purposes.

Similarly, additional BTAs were added to the Rand McNally list. These additional BTAs are American Samoa, Guam, Northern Mariana Islands, Mayagüez-Aguadilla-Ponce, San Juan, and U.S. Virgin Islands. The addition of these six BTAs to the original 487 BTAs defined by Rand McNally leads to a total of 493 BTAs for licensing purposes.

Blocks A and B were licensed by an auction process. Blocks C and F are reserved for small businesses with less than $125 million in gross revenue and $500 million in total assets. Winning bidders for these blocks may pay for their licenses in installments. Small businesses, under $40 million in gross revenue, are eligible for bidding credits and enhanced installment payments.

In order to prevent an unusual concentration of spectrum and to promote competition, the commission has imposed spectrum caps that limit the amount of spectrum that any one entity may control within the same area. There is a 10-MHz limit on broadband PCS spectrum for cellular carriers within their cellular service areas; a 40-MHz limit on all entities for broadband PCS; and a 45-MHz limit on all entities for broadband PCS, cellular, and SMR spectrum.

The spectrum-cap restriction for cellular carriers becomes effective if an entity has a 20 percent or greater interest in a cellular carrier, and the cellular carrier's CGSA covers 10 percent or more of the PCS service area population. If an entity has an interest of 20 percent or greater in more than one cellular system overlapping into the PCS service area, its coverage is aggregated to determine the extent of overlap of the cellular and PCS licenses. However, if an entity owns less than 20 percent in a cellular license, then the population covered by that license is not counted toward the 10 percent population requirement.

Broadband PCS licenses cannot acquire more than 40 MHz of broadband PCS spectrum in any geographic area. PCS licenses are entities having an ownership interest of 5 or more percent or an attributable management agreement in a PCS license. The attribution rules are described in detail in Part 47, Section 24.204(d) in the Code of Federal Regulations.

A single entity cannot hold more than 45 MHz of combined spectrum in PCS, cellular, and SMR licenses in a given geographic area. The attribution rules are similar to that defined for PCS, cellular, and SMR carriers. Thus, any interest of 20 percent or greater in a cellular or SMR license is attributable and any interest of 5 percent or greater in a PCS license is generally attributable.

Based on the FCC regulation, each area in the United States will have at least six PCS operators. This is in addition to the existing two cellular operators and one SMR operator. Thus, consumers can potentially have a choice of nine operators for their wireless telecommunications needs. The arrival of PCS will fundamentally change the telecommunications landscape. With the release of spectrum for PCS, the FCC has accelerated the evolution of the telecommunications industry from the monopoly structure of LEC to the duopoly structure of cellular operators to the eventual oligopoly structure as PCS networks become operational.

The FCC's release of spectrum and the efficiency of spectrum utilization by digital technologies has essentially changed the industry from being spectrum scarce to spectrum abundant. Due to the change in the spectrum availability, the PCS operators need to have different strategies. The strategies of the past developed in an environment of scarce spectrum may not be relevant, as shown in Fig. 5.4. In an environment of scarce spectrum the operators could charge high prices to maximize revenue. With the introduction of PCS and digital technologies the spectrum is no longer scarce. In an environment of abundant spectrum, as in the manufacturing industries with excess capacity, price wars will occur. In order to compete, the operators will have to focus on lowering their cost structure. In such an industry, marketing and brand image become crucial.

## 5.3.2  Subscribers

Subscribers have the desire for untethered communication. Subscribers want terminal and personal mobility, as discussed in the previous section. PCS, like cellular services, may not even provide full terminal mobility because of several incompatible wireless technologies being implemented by the different operators. In fact, it is argued that there is no need for personal mobility if there is ubiquitous wireless coverage. If wireless coverage exists everywhere, then subscribers will be able to use their portable terminals everywhere and the sub-

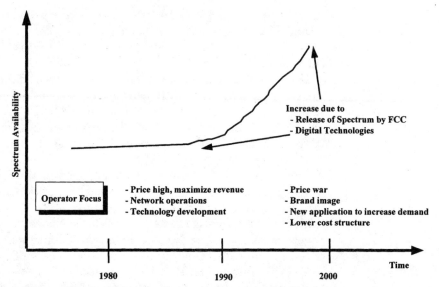

**Figure 5.4**   Changes in operator focus as spectrum availability changes.

scriber will become associated with the terminal. Thus, the personal number is the terminal number. Even though it is quite possible that this may be how the PCS networks will operate in the distant future, there will be coexistence of wireless and wire-line networks. This is for several reasons, such as radio coverage, services, and cost.

Radio coverage is not ubiquitous. For mobile terminals to be used everywhere, it means that there has to be radio coverage in buildings, in malls, outdoors, and so on. Moreover, with today's technology and limited radio spectrum, it is just not possible for wireless to support the subscriber-penetration levels needed to completely displace wire-line networks.

The wireless link is low bandwidth and optimized for speech. Enhanced services requiring very high bandwidth, such as high-speed data, video, and broadband ISDN services, cannot be supported on the wireless link.

The wire-line local loop has been in service for several decades and has been cost optimized; whereas the wireless networks are just implemented and have very high infrastructure and terminal costs associated with them. Until the wireless networks are fully deployed and mature, it will always cost less to communicate over wire than over the RF. This will translate to lower costs for the users. Whenever mobility is not a concern, users will attempt to use wire-line services because it is less costly.

PCS networks are attempting to provide enhanced services and full terminal mobility. Wherever there are incompatible wireless technologies vendors are striving to provide dual-mode, dual-frequency portable terminals to the users. These terminals are able to access different networks operating in different frequency ranges. Also, some mobile phones now come with smart cards, which store the subscriber's profile. It is possible to remove these cards and use them in different mobile terminals such as a rental phone. More and more wire-line services are being offered on wireless networks, such as ISDN services or PBX type of services. The similarities between the wire-line network and wireless network are becoming greater, with new service offerings being developed by both. The PCS network is just now beginning operation and will eventually migrate toward the PCS vision of the subscriber with terminal and personal mobility.

The implementation of PCS networks will provide subscribers with multiple choice for services from potentially six new PCS operators. In addition, alternatives such as cellular-product offerings, essentially offer the same type of service. For the most part the services offered by PCS and cellular carriers are standard and undifferentiated. As in the long-distance market, excess capacity will lead to price erosion. The subscribers' costs of switching from one carrier to another are relatively low. Therefore, buyers can easily switch from one carrier to another if they are unhappy with their carrier. Cellular subscribers, in spite of the lack of a trade association to increase their bargaining power, will benefit from the intense competition among cellular and PCS operators, thereby extracting a piece of the revenue and limiting the profitability of the PCS industry.

### 5.3.3  Equipment vendors

As discussed in the technology section, there are multiple choices for PCS technology. There are currently seven different standards for PCS deployment. Each of these standards has its merits and demerits. The three key technologies that appear to be attracting the most attention are GSM, TDMA, and CDMA. Vendors are providing products based on one or more of the different technologies of their current products, R&D capability, and current and future market position.

It is natural that the equipment vendors who have a strong GSM portfolio are pushing for a GSM equipment standard in the United States. The major vendors supplying GSM equipment are Ericsson, Nokia, and Nortel. Of the 500+ million pops (2 licenses per MTA), about 125 million pops are potentially planned for GSM coverage so roughly 50 percent of the population will be covered by GSM. Due to this partial coverage there could be problems with nationwide roaming. GSM uses a different networking protocol, GSM MAP, than that used by the cellular network, IS41. The different networking protocols and lack of GSM/AMPS dual-mode phone will not allow users to have access to wireless services everywhere in the United States. GSM vendors and operators are hoping that future auctions will allow the possibility of a national GSM footprint, thereby mitigating the nationwide roaming issues. Ericsson and Nortel have the majority share of GSM infrastructure contracts.

The CDMA infrastructure equipment providers are AT&T, Motorola, and Nortel/Qualcomm. Of the 500+ million pops, CDMA technology will be deployed to cover 265 million pops. Incumbent cellular carriers planning to deploy CDMA as their digital technology have naturally opted for CDMA implementation with their PCS licenses. This provides for nationwide coverage. AT&T, Nortel, and Motorola have shared the CDMA infrastructure contracts.

TDMA is adopted by PCS carriers, AT&T Wireless, and Southwestern Bell. These two carriers have deployed TDMA networks for their cellular properties. To gain operational efficiency it is natural that these companies plan to deploy TDMA for the PCS licenses. The PCS properties of these two operators collectively cover 114 million pops. These companies' PCS licenses are complementary to their cellular licenses; therefore, TDMA has practically nationwide coverage, albeit at different frequency. The IS41 network protocol for TDMA and the current analog network are the same. This allows AMPS/TDMA dual-mode phones to roam anywhere in the United States. AT&T and Ericsson are the major vendors that have TDMA infrastructure equipment contracts.

Due to intense competition among the vendors, operators have been able to obtain attractive pricing from the equipment manufacturers. In many cases, the equipment vendors are providing up to 120 percent financing of their infrastructure equipment. The operators are in a better position to negotiate better terms and arrangements with the vendors because of the intense competition for infrastructure equipment. AT&T has obtained the largest share of infrastructure equipment contracts, followed by Nortel, Ericsson, and Motorola.

### 5.3.4   Operators

The major players in the PCS market are the local exchange carriers (LECs), interexchange carriers (IECs), cable operators (CATVs), cellular operators, and newer entrants. The driving force and motivation for each of these groups to enter into the PCS market are summarized in Fig. 5.5.

The IECs pay an average of 40 percent of their long-distance revenue to the LECs for access fees. By increasing local access competition as

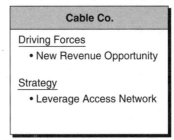

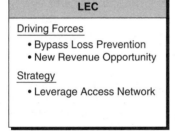

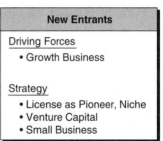

**Figure 5.5**   Major group of PCS players.

promised by PCS, the IECs increase their bargaining power to lower access fees. In addition, PCS provides an opportunity to leverage the IECs' existing brand name and to use their backbone intelligent network as the PCS backbone to develop a separate access network and bypass the LECs'. For example, AT&T's revenue would have increased 40 percent in 1992 if access costs were reduced by 10 percent.

The LECs have considered PCS as their competition since its inception and have tried to delay PCS through legal and political avenues. The LECs now realize that PCS is inevitable and have jumped on the PCS bandwagon. The LECs now want to leverage their existing network and access to subscribers.

Similarly, the cable operators plan to leverage their existing cable networks and access to subscribers. Furthermore, PCS provides a powerful counterthrust strategy for the cable operators to thwart the attempts of the LECs to get into the business of video delivery.

The current cellular operators consider PCS as a competitive substitute service. Furthermore, the existence of multiple operators will lead to lower profit margins and increased competition. The cellular operators plan to upgrade their network to digital in order to lower their infrastructure cost. Cellular operators are currently trying several PCS-like services. In addition, many cellular operators do not have a national/regional footprint. They will use their interest in their PCS operations to complement their existing cellular coverage.

The new entrants into the PCS market are aided by the FCC and Congressional mandates to include smaller players in the telecommunications sector. These new players are given the C and F band exclusively to operate their network. In addition, the FCC has encouraged the new entrants by providing preferential payment terms. The biggest challenge for the new entrant is to establish a nationwide presence and brand name and to obtain financing in order to compete with the bigger players.

All these players realize the need for nationwide coverage and brand recognition. The potential for price wars has led to the merging of several operators. These mergers, acquisitions, and alliances will help lower the unit costs of operation and marketing, provide an opportunity to obtain larger discounts by volume purchases, and have greater capital resources with a larger subscriber base. The two key alliances in the PCS industry are (1) the alliance between Bell Atlantic, NYNEX, Airtouch, and US West to form PCS PrimeCo, (2) the alliance between U.S. Sprint, Cox Cable, Comcast and other cable operators to form Sprint Telecommunications Venture, and (3) the purchase of McCaw Cellular by AT&T.

There are 20 PCS operators (based on A and B MTA licenses) that have licenses to provide PCS services. Unlike the cellular industry, which was initially fragmented and evolved by consolidation and merger into a small number of players with large footprints, the PCS industry is starting with operators that have almost nationwide coverage. The five largest PCS operators are Sprint Telecommunications Venture, AT&T Wireless, PCS PrimeCo, Pacific Bell Mobile Services, and American Portable Telecommunications Incorporated. Table 5.3 gives a list of their PCS coverage areas.

These five operators cover approximately 75 percent of the population with AT&T Wireless and PCS PrimeCo having the most significant cellular presence. Table 5.4 shows the population covered by their combined cellular and PCS licenses.

These companies, because of their nationwide presence and their name awareness, will be formidable and potentially attract the majority of wireless subscribers. In addition, they offer other telecommunications services, such as long distance and/or local services that could

TABLE 5.3   Licenses of the Top Five PCS Operators

| Operator | MTA areas | Total population (M) |
|---|---|---|
| Sprint Telecommunications Venture | New York, San Francisco–Oakland–San Jose, Detroit, Dallas–Fort Worth, Boston-Providence, Minneapolis–St. Paul, Miami–Ft. Lauderdale, New Orleans–Baton Rouge, St. Louis, Milwaukee, Pittsburgh, Denver, Seattle, Louisville, Phoenix, Birmingham, Portland, Indianapolis, Des Moines, San Antonio, Kansas City, Buffalo-Rochester, Salt Lake City, Little Rock, Oklahoma City, Spokane-Billings, Nashville, Wichita, Tulsa, and Philadelphia | 154 |
| AT&T Wireless | Chicago, Boston-Providence, Cleveland, Detroit, Charlotte-Greensboro-Greenville-Raleigh, Philadelphia, Washington-Baltimore, Atlanta, Cincinnati-Dayton, St. Louis, Richmond-Norfolk, Puerto Rico–U.S. Virgin Islands, Louisville-Lexington, Phoenix, Buffalo-Rochester, Columbus, El Paso–Albuquerque, Nashville, Knoxville, Omaha, and Wichita | 107 |
| Pacific Bell Mobile Services | Los Angeles–San Diego and San Francisco | 31 |
| American Portable Telecommunications, Inc. | Minneapolis–St. Paul, Kansas City, Columbus, Pittsburgh, Houston, Tampa–St. Petersburg–Orlando, Alaska and Guam–Northern Mariana Islands. | 27 |

be bundled together with wireless services. These companies have well-established brand names and can leverage their existing product offerings to entice subscribers and also obtain operational efficiency.

Some of the MTA license winners have already started selling and exchanging MTA licenses, based on the need for a regional footprint. Recently, GTE sold the Atlanta and Denver license to Intercell and Western Wireless. In turn, GTE bought the Spokane-Washington-Billings license from Poka Lambro Telephone Cooperative, Incorporated. This allows GTE to concentrate on clustering wireless services in markets where it already has a local telephone presence. Intercell, by acquiring the Atlanta license, will have four contiguous MTAs: Atlanta, Birmingham, Jacksonville, and Memphis-Jackson.

There are other companies, such as MCI, who plan to be major players in the wireless services by buying airtime wholesale and reselling

TABLE 5.4   Combined Population Coverage of Two Operators

| Operators | Population covered (M) | Percent covered (%) |
|---|---|---|
| AT&T Wireless | 189 | 74 |
| PCS PrimeCo | 168 | 66 |

it to their subscribers rather than building their own network. With a large number of wireless operators and abundant, new spectrum due to digital technology, there is potential for excess capacity. This excess capacity will allow companies to buy airtime at wholesale and sell it at retail and bundle special packages for larger corporate users at specially discounted prices. This way the company focuses on marketing its services rather than diffusing its energy by building and operating a network.

## 5.4  Technology

The FCC did not mandate a technology standard for the PCS market. As a result, there are currently seven standards, with different claims and counter claims, being peddled to the operators. Market forces such as the technology selected by the big carriers and consortia, system interoperability, equipment availability, voice quality, security, and price will determine the ultimate winners. However, there will be no single victor, but perhaps two or three winners. The seven PCS technology standards are as follows:

1. GSM (PCS 1900)—A derivative of the GSM/DCS 1800 standard with an 8-time-slot air interface
2. CDMA—A 1.25-MHz spread-spectrum-based air interface
3. DAMPS—A 3-time-slot TDMA air interface
4. PACS—An 8-time-slot air interface for pedestrian application
5. CDMA/TDMA—a composite hybrid that uses TDMA within cells and CDMA between cells
6. DCT-based TDMA—Based on the digital European cordless telephone with a 12-time-slot air interface
7. Wideband CDMA—A 5-MHz CDMA air interface

The four most widely deployed technologies, GSM, CDMA, TDMA, and PACS, are discussed in detail in subsequent chapters.

## 5.5  Summary

The concept of PCS involves terminal and personal mobility. Cellular and PCS both currently offer terminal mobility. It is expected that as technology and network evolve, personal mobility will also become universally available. The industry structure for the PCS market is essentially an oligopoly structure, which is largely influenced by regulatory environment. Regulation created the monopoly and duopoly structures in the local wire-line and cellular industry. The oligopoly structure in PCS is also created by regulatory rules on the number of service areas, the number of operators within each area, and the spectrum caps for the participants within each area.

The rest of this book focuses, first, on the different technology that will be deployed in PCS networks, and second, on the economic and operational considerations of a PCS operator.

# 6

# GSM

## 6.1 Introduction

Europe during the 1980s was experiencing a rapid growth of wireless communications. Subscribers were being added every day and network coverage was expanding. This explosive growth of wireless networks, though beneficial to the operators, was causing problems due to the proliferation of different, incompatible technologies and lack of a central body to coordinate the growth. Most of the existing technologies operated at different frequencies and all of them were analog. European operators realized that these analog networks would soon run out of capacity to support new subscribers. All of the following factors lead to the state of the wireless networks in Europe:

- There could be no roaming across a country's sovereign boundary due to incompatible wireless technologies that operated at different frequencies.

- Most of the networks were running out of capacity and looking for solutions to alleviate the problem.

- Economies of scale could not be reached for infrastructure and subscriber equipment due to a limited market for each technology. Further growth was being hampered by the high subscriber equipment costs.

Table 1.5 refers to the different cellular networks in Europe as of 1993. In contrast to Europe, the North American market had decided on a single technology called Advanced Mobile Phone Service (AMPS). It enabled subscribers to roam across most of the country, had less expensive handsets, and charged lower prices for service to the subscriber. To provide the advantages that come with a single technology and to overcome the deficiencies in their analog networks, telecommunication administrators of 26 countries met in 1982 under the auspices of the European Conference of Posts and Telecommunications Administrations (CEPT) and decided to establish a team with the title Groupe Speciale Mobile (GSM). The mandate of this team was to develop a set of common standards for a future Pan-European wireless network. The team decided to recommend that two blocks of frequencies in the 900-MHz band be set aside for the system. Later on, as the capacity of the analog systems became more of a driving factor, the GSM team decided that the new standard would be based on digital technology. Digital technology would offer better spectrum efficiency and quality, enhanced services, and would also allow lightweight, portable, and cheaper handsets to be manufactured. The United Kingdom later requested a version of GSM adapted to the 1800-MHz frequency band with a frequency allocation of 75 MHz each way. This was approved and was known as Digital Cellular System at 1800 MHz (DCS 1800). It was primarily for deployment in dense urban areas such as a downtown. This type of network was known as Personal Communications Network (PCN) and was very similar to the PCS network in North America.

The GSM team, from its inception, was strongly influenced by the migration of the wire-line networks toward ISDN. To enable the developing wireless digital standard to be compatible with the wire-line standard, the team decided that the GSM standard should closely follow that of the ISDN. This meant the use of similar signaling schemes and feature transparency across GSM and ISDN networks. This decision enabled a consistent access platform across both wire-line and wireless networks and facilitated commonality of features and services. Some of the main objectives of the GSM team were that the GSM standard should support the following:

- Vehicle-mounted, transportable, handhelds and other categories of mobile stations.

- International roaming, that is, roaming within all GSM countries.

- Quality of service as good as or better than the analog systems.

- Various ISDN services.

- Encryption (we will refer to it by its GSM term, ciphering) cost effectiveness.

- Coexistence with earlier systems in the same frequency band.

- Spectrum efficiency.

- Operation in the 890–915 and 935–960 frequency bands.

- CCITT-based numbering plan and signaling systems.

- Low-cost systems including mobile terminals.

Until 1986 GSM received support only from the equipment manufacturers, but this situation changed in 1987 when a memorandum of understanding (MoU) was signed by thirteen network operators from twelve countries. This ensured that when a GSM product became available, there would be customers ready to implement it into their networks. The GSM team decided to have two phases due to a delay in the completion of the specifications: phase 1 and phase 2. These phases were to be compatible yet specify different services.

Presently, GSM networks support close to 6 million subscribers in over 102 operator networks spread over 60 countries. The forecast is for close to 10 million subscribers by mid-1996. GSM has been adopted in Europe, the Middle East, Asia and Pacific, and Africa. The major GSM switching and radio equipment vendors are Alcatel, AT&T, Ericsson, Nokia, Northern Telecom, Motorola, and Siemens.

TABLE 6.1    GSM Systems (1995)

| Country | System | Status | Operator |
|---|---|---|---|
| ANDORRA | GSM | Licensed | STA |
| AUSTRALIA | GSM | Commercial | Optus |
| | GSM | Commercial | Telecom Australia |
| | GSM | Commercial | Vodafone |
| AUSTRIA | GSM | Commercial | PTV Austria |
| BAHRAIN | GSM | Licensed | Batelco |
| BELGIUM | GSM | Commercial | Belgacom Mobile |
| | GSM | Proposed | Second Operator |
| BRUNEI | GSM | Licensed | Jabatan Telekom |
| BULGARIA | GSM | Licensed | Mobile TEL |
| CAMEROON | GSM | Commercial | PM |
| CHILE | DCS 1800 | Proposed | Telex Chile |
| CHINA Beijing | GSM | Licensed | Beijing Telecoms Administration |
| CHINA Guangdong | GSM | Licensed | Guangdong Machinery Import & Export Corp. |
| | GSM | Licensed | Guangdong Mobile Comms Corp |
| CHINA Jiaxing | GSM | Licensed | MPT |
| CHINA Shanghai | GSM | Licensed | PTA |
| CHINA Shenzen | GSM | Licensed | PTA |
| CHINA Zhuhai | GSM | Licensed | Zhuhai Comms |
| CROATIA | GSM | Proposed | Croatian Post & Telecommunications |
| CYPRUS | GSM | Licensed | Cyprus Telecom. Authority |
| CZECH REPUBLIC | GSM | Proposed | Eurotel |
| | GSM | Proposed | Testcom Prague |
| DENMARK | GSM | Commercial | Dansk MobilTelefon |
| | GSM | Commercial | Tele Danmark Mobil |
| EGYPT | GSM | Proposed | Arento |

**TABLE 6.1    GSM Systems (1995) (Continued)**

| | | | |
|---|---|---|---|
| ESTONIA | GSM | Commercial | Eesti Mobiltelefon |
| | GSM | Licensed | Radiolinja Estonia |
| FIJI | GSM | Commercial | Vodafone Fiji |
| FINLAND | GSM | Commercial | Radiolinja |
| | | Commercial | Telecom Finland |
| | | Licensed | Alands Mobiltelefon |
| FRANCE | GSM | Commercial | France Telecom |
| | GSM | Commercial | SFR |
| | DCS 1800 | Licensed | Bouygues Telecom |
| | DCS 1800 | Proposed | SFR |
| FRENCH POLYNESIA | GSM | Licensed | Tikiphone |
| GERMANY | GSM | Commercial | DeTeMobil |
| | GSM | Commercial | Mannesmann Mobilfunk |
| | DCS 1800 | Commercial | E-Plus Mobilfunk |
| GIBRALTAR | GSM | Licensed | Gibtel |
| GREECE | GSM | Commercial | Panafon |
| | | Commercial | STET Hellas |
| GUERNSEY | GSM | Licensed | Guernsey Telecoms |
| HONG KONG | GSM | Commercial | Hongkong Telecom CSL |
| | GSM | Commercial | SmarTone Mobile Comms |
| | GSM | Licensed | Hutchinson |
| | DCS 1800 | Proposed | Hong Kong DCS 1800 1 |
| | DCS 1800 | Proposed | Hong Kong DCS 1800 2 |
| | DCS 1800 | Proposed | Hong Kong DCS 1800 3 |
| | DCS 1800 | Proposed | Hong Kong DCS 1800 4 |
| | DCS 1800 | Proposed | Hong Kong DCS 1800 5 |
| | DCS 1800 | Proposed | Hong Kong DCS 1800 6 |
| HUNGARY | GSM | Commercial | Pannon GSM |
| | GSM | Commercial | Westel 900 |
| ICELAND | GSM | Commercial | Iceland PTT |
| INDIA | GSM | Proposed | 36 regional licenses |
| INDIA Bombay | GSM | Licensed | Bharati Telecom |
| | GSM | Licensed | Max India |
| INDIA Calcutta | GSM | Licensed | India Telecom |
| | GSM | Licensed | Usha Martin |
| INDIA Delhi | GSM | Licensed | BPL |
| | GSM | Licensed | Sterling Cellular |
| INDIA Madras | GSM | Licensed | Mobile Telecom Services |
| | GSM | Licensed | Skycell |
| INDONESIA | GSM | Commercial | PT Telekom Indonesia |
| | GSM | Commercial | Satelindo |
| | GSM | Proposed | PT Kartika Ekamas Nusantara |
| IRAN | GSM | Commercial | TCI |
| IRELAND | GSM | Commercial | Telecom Eireann Eircell |
| | GSM | Proposed | Second Operator |
| ITALY | GSM | Commercial | SIP Mobile |
| | | Licensed | Omnitel |
| JERSEY | GSM | Commercial | Jersey Telecoms |
| JORDAN | GSM | Proposed | Jordan Mobile Telephone Services |
| KUWAIT | GSM | Commercial | Mobile Telecommunications Comp |
| LAOS | GSM | Licensed | Lao Shinawatra Telecom |

**TABLE 6.1    GSM Systems (1995) (Continued)**

| | | | |
|---|---|---|---|
| LATVIA | GSM | Commercial | Latvian Mobile Telephone |
| LEBANON | GSM | Licensed | Libancell |
| | GSM | Proposed | Lebancell |
| LITHUANIA | GSM | Proposed | Litcom |
| LUXEMBOURG | GSM | Commercial | Luxembourg P & T |
| MACAU | GSM | Licensed | CTM |
| MALAYSIA | GSM | Licensed | Binariang |
| | GSM | Licensed | Celcom |
| | DCS 1800 | Licensed | Berjaya |
| | DCS 1800 | Licensed | Malaysia Resources Corporation |
| | DCS 1800 | Licensed | Sapura |
| MAURITIUS | GSM | Licensed | Mauritius Telecom |
| MOROCCO | GSM | Commercial | Morocco PTT |
| NAMIBIA | GSM | Licensed | Mobile Telecommunications Ltd. |
| NETHERLANDS | GSM | Commercial | PTT Telecom |
| | GSM | Proposed | Second Operator |
| NEW ZEALAND | GSM | Commercial | BellSouth New Zealand |
| | GSM | Licensed | Telstra |
| NIGERIA | GSM | Proposed | EMIS Nigeria |
| NORWAY | GSM | Commercial | NetCom GSM |
| | GSM | Commercial | Tele-mobil |
| OMAN | GSM | Licensed | General Telecommunications Org. |
| PAKISTAN | GSM | Licensed | Pakcom |
| PHILIPPINES | GSM | Commercial | Globe Telecom |
| | GSM | Commercial | Isla Communications |
| POLAND | GSM | Proposed | PTK-Centertel |
| PORTUGAL | GSM | Commercial | Telecel |
| | GSM | Commercial | TMN |
| QATAR | GSM | Commercial | Q-Tel |
| REUNION | GSM | Licensed | SRR |
| ROMANIA | GSM | Proposed | Romanian Telecoms |
| RUSSIA | | | |
| Moscow | GSM | Commercial | Mobil Tele Systems |
| Bashkorstan | GSM | Licensed | United Telecom Bashkorstan |
| Blagoveshchensk | GSM | Licensed | United Telecom Far East |
| Kaliningrad | GSM | Licensed | DeTeMobil |
| Khabarovsk | GSM | Licensed | United Telecom Far East |
| Kazakhstan | GSM | Licensed | Wireless Technology Corporation |
| Krasnodar | GSM | Licensed | United Telecom |
| Nizhni Novgorod | GSM | Licensed | United Telecom Nizhni Novgorod |
| Novgorod | GSM | Licensed | DeTeMobil |
| Perm | GSM | Licensed | United Telecom Ural |
| Petropavlovsk | GSM | Licensed | United Telecom Far East |
| Pskov | GSM | Licensed | DeTeMobil |
| Rostov | GSM | Licensed | United Telecom |
| Rostov Oblast | GSM | Licensed | RTDC |
| St. Petersburg | GSM | Commercial | North West GSM |
| Tver | GSM | Proposed | DeTeMobil |
| Udmurtiya | GSM | Licensed | United Telecom |
| Vladivostok | GSM | Licensed | United Telecom Pacific |
| Volgograd | GSM | Licensed | United Telecom |
| SAUDI ARABIA | GSM | Licensed | MoPTT |
| | GSM | Licensed | Royal Palace of Saudi Arabia |

**TABLE 6.1    GSM Systems (1995) (Continued)**

| | | | |
|---|---|---|---|
| SINGAPORE | GSM | Commercial | Singapore Telecom |
| | GSM | Proposed | Second Operator |
| | DCS 1800 | Proposed | Singapore Telecom |
| SLOVENIA | GSM | Licensed | Slovenian PTT |
| SOUTH AFRICA | GSM | Commercial | Mobile Telephone Network |
| | GSM | Commercial | Vodacom |
| SPAIN | GSM | Commercial | Telefonica |
| | GSM | Proposed | Second Operator |
| SRI LANKA | GSM | Licensed | MTN |
| SWEDEN | GSM | Commercial | Comviq GSM |
| | GSM | Commercial | NordicTel |
| | GSM | Commercial | Telia Mobitel |
| SWITZERLAND | GSM | Commercial | Swiss Telecom PTT |
| | DCS 1800 | Licensed | PTT Telecom |
| SYRIA | GSM | Licensed | Syrian Telecom |
| TAIWAN | GSM | Licensed | DGPT |
| THAILAND | GSM | Commercial | Advanced Information Service |
| | DCS 1800 | Commercial | Total Access Communications |
| TUNISIA | GSM | Proposed | Tunisian PTT |
| TURKEY | GSM | Commercial | Turkish PTT |
| | GSM | Commercial | Turkish PTT |
| UGANDA | GSM | Licensed | MTN Uganda |
| UAE | GSM | Commercial | Etisalat |
| | GSM | Commercial | Etisalat |
| UK | GSM | Commercial | Telecom Securicor Cellular Radio |
| | GSM | Commercial | Vodafone |
| | DCS 1800 | Commercial | Hutchison Telecom |
| | DCS 1800 | Commercial | Mercury |
| VIETNAM | GSM | Commercial | Vietnam Telecom Services Co |
| | GSM | Proposed | DGPT Singapore Telecom |
| ZAIRE | GSM | Proposed | African Telecom Network |

## 6.2    Telecommunication Services in GSM

Telecommunication services supported by a GSM network are the communication capabilities made available to a subscriber by network operators. The GSM network, in collaboration with other networks such as the PSTN, provides a set of network capabilities that are defined by standard protocols and functions and enables telecommunication services to be offered to customers. The GSM team was very strongly influenced by the telecommunication services being provided by ISDN. At the very minimum they wanted to provide some of the basic telecommunication services supported by ISDN. They were hampered by not being able to support all of the ISDN services because of the bandwidth limitation on the air interface. ISDN supports speech at 64 kbps as a basic service, which is not possible for GSM since the air interface cannot support this rate. The telecommunication services supported by GSM are derived from the wire-line ISDN, current analog-cellular, and paging networks.

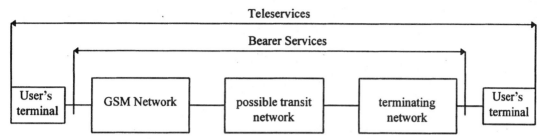

**Figure 6.1**    Teleservices and bearer services.

Figure 6.1 shows the scope of teleservices and bearer services. Telecommunication services in GSM are divided into two main groups: basic services and supplementary services. Basic services are further divided into the following two main categories:

1. Bearer services—These are telecommunication services that give the user the capacity needed to transmit appropriate signals between certain access points. They provide the means to convey information (speech, data, etc.) between users in real time and without alteration of the content of the message. These services correspond to the lower three layers of the OSI model. Bearer services define requirements for network functions.

2. Teleservices—These are telecommunication services that provide the user with necessary capacities, including terminal-equipment functions, to communicate with any other users. They combine the transportation function with the information-processing function, employ bearer services to transport data, and provide a set of higher layer functions. These higher layer functions correspond to layers four through seven. Teleservices include network as well as terminal capabilities. Examples include telephony, facsimile, and so on. While bearer services will be used to bring the digital bit stream containing speech to the terminal, teleservice will allow it to be converted to sound the user can hear.

Several criteria have to be fulfilled for a customer to be able to use any of these services. First, the GSM network has to support the service. As mentioned, the various services have been categorized as Phase 1 and Phase 2 services. The rollout of these services is operator dependent. In order to be able to use a particular service, it has to be first implemented and supported by the GSM network. Support of the service by the network indicates that it has to be supported not only by the subscriber's home GSM network, but also by the visiting GSM network if the subscriber wishes to use the service in the visiting network.

Once a service is supported by the network, the customer can subscribe to that particular service. This is the second criterion that has to be met for a customer to be able to use a service. *Subscription* means that the subscriber is authorized to use the service by the network. Depending on the type of service subscribed to, the subscriber may be given the option to activate or deactivate the service for a certain time period. Additionally, there may be different charges for various services. Another criterion that has to be met is that the subscriber's mobile terminal must be able to support the service. For example, a speech-only mobile phone cannot support facsimile even though the network supports it and the subscriber may have subscribed to it. Finally, there has to be terminal compatibility between both the calling and called party's equipment.

### 6.2.1  Bearer services

A variety of bearer services are offered. GSM users can send and receive data at rates up to 9600 bps, to users on PSTN, ISDN, Packet Switched Public Data Networks, and Circuit Switched Public Data Networks using a variety of access methods and protocols, such as X.25 or X.32. Since GSM is a digital network, a modem is not required between the user and GSM network, although an audio modem is required inside the GSM network to interwork with PSTN. An important observation related to bearer services is that all of them are point-to-point services. Table 6.2 lists the bearer services supported by GSM.

**TABLE 6.2   Bearer Services**

| Service | Details |
|---|---|
| Asynchronous data | 300–9600 bps |
| Synchronous data | 1200–9600 bps |
| PAD access | 300–9600 bps. Provides an asynchronous connection to a Packet assembler/dissembler. This enables GSM subscribers to access a Packet network. This service is applicable to mobile-originated calls only. |
| Packet access | 2400–9600 bps. Provides a synchronous connection that enables a GSM subscriber to access a Packet network. This service is applicable to mobile-originated calls only. |
| Alternate speech/data | The data phase could be asynchronous/synchronous up to 9600 bps. This service provides the capability to swap between speech and data during a call. For instance, a user can be talking and then send a fax and resume the conversation after the fax has been received at the other end. Some means has to be provided at the mobile station to select speech/data capability. |
| Speech followed by data | The data phase could be asynchronous/synchronous up to 9600 bps. This service provides a speech connection first and then at some time while the call is in progress, the user can switch to a data connection. The user cannot switch back to speech after the data portion. |

### 6.2.2  Teleservices

**6.2.2.1  Telephony and emergency calls.**  Support for speech is provided by encoding the speech and digitally transmitting it to the network. A speech encoder is essential, since without it speech would have to be transmitted digitally at 64 kbps. Speech can be transmitted at 64 kbps, but it would be spectrally inefficient, which is one of the reasons GSM is standardized for 13 kbps speech transmission. There are low-bit-rate speech encoders available that can provide acceptable speech quality at close to 16 kbps. GSM makes use of a 13-kbps (full rate) speech coder to encode the speech. There is also provision for use of still lower bit rates of close to 6.5 kbps (half rate) when they become commercially available. GSM also provides for emergency calls by providing a three-digit code that would connect the user to the nearest emergency call center. Table 6.3 lists the teleservices supported by GSM.

**6.2.2.2  Short message services (SMS).**  Short message services, offered to subscribers, have their origin in the paging operators' services. Paging subscribers are able to receive short, alphanumeric messages indicating the caller and the reason for the call. The paging subscriber then has to locate a phone, either a fixed wire-line phone or a mobile phone, and call the sender of the message to get the details. GSM has provided for this simple action by incorporating this paging functionality within

**TABLE 6.3   Teleservices**

| Service | Details |
| --- | --- |
| Telephony (speech) | Full rate (13 kbps), half rate (6.5 kbps). This service provides for the transmission of speech information and audible signaling tones to the network. |
| Emergency calls (speech) | Typically, emergency calls will supersede any constraints placed by other services. It is a mobile-originated service only. |
| Short message service (mobile terminated point-to-point, MT/PP) | Short alphanumeric message, ≤160 characters. This service provides for the transmission of a short message from a message-handling system (service center) to a mobile station. The service center is functionally separate from a GSM network. |
| Short message service (mobile originated point-to-point, MO/PP) | Short alphanumeric message, ≤160 characters. This service provides for the transmission of a short message from a mobile station to a message-handling system (service center). Information from the following sources at the mobile station might be transmitted: a prerecorded message in a store, a number from the dialing keypad, information from an external keyboard or terminal equipment connected to the mobile station. |
| Short message transmission (cell broadcast) | Short alphanumeric message, ≤93 characters. This service provides for the transmission of a short message from a message-handling system to all mobile stations in a cell. This is a point-to-multipoint service. |
| Automatic facsimile | Group 3 fax. This service supports facsimile group 3 autocalling/autoanswering mode only. |

the GSM network and the mobile stations. This alleviates the need for GSM subscribers to carry two wireless devices: the GSM mobile station and a pager. GSM designers have even allowed for mobile-originated short messages. This can allow for a potentially bidirectional short message conversation between users.

For both MO/PP and MT/PP short message services the service center acts as a store-and-forward center. The service center is functionally separate from the GSM network. Messages may be input to the service center either from a fixed network customer by speech, telex, and facsimile or from a mobile customer. All GSM point-to-point short messages are either to or from the service center. A message from one mobile station to another must pass through a service center. GSM allows for short messages to be sent either when the subscriber is involved in a call (speech or data) or when idle.

Cell broadcast messages are sent cyclically in a given area of the GSM network. The mobile stations can receive cell broadcast messages only when they are in the idle mode. Typically, all mobile stations equipped with this functionality will be able to receive the broadcast messages, hence this is not a service that can be subscribed to on an individual basis. Information that is transmitted in a cell broadcast message is of relevance to the mobile stations in that geographical area, not to a specific user. The message information may be related to weather, traffic congestion, accidents, and so on.

### 6.2.3  Supplementary services

Supplementary services may enhance both bearer services and teleservices. A supplementary service is one that may be used in conjunction with one or more of the bearer or teleservices. It cannot be used alone. It must be offered to the customer together with a basic telecommunication service. The same supplementary service may be applicable to a number of telecommunication services. An example of a supplementary service is call forwarding. The prerequisite for this service is either the speech or facsimile teleservice. If so desired by the subscriber, the call forwarding supplementary service can be applied to both speech and fax calls. Table 6.4 lists the supplementary services supported by GSM.

#### 6.2.3.1  Calling line identification presentation (CLIP).   The CLIP supplementary service provides the called party with the possibility of receiving the identity of the calling party. Identity includes the subscriber's number along with the country code. For instance, assume that a caller, Ann, is calling another user, Bob. When Ann calls Bob, if Bob has subscribed to CLIP, then Ann's phone number will be displayed to Bob when he is alerted. This service provides Bob the information about the

**TABLE 6.4    Supplementary Services (SS)**

| Service | Details | |
| --- | --- | --- |
| Number identification SS | CLIP | calling line identification presentation |
| | CLIR | calling line identification restriction |
| | CoLP | connected line identification presentation |
| | CoLR | connected line identification restriction |
| Call offering SS | CFU | call forwarding unconditional |
| | CFB | call forwarding on mobile subscriber busy |
| | CFNRy | call forwarding on no reply |
| | CFNRc | call forwarding on mobile subscriber not reachable |
| Call completion SS | CW | call-waiting |
| | HOLD | call hold |
| Multiparty SS | MPTY | multiparty service |
| Community of interest SS | CUG | closed user group |
| Charging SS | AoCI | advice of charge (information) |
| | AoCC | advice of charge (charging) |
| Call restriction SS | BAOC | barring of all outgoing calls |
| | BOIC | barring of outgoing international calls |
| | BOIC-exHC | barring of outgoing international calls except those directed to the home GSM country |
| | BAIC | barring of all incoming calls |
| | BIC-Roam | barring of incoming calls when roaming outside the home GSM country |
| Unstructured SS | | Unstructured supplementary services data |
| Operator-determined barring | | Restriction of different calls/services by operator |

calling party that could be used by him to either accept or reject the call. If Bob is waiting for an important call from a customer, he might choose to ignore Ann's call and wait for his customer's call. Without this service, Bob does not have any indication about the identity of the calling party. This service is applicable to all mobile terminated telecommunication services, except for the short message service. Typically, when provisioned, this service is permanent; that is, all calls to the user will have the calling party's identity presented. Invocation of the service is done automatically by the network during the call setup phase.

**6.2.3.2   Calling line identification restriction (CLIR).**   The CLIR supplementary service enables the calling party to prevent presentation of its identity to the called party. In the previous instance, Ann could have prevented the presentation of her phone identity to Bob by having subscribed to CLIR. This service would be useful whenever the calling party needs to maintain user confidentiality. For instance, whenever a user is calling a direct-marketing firm for information, the user might wish to maintain confidentiality to prevent unwanted sales calls from telemarketers. The CLIR supplementary service is applicable to all mobile-originated telecommunication services, except short message service and emergency calls. CLIR can be offered on a permanent basis (all outgoing calls) or on a temporary basis (suppressed by user on a per call basis). If the temporary basis is chosen, the user has to explicitly suppress the service for each outgoing call. CLIR takes precedence over CLIP; for example, if Ann, with the CLIR provision, is calling Bob, who has the CLIP service, then Ann's identity will not be presented to Bob.

**6.2.3.3  Call forward unconditional (CFU).**  This service permits a called mobile subscriber to direct the network to send all incoming calls to another directory number. *Unconditional* indicates that the type of termination is not a consideration during forwarding. This means that irrespective of the mobile station being turned off, idle, or involved in a call, the forwarding takes place. The ability of the served mobile subscriber to originate calls is unaffected (i.e., this is a mobile termination service only). The subscribed user can request a different forwarded-to number for each basic service group, such as speech and facsimile. For example, a user can have a forwarded-to number to a voice mail system for all speech calls and a different forwarded-to number to an office fax machine for all facsimile calls. CFU is applicable to all telecommunication services except emergency calls and short message service. For CFU to function correctly, the user has to register the forwarded-to number in the network. This is typically done by the user, who dials a digit string via the mobile phone keypad. The forwarding subscriber is charged for the forwarded portion of the call.

**6.2.3.4  Call forwarding on mobile subscriber busy (CFB).**  This service permits a called mobile subscriber to have the network send all incoming calls to another directory number when that subscriber is currently involved in a call. This service is similar to CFU, with the difference that CFB is instigated only when the called user is already involved in another call. Similar to CFU, the user can have different forwarding numbers for each basic service group. CFB is applicable to all telecommunication services except emergency calls and short message service. A CFB service option allows the forwarding subscriber to be notified that a call has been forwarded.

**6.2.3.5  Call forwarding on no reply (CFNRy).**  This service is similar to CFU, with the difference that the forwarding is done only on not getting a reply from the called user. In CFU, the call is forwarded without the user being alerted. In CFNRy, the user is first given an opportunity to answer the call by being alerted. This phase of the call is no different from that of regular call termination. CFNRy comes into play when the user does not answer the call before the time-out period. Apart from the common registration elements, a CFNRy subscriber also has to register the time-out period with the network. This period usually ranges from 5 to 30 seconds.

**6.2.3.6 Call forwarding on mobile subscriber not reachable (CFNRc).** This service is similar to CFU, with the difference that the forwarding is done only when the mobile station is not reachable. *Not reachable* means that the network is unable to contact the mobile station for call termination. This could occur due to several reasons, such as the mobile station is turned off, out of radio coverage, or not registered in the system. The network first attempts to make contact with the mobile station. If it is unsuccessful the call will be forwarded to the predetermined number.

**6.2.3.7 Connected line identification presentation (CoLP).** Figure 6.2 describes the differences between CLIP and CoLP service. The connected line identification presentation (CoLP) supplementary service provides the calling party with the identity of the connected party. This service differs from the CLIP in that CLIP provides identity of the calling party to the called party, whereas CoLP provides identity of the connected party to the calling party. For a simple call termination the called party address is the same as the connected party address, but it will be different if there is forwarding involved. For example, assume that Ann calls Bob. If Ann has subscribed to CoLP, then Bob's number will be displayed on Ann's terminal. If Bob has CFU activated and his phone is forwarded to Cathy, then the call is forwarded to Cathy when Ann calls Bob and Cathy's number is now displayed on Ann's terminal, since Cathy is the connected party. This enables Ann to terminate the call if she wishes to communicate only with Bob, not Cathy. This service is applicable to all telecommunication services except SMS.

**6.2.3.8 Connected line identification restriction (CoLR).** This supplementary service enables the connected party to prevent presentation of its line identity to the calling party. In the previous example, if Bob or Cathy had subscribed to CoLR, then their numbers would not have been displayed on Ann's terminal.

**6.2.3.9 Call hold (HOLD).** The call hold service allows a served mobile subscriber having subscription to this service to interrupt communication on an existing active call and then, subsequently, if desired, reestablish communication. The communication channel remains assigned to the mobile subscriber after the communication is interrupted to allow the origination or termination of other calls. Typically, if user Ann is on an active call with Bob and Bob wants to consult with Cathy, Bob has to invoke the call hold service. Bob would place Ann on

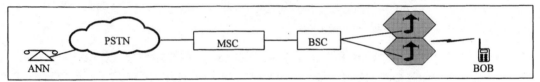

Bob has subscribed to CLIP service
When Ann calls Bob, Ann's phone number is displayed at Bob's terminal

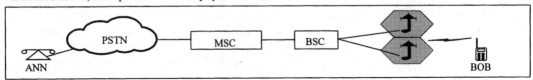

Ann has subscribed to CLIR service and Bob has subscribed to CLIP service
When Ann calls Bob, Ann's phone number will not be displayed at Bob's terminal

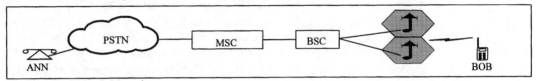

Ann has subscribed to CoLP service
When Ann calls Bob, Bob's phone number will be displayed at Ann's terminal

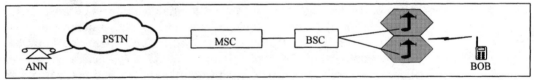

Ann has subscribed to CoLP service , Bob has subscribed to CoLR
When Ann calls Bob, Bob's phone number will not be displayed at Ann's terminal

**Figure 6.2**   CLIP/CLIR and CoLP/CoLR service.

hold, dial Cathy's number and get connected to Cathy. Now Bob can toggle between Ann and Cathy, but cannot talk to both at the same time. GSM man-machine interfaces to the handsets to support supplementary services are very different from the analog protocol. GSM is very feature rich and allows a lot of flexibility in the handling of supplementary services, but with it comes an increase in complexity and loss of user-friendliness. For invocation and support of call hold, the procedures involved are complicated and the reader is referred to the GSM specification 2.30 that deals with the MMI of the mobile station for details. Suffice it to say the GSM allows users to hold calls and make new calls, toggle between the active and held calls, terminate either one and still be on the remaining call.

**6.2.3.10   Call-waiting (CW).**   The call-waiting service permits a mobile subscriber to be notified of an incoming call while being engaged in another call. The subscriber then has the option of either accepting, rejecting, or ignoring the incoming call. The notification to the subscriber is in the form of a call-waiting tone. If the subscriber has subscribed to call hold also, then it is possible to toggle between the two calls without releasing either one. This service is applicable to all GSM telecommunication services, except emergency calls, using a circuit switched connection. The call-waiting call can be of any type.

**6.2.3.11 Multiparty (MPTY).** This supplementary service provides for the mobile subscriber to have a conference call (i.e., a simultaneous communication with more than one party). A condition for multiparty is that the subscriber currently is involved in two calls—one active and one on hold. The subscriber can then conference these two calls into a single call. GSM allows for a maximum of 5 parties to conference in, along with the user. Once a multiparty call is established, the user can receive or initiate additional calls.

**6.2.3.12 Closed user group (CUG).** This service enables a group of users to customize the use of telecommunication services. Unlike most other supplementary services, CUG service does not describe a single service but a group of services that are specific to private user groups. A *user group* is a list of individual subscribers who are bound together by a common interest (usually business). Any subscriber can become part of a user group. The user groups are typically used by companies who are in the service, construction, or utility industries. It allows these companies to tailor the service to the function of the user. Some of the common services available in CUG are: access to outgoing calls only, incoming calls only, or restriction of calls to members of the group only, or any other combination. For example, it is possible to provide a CUG service where the user can make and receive any type of call to and from users who are part of the designated group, but only receive incoming calls from people outside the group. Thus, this user cannot make a call to anyone who is not a member of his or her user group. Provisioned in this way CUG allows, for instance, a salesperson of a company to contact his or her colleague but prevents misuse of the phone for personal use. This service is traditionally used in specialized mobile radio markets (SMR) but GSM has adopted these services to be offered to any cellular user. Any individual member is at all times limited to a maximum of 10 CUGs.

**6.2.3.13 Advice of charge (AoC).** This service permits a mobile station to display an accurate estimate of the size of the bill that will be levied in the home GSM network. It provides the MS with the information to produce an estimate of the cost of the service used. Charges are indicated for the call in progress when the call is mobile originated, or for the roaming leg only when the call is mobile terminated. Any charges for non-call-related transactions and for certain supplementary services, such as call forwarding, are not indicated. This service can be requested by both the calling and called party. The total, accumulated charge for each call is stored in the mobile station.

**6.2.3.14    Call barring services.**    These services allow restrictions to be placed on the subscriber on the type of calls allowed. Typically, the restrictions are categorized into incoming and outgoing calls. Barring of outgoing calls can further be customized to specific types of calls, such as international, all outgoing calls, or those directed toward the home GSM network country. The last one requires special mention. If a subscriber has subscribed to call barring except to those directed toward the home GSM network country then, essentially, when the user is roaming outside of the home country, all international calls except to those to his home country are denied. In all types of outgoing call barring, the ability of the user to receive incoming calls remains unaffected. Similarly, there are also restrictions on incoming calls, such as all incoming calls or incoming calls when roaming outside the home GSM network country.

**6.2.3.15    Unstructured supplementary services.**    These services provide a way for operators to differentiate themselves from other networks. They allow operators to customize specific services or features for their subscribers. There is no way to ensure that these services will work similarly outside the home network. A particular operator may have a service where subscribers can dial a specific number to get stock quotes, or maybe pizza, or to get connected to a radio station. These numbers may differ from network to network. However, operators have the flexibility to offer these kinds of services to their subscribers due to the support of unstructured supplementary services by GSM.

## 6.3    GSM Architecture

GSM has a modular network architecture with standardized interfaces between segments. This allows an operator to mix-and-match any vendor's equipment into the system. Figure 6.3 shows the GSM network architecture, its elements, and its standard interfaces.

The following are the network elements.

### 6.3.1    Mobile station (MS)

The mobile station usually represents the only equipment the user ever sees in the GSM system. Mobile stations are used by the sub-

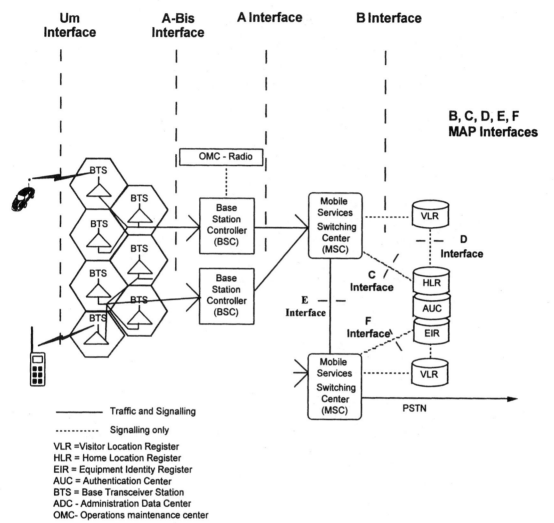

**Figure 6.3** GSM network architecture.

scriber to access the GSM network and interface to it via the air interface. Types of mobile stations include vehicle-mounted, portable equipment, and handheld stations that make up most of the market. The GSM handset from Northern Telecom is shown in Fig. 6.4.

Different types of stations are distinguished by power and application. Mobile stations used in the GSM 900-MHz band have different power requirements from the ones used in the DCS 1800-MHz band. Tables 6.5 and 6.6 list the different power classes of the mobiles by application.

**Figure 6.4**  GSM mobile handset. (*Courtesy: Northern Telecom.*)

The DCS 1800 mobile stations have lower power levels since at that frequency the DCS system has a smaller cell size compared to that of the GSM 900 system. Smaller cells means that the mobile stations need to transmit with lower power, since they are relatively closer to the cell transmitter.

**TABLE 6.5    Mobile Power Levels in the GSM 900 System**

| Power class | Maximum power of a mobile station |
|---|---|
| 1 | 20 W |
| 2 | 8 W |
| 3 | 5 W |
| 4 | 2 W |
| 5 | 0.8 W |

**TABLE 6.6    Mobile Power Levels in the DCS 1800 System**

| Power class | Maximum power of a mobile station |
|---|---|
| 1 | 1 W |
| 2 | 0.25 W |

Besides generic radio and processing functions to access the network through the radio interface, a GSM mobile station offers an interface to the human user and to some other terminal equipment (such as an interface to a personal computer or a facsimile machine). GSM specifications describe several features that are supported by the mobile station. These can be classified into basic features, supplementary features, and additional features. *Basic features* are those related directly to the operation of the basic telecommunication services. Some of the mandatory basic features are: display of a called number, indication of call progress signals, emergency call capabilities, DTMF tones, and subscription management. Support of supplementary services, such as the display of a calling line number (CLIP) is optional. Support of the call barring and call forwarding supplementary services is mandatory. In addition to basic and supplementary services support, the mobile station manufacturer can support additional features, such as abbreviated dialing or last number redial. All additional features are optional and vary from manufacturer to manufacturer.

A GSM mobile station is split in two parts, one of which contains the hardware and software specific to the radio interface, and another which contains the subscriber-specific data: the subscriber identity module or SIM. The SIM is one of the most attractive and user-friendly features of GSM. The SIM can be either a smart card, having the well-known size of smart cards (similar to a credit card), called ISO SIM, or it can be a much smaller format, known as a plug-in SIM. This smaller format is the one used in most of the hand portables. The SIM is a removable module and provides some personal mobility. The subscribers can remove the SIM and carry it with them when they are traveling. At their new destination, they can plug their SIM into a GSM mobile station and get service. This portability feature allows GSM subscribers to roam (called SIM-roaming) into GSM networks that do not have a signaling connection to the subscriber's home HLR. It also allows users to easily rent or lease GSM mobile stations without a lot of administrative overhead. The SIM contains information related to the mobile subscriber, subscribed GSM services, and network. It stores data, such as the subscriber's identity, location information, and security-related data, such as ciphering keys, forbidden GSM networks, and language preference. Optionally, the SIM can store abbreviated dialing numbers, last number dialed, short messages and associated parameters, call meter, cell broadcast message identifier and more. SIM supports the use of a personal identification number (PIN) that authenticates the user of the card and provides protection against the use of stolen cards. The PIN is a 4- to 8-digit number and is initially loaded during provisioning of the SIM. If the SIM is removed from the mobile station, the mobile station is capable of making only emergency calls.

GSM mobile stations also differ from traditional, analog mobile stations in the manner in which they handle supplementary services. Due to the specific type of signaling used in GSM to invoke supplementary services, not all of the GSM mobile stations support supplementary services. Call-waiting/call hold, short message service (SMS), data, cell broadcast, calling line identity presentation (CLIP), and advice of charge (AoC) are some of the services that many GSM mobile stations may or may not support.

### 6.3.2 Base station or base transceiver station (BTS)

The mobile's interface to the network is the base transceiver station. A BTS comprises radio transmission and reception devices, up to and including the antennae, and also all the signal processing specific to the radio interface. Toward the BSC, the BTS separates the speech and control signaling associated with a mobile station and sends them to the BSC on separate channels. Toward the MS, the BTS does the reverse—it combines the signaling and speech onto a single carrier. A BTS is usually located in the center of a cell and is responsible for one cell. The transmitting power of the BTS determines the absolute cell size. A BTS will typically range from 1 to 24 transceivers (TRX), each of which represents a separate RF channel. The GSM BTS from Northern Telecom is shown in Fig. 6.5.

Table 6.7 lists the base-station power levels.

Base stations come in various configurations depending on the application. Some base stations are made more rugged for outdoor use where

**Figure 6.5** GSM BTS. (*Courtesy: Northern Telecom.*)

TABLE 6.7    Base Station Power Levels

| Power class | Maximum power of a GSM 900 base station | Maximum power of a DCS 1800 base station |
|---|---|---|
| 1 | 320 W | 20 W |
| 2 | 160 W | 10 W |
| 3 | 80 W | 5 W |
| 4 | 40 W | 2.5 W |
| 5 | 20 W | |
| 6 | 10 W | |
| 7 | 5 W | |
| 8 | 2.5 W | |

both environmental and mechanical considerations are taken into account by providing year-round weatherproofing. Others are for indoor use where aesthetics are more important. One of the critical factors of indoor base stations is minimized size. Base stations can be configured as sectored or omnidirectional. Typically, 1+1+1 TRX sectored BTS are common. This indicates that the BTS can support three sectors in the same cell with each sector having 1 TRX. Other required equipment includes a power supply and backup power in case of power failure.

### 6.3.3    Base station controller (BSC)

A base-station controller monitors and controls several base stations. The chief tasks of a BSC are frequency administration, control of BTSs, and exchange functions. The BSC is in charge of all the radio interface management through the remote command of the BTS and the MS, mainly the allocation and release of radio channels and handover management. The BSC is connected on one side to several BTSs and on the other side to the mobile switching center (MSC). A BSC along with its associated BTSs is known as a BSS—base-station subsystem. The GSM BSC from Northern Telecom is shown in Fig. 6.6.

### 6.3.4    Transcoder and rate adaptor unit (TRAU)

This network element is responsible for transcoding the user data from 16 kbps to 64 kbps. Physically the TRAU can reside either at the BTS, between the BTS and BSC, at the BSC, between the BSC and the MSC or at the MSC. The GSM specifications allow for the TRAU to be placed at any of these places.

### 6.3.5    Mobile services switching center (MSC)

The main function of an MSC is to coordinate the setting up of calls to and from GSM users. It is primarily a switch and has access to several

**Figure 6.6**   GSM BSC. (*Courtesy: Northern Telecom.*)

databases to assist in the task of setting up calls. It is the interface between the GSM network and the PSTN. The MSC is an exchange that performs all the switching and signaling functions for mobile stations located in a geographical area designated as the MSC area. An MSC can connect to several BSCs. Apart from supporting the BSCs, the MSC also handles inter-BSC/MSC handover, interworking for data calls, DTMF digit collection, call routing, billing, paging function for mobile terminated calls, location registration update and management of radio resource during a call (what kind of channel is needed at each stage of the call). The GSM MSC from Northern Telecom is shown in Fig. 6.7.

### 6.3.6   Home location register (HLR)

HLR is a database in charge of the management of mobile subscribers. Logically, there is one HLR per mobile network. It contains information on teleservices and bearer services subscription, service restrictions, and supplementary services. The data stored in a HLR is of a semipermanent nature and does not usually change from call to call. Some of the data stored is the IMSI, service subscription information, location information (the identity of the currently serving VLR to enable routing of mobile-terminated calls), service restrictions, and supplementary services information. It handles transactions with both the MSC and the VLR nodes, which either request information from the HLR or update the information currently held in the HLR. The HLR also initiates transactions with the VLR to complete incoming calls and update

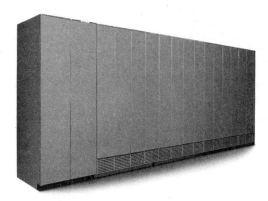

**Figure 6.7**  GSM  MSC.  (*Courtesy: Northern Telecom.*)

subscriber data. The subscriber data stored in the HLR is configurable by authorized operator personnel.

### 6.3.7  Visitor location register (VLR)

The VLR contains the relevant data of all mobiles currently located in a serving MSC. Usually there is one VLR per mobile switch. The permanent data is the same as data found in the HLR; the temporary data differs slightly. This data includes features currently activated, temporary subscriber identity (TMSI), and precise location of the mobile in the network (location area identity). The VLR also allocates mobile subscriber roaming numbers (MSRNs) for the incoming call setup. The data stored in the VLR is not for administrative use, unlike the data in the HLR.

### 6.3.8  Authentication center (AuC)

The authentication center is associated with an HLR and stores an identity key for each mobile subscriber registered with the associated HLR. The key is used to generate data that is used to authenticate the mobile subscriber identity, and this key in turn is used to generate another key used to encrypt the communication over the radio path between the mobile station and the network. It also stores the authentication (A3) and ciphering (A8) algorithms. The AuC communicates only with the HLR.

### 6.3.9  Equipment identity register (EIR)

EIR is essentially a database used to store the identity of the mobile station equipment. Its purpose is to help the network deny service to stolen or fraudulent mobile stations. To facilitate this the GSM specifications recommend three lists—white, black, and gray. A black list will contain the equipment identities of all mobiles that have been flagged as stolen, fraudulent, or severely malfunctioning. These mobiles will be denied service when they try to access the network. The white list contains the equipment identities of all approved mobile stations. These mobiles will be allowed service when they access the network. A gray list is a list of mobile equipment identities that are slightly malfunctioning but not severely enough to justify being denied service.

### 6.3.10   Operation and maintenance center (OMC)

The operation and maintenance center (OMC) has access to both the MSC and the BSC, handles error messages coming from the network, and controls the traffic load of the BSC and the BTS. The OMC configures the BTS via the BSC and allows the operator to check the attached components of the system.

## 6.4   GSM Network Interfaces

GSM has created a set of standard interfaces which allows an open system architecture. An operator can mix and match different vendors' equipment as elements in the network. Previously, each vendor had a closed system and each element was proprietary and restricted to the vendors equipment. In GSM it is possible for an operator to choose the BSS (BSC and BTS) from one vendor, the MSC and VLR from another, and the HLR from still another. Interworking is simpler due to the standardized interfaces among all of these entities.

### 6.4.1   Air interface (U$_m$)

Figure 6.8 shows the protocol used on the air interface. The radio interface between the BTS and the mobile station is known as the air interface or U$_m$ (user interface—mobile). The radio interface uses RF signaling as the layer one and modification of integrated digital services network (ISDN) protocol as layers two and three. This interface has been very well documented in the GSM standards and all mobile station and BTS vendors adhere to it strictly. Each RF channel on the air interface is broken down into time slots wherein mobile subscribers can transmit information. Each RF channel supports eight time slots (i.e., eight users/RF channel). Since the available radio spectrum is first divided into individual RF channels and each channel then further subdivided into time slots, the scheme used in GSM is a frequency-division duplex (FDD) time-division multiple access (TDMA) scheme. FDD indicates that two different RF channels are used for uplink and downlink communications.

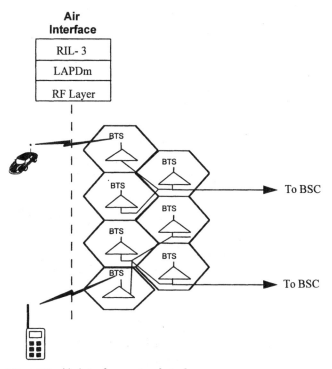

**Figure 6.8**   Air interface protocol stack.

## 6.4.2   A-bis interface

A-bis interface is the interface between the BTS and the BSC (see Fig. 6.9 for details about the protocol stack). This interface has been well documented in the GSM standards but most vendors have their own proprietary versions. Most of them differ in how they support some of the optional elements of this interface as well as in how they do operations, administration, and maintenance (OA&M). All the connections from the BSC to the BTS utilize a modification of ISDN signaling for layer three and use ISDN signaling for layer two. The physical interface is an E1. Since speech is compressed in GSM, each 64-kbps channel on the E1 supports four TDMA time slots (i.e., four users). There is a separate signaling channel used for control of the BTS that is also transported via an E1 time slot.

## 6.4.3   A interface

The A interface occurs between the MSC and the BSS (see Fig. 6.10). This interface is well documented in the GSM standards and is supported by all the MSC vendors. The A interface uses CCITT signaling system number seven (CCS7) for the lower three layers to trans-

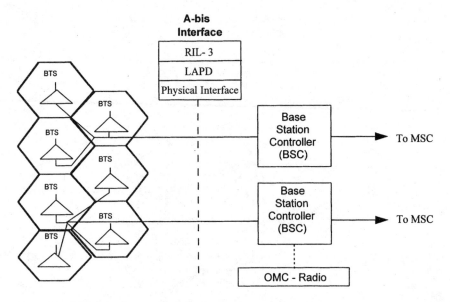

**Figure 6.9**   A-bis interface protocol stack.

port modified ISDN call-control signaling. The information carried on this interface pertains to management of the BSS, call handling, and mobility management. The SCCP and MTP layers provide for data transport. SCCP is implemented in two classes—0 and 2. Class 0 (connectionless) is for messages for the BSC, while class 2 (connection oriented) is for messages to a particular mobile station or logical connection. BSSMAP controls base-station functions and manages the physical connection between the BSS and the MSC. If also controls allocation of radio channels and intra-BSS handover.

### 6.4.4   PSTN interfaces

These are the interfaces between the MSC and the PSTN. All of these protocols are grouped under call-associated signaling. They are not specific to GSM and are commonly used in PSTNs for call setup. The GSM architecture is based on ISDN access and as such the MSC is based on an ISDN switch. To take full advantage of all the ISDN services the MSC should be connected to the PSTN via CCS7-based protocols such as ISUP. The MSC is forced to use whatever signaling protocol is in use by the connected PSTN switch. In many countries this may not be ISUP. There is nothing in GSM which precludes using a non-ISUP signaling scheme to connect to the PSTN to support basic calls. As mentioned, the drawback of this approach would be that some of the ISDN features, especially supplementary services, may be difficult to support.

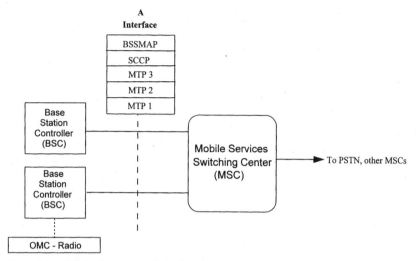

**Figure 6.10**    A interface protocol stack.

### 6.4.5    Mobile application part (MAP)

All non-call-associated signaling in GSM is grouped under MAP. Non-call-associated signaling implies all signaling dealing with mobility management, security, activation/deactivation of supplementary services, and so on. All MAP protocols use CCS7 lower three layers (i.e., MTP 1,2,3, SCCP layer, and TCAP layer). These protocols are used primarily for database queries and responses. Following are descriptions of specific MAP protocols.

**6.4.5.1    MAP-B.**  MAP-B is the interface between the MSC and its associated VLR. Whenever the MSC needs data related to a given mobile station currently located in its area, it interrogates the VLR. When a subscriber activates a specific supplementary service or modifies some data attached to a service, the MSC informs (via the VLR) the HLR that stores these modifications and updates the VLR if required. This interface between the MSC and the VLR is very heavily used, and hence the decision by several manufacturers to integrate the VLR functionality with the MSC.

**6.4.5.2    MAP-C.**  MAP-C is the interface between the MSC and the HLR. There is a specific function in GSM known as the gateway functionality. Due to the numbering and routing scheme used in GSM all calls to mobile subscriber are first handled by a gateway MSC. The term *gateway* refers to a path to the mobile network from the PSTN. This gateway MSC then queries the corresponding subscriber HLR to determine the routing information for a call or a short message directed toward the user. This messaging is handled by the MAP-C protocol. Additional SMS and charging messages also form part of this interface message set.

**6.4.5.3  MAP-D.**  MAP-D is the interface between the HLR and the VLR. It is used to exchange data related to the location of the mobile station and for the management of the subscriber. The VLR informs the HLR of the location of a mobile station managed by the latter and provides it with the roaming information for that subscriber. Exchanges of data may occur when the mobile subscriber requires a particular service, when changes to the subscription have to be done, or when some parameters of the subscription are modified by administrative means.

**6.4.5.4  MAP-E.**  When a mobile station moves from one MSC area to another during a call, a handover procedure has to be performed in order to continue the communication. For that purpose the MSCs have to exchange data to initiate and then to realize the operation. This interface supports the necessary signaling support for the handover function. When a short message is to be transferred between a mobile station and short message service center, this interface is used to transfer the message between the MSC serving the mobile station and the MSC acting as the interface to the message center.

**6.4.5.5  MAP-F.**  MAP-F is the interface between the MSC and the equipment identity register (EIR). It is used to exchange data to enable the EIR to verify the mobile subscriber equipment.

## 6.5  Numbering, Identification, and Security in GSM

### 6.5.1  Numbering and routing

**6.5.1.1  Mobile subscriber ISDN number (MSISDN).**  The MSISDN is the number dialed by a user in order to reach a GSM subscriber. The PSTN routes this call based on the MSISDN to the gateway MSC. The GMSC, based on its internal tables, correlates the MSISDN to the specific HLR, which has to be queried to get subscriber information. The HLR replies with specific information about the identity of the MSC where the subscriber is currently parked. It essentially provides a number at which the subscriber can be reached. This number is known as the mobile station roaming number (MSRN). It's structure is similar to MSISDN. The HLR may, in case of call forwarding, return a forwarded-to number. The GMSC then, using the MSRN, reroutes the call to that particular MSC. MSISDN is usually similar to the current dialing plan

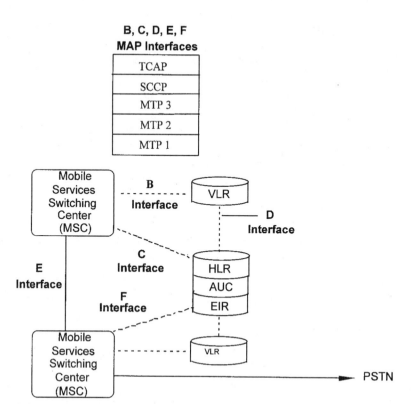

**Figure 6.11**   MAP interface protocol stack.

in use in wire-line networks. MSISDN has relevance only in the network and only for terminating calls. Its only interaction with the subscriber is that it supplies the number used by the subscriber to receive calls. In essence it is the directory number. It is not used on the air interface. MSRN's use is even more restrictive. It is used solely between network entities and no user has access to it. Moreover, unlike MSISDN, it is not permanently associated with a user and has significance only for that particular call.

MSISDN consists of a country code (CC), national destination code (NDC), and subscriber number (SN). The country code is used to identify the destination country and varies in length. For instance, the country code for the United States is 1 while it is 44 for the United Kingdom. The country code is administered by CCITT. The NDC can be used in two ways, either to identify the destination network or to identify the geographical area to which the called subscriber belongs. In the United States, it is used in the latter sense where the NDC is essentially the numbering plan area (NPA) or area code. It does not identify the destination network, but the geographical area. The subscriber number is the number to be dialed or called to reach a subscriber in the same numbering area. This number is the one usually listed in the directory beside the name of the subscriber. The NDC and SN are administered by the country-specific administration bodies.

**6.5.1.2    International mobile subscriber identity (IMSI).**    This is a unique number allocated to each mobile subscriber in the GSM system. It identifies each individual mobile subscriber uniquely, on an international basis. This number resides in the SIM and is thus transportable across mobile station equipment. It identifies the subscriber and the subscription which that subscriber has with the network. IMSI is used for internal purposes within the GSM network, such as accessing, identifying, and billing. As can be seen, the IMSI plays a critical role in GSM networks and there are elaborate procedures to ensure that it cannot be duplicated or used fraudulently. Figure 6.12 shows the structure of IMSI.

IMSI is composed of three parts: (1) the mobile country code (MCC) consisting of three digits, (2) the mobile network code (MNC) consisting of two digits, and (3) the mobile subscriber identification number (MSIN). The MCC uniquely identifies the home country of the mobile subscriber. MCC differs from the CC of the MSISDN in that it is of fixed length (i.e., 3 digits—unlike the CC which can be variable). Also the MCC differs from the CC in each country. For instance, the CC for the United States is 1 whereas the MCC ranges from 310–316. The reason for this change is to avoid ambiguity and to allow more numbers to be brought into service to support mobile subscribers. The MCC is administered by CCITT. The MNC uniquely identifies the network within the country. This means that the MNC does not have any geographical significance unlike the NDC of the MSISDN. The MSIN identifies the subscriber within the particular network.

**6.5.1.3    Temporary mobile subscriber identity (TMSI).**    Figure 6.13 shows the different numbering and identification parameters used in GSM. GSM provides for user confidentiality. To ensure that subscribers do

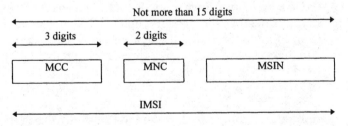

**Figure 6.12**    IMSI structure.

not transmit IMSI numbers in the clear where they can be intercepted, GSM provides for use of an alias over-the-air interface for IMSI (i.e., TMSI). The TMSI has only local significance (i.e., within the area controlled by a VLR). Whenever a mobile station has a TMSI available it uses it in place of IMSI to communicate to the network. TMSIs are allocated from a pool of numbers earmarked for that purpose and new mobiles are given the next available number. This scheme effectively separates the identity of the user from the number being used. Since the number varies over time and has only local significance (i.e., it is valid only so long as the subscriber is parked at that VLR), it is not of much use if intercepted.

**6.5.1.4  Location areas.** As stated before, the primary difference between wireless and wire-line networks is the issue of mobility and hence that of determining the current location of the subscriber. There are two ways in which this can be achieved. In the simplest nonlocation scheme the network does not attempt to keep track of the mobile and has no fore-knowledge of where the subscriber is currently located. Whenever an incoming call is placed for the subscriber the network does a networkwide page for the user. When the user responds to the page the network determines the location and connects the call. As we can quickly see, this scheme is not practical for large networks that support national and international roaming. Another way of terminating calls would be for the network to have some fore-knowledge of the location of the user and when a call comes in, page only in that location instead of a networkwide page. This introduces the concept of location areas. Location areas at the simplest level could consist of a single cell. Whenever a subscriber enters a new cell the network is informed of the new location area. This scheme lays undue burden on the network since a lot of signaling is involved in a location update and it is very

Information resident in the Mobile Station and SIM Card          Routing information used by the Network

**IMSI = International Mobile Subscriber Identity**
Nondialable number which acts as a unique identification of the subscriber.

**TMSI = Temporary Mobile Subscriber Identity**
Allocated by the VLR. Used on the air interface for security purposes. It provides user confidentiality.

**IMEI = International Mobile Equipment Identity**
Used on the air interface to provide equipment identity. Uniquely identifies the handset.

**Misc Information**
Authentication and ciphering parameters and algorithm, PIN etc.

**MSISDN = Mobile Station ISDN Number**
Dialable subscriber mobile phone number.

**MSRN = Mobile Station Roaming Number**
Used internally by GSM network elements. Provides routing to the visited MSC.

**Figure 6.13**  Parameters used in GSM for numbering and identification.

expensive to do it for every cell change for every subscriber. The compromise is to define a location area as an aggregate of cells. Each cell transmits the identity of the location area it is a part of to the mobile. Whenever the mobile realizes that the location areas have changed as a result of a change in cells, it sends in a location area update to the network. This is the scheme which is used in GSM. Location areas have to be very carefully engineered since they are a product of two conflicting requirements, that of reducing the paging traffic and that of reducing the location update traffic. If the location areas are very large, then the paging traffic will be very heavy and if they are very small then the location update traffic is high.

Refer to Fig. 6.14 for the structure of the location area identity. The location area identity and the TMSI is stored in the SIM. In GSM the mobile can initiate a location update either on its own or on command from the network (periodic location update). The location areas are identified by a location area identification (LAI). Each cell within a location area has its own identity known as *cell identity* (CI). An LAI and a CI uniquely identify each cell in the network. LAI's structure is similar to that of IMSI in that it has an MCC, MNC, and LAC (location area code). The MCC and MNC have the same significance as that of the IMSI and LAC—a fixed length code that identifies a location area within the GSM network.

MCC - Mobile Country Code
Uniquely identifies the country of the GSM subscriber.

MNC - Mobile Network Code
Identifies the GSM operator within the country. Each country can have several GSM operators each having a unique MNC.

LAC - Location Area Code
Defines a location area, which consists of a group of cells. Each MNC can have several LACs.

CI - Cell Identity
Uniquely identifies a cell in a location area.

LAI - Location Area Identity
Uniquely identifies a location area in the network. Made up of MCC + MNC + LAC.

CGI - Cell Global Identifier
Uniquely identifies the cell within the network. Made up of LAI + CI.

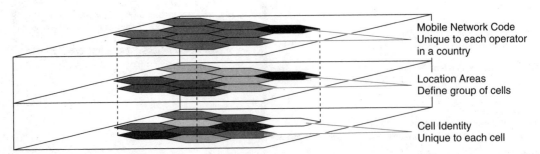

**Figure 6.14** Location area and cell identification parameters.

### 6.5.2 Authentication and ciphering

Figure 6.15 describes the authentication process in GSM. GSM provides for very strong ciphering and authentication procedures. As explained previously, authentication is necessary to prevent unauthorized or fraudulent use of the network, and ciphering is necessary for user communication confidentiality. The other issue related to security is that of user identity confidentiality and is resolved by the use of TMSIs, as explained in the previous section. GSM security procedures make use of the trapdoor functions to define the authentication and ciphering procedures. These functions make it computationally very difficult to determine the key, even when the plaintext, the corresponding encrypted message, and the algorithm used to encrypt the message are known. In GSM the authentication procedure is based on what is known as the unique challenge-response scheme. Whenever the network has to authenticate a mobile subscriber, it does several things. It has the user's secret key ($K_i$), the authentication algorithm (A3), and a list of challenge-response pairs. The challenge is a random number (RAND) which the network generates and uses it along with $K_i$ as input to the A3 algorithm. The output of the algorithm is the response and is known as signed response (SRES) in the GSM specification. On the other hand, the mobile station also has $K_i$, which is unique to each mobile station, and

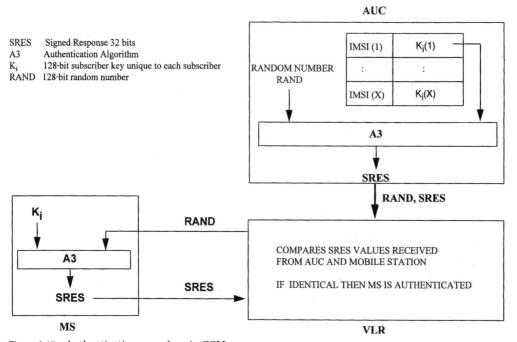

**Figure 6.15** Authentication procedure in GSM.

the A3 algorithm, which is the same for all mobile stations. The network, in order to authenticate the mobile subscriber, sends it a random number (RAND) picked from the list corresponding to the particular subscriber. The mobile station, on receipt of the random number (challenge), computes the signed response (SRES) exactly as is done in the network and sends the SRES to the network. The network then compares the SRES received with the SRES already computed. If they match the network permits or denies service to the user. Thus, even though the random number and the SRES can be intercepted on the air interface and the A3 algorithm is identical, $K_i$ cannot be known due to the nature of the trapdoor function. GSM provides for additional security by restricting the use of the A3 algorithm to operators only. Furthermore, $K_i$ is stored in the SIM in a very protected mode; even the mobile subscriber does not know what it is. The choice of different A3 algorithms is left up to the operators who have maximum flexibility in choosing the algorithm they need. Since authentication computations are very processor intensive, there is provision in GSM for an authentication center (AuC). Typically, the AuC computes several challenge-response pairs for each subscriber and sends them to the HLR, which stores them and makes

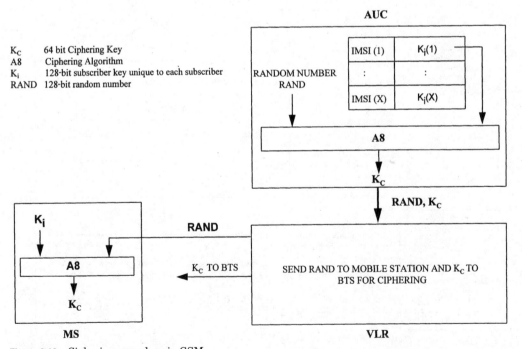

| | |
|---|---|
| $K_C$ | 64 bit Ciphering Key |
| A8 | Ciphering Algorithm |
| $K_i$ | 128-bit subscriber key unique to each subscriber |
| RAND | 128-bit random number |

**Figure 6.16**  Ciphering procedure in GSM.

**TABLE 6.8    Comparison of Authentication Procedures in IS-41 Rev. C and GSM**

| | GSM | IS-41 Rev. C |
|---|---|---|
| Provisioning of authentication key | The 128-bit authentication key $K_i$ resides in the SIM. By providing an SIM to the user, the authentication key is provisioned. | The 64-bit authentication key A-key is sent to the user via mail. The user, on receiving the key, has to enter it into the phone. Optionally, the key can be entered into the phone using over-the-air activation procedures. This does not require any user intervention. Shared, secret data is derived from the A-key. |
| Authentication mechanism | Unique challenge process wherein each user is provided a different random number (RAND) to authenticate. | Global challenge process wherein all users in the same cell use the same RAND. Unique challenge process is also supported, but it is not the primary authentication process. |
| Authentication procedure | Unique RAND along with $K_i$ is used as input to the A3 algorithm to generate SRES, which is sent to the network for verification. The input parameters to the A3 algorithm are independent of the type of access. | Broadcast RAND, ESN, MIN1 and SSD-A are used as input to the CAVE algorithm to generate AUTHR, which is sent to the network for verification. The input parameters to the CAVE algorithm vary on the type of access (e.g., registration, origination, etc.). |
| Authentication of roaming subscribers | Triplets are provided to the VLR. Each triplet contains the RAND, SRES, and ciphering key $K_C$. The VLR chooses a RAND from a triplet and requests the MS to authenticate it. It compares the SRES sent from the MS with the corresponding SRES in the triplet to authenticate the MS. Typically, the number of triplets sent at a time is five. Once the VLR uses them all, it has to request additional triplets from the HLR/AuC. The A3 algorithm resides at the AuC. | SSD is provided to the VLR. The VLR also has the CAVE algorithm. On receiving the RAND, ESN, and MIN1 from the MS, the VLR is able to use its stored copy of SSD-A as input to the CAVE algorithm to autonomously authenticate the MS. |
| Transmission of authentication parameters | Authentication parameters are transmitted once a channel has been established. This is more bandwidth efficient. | Authentication parameters can be sent as part of the initial service access message. |
| Protection against interception of authentication data | If the triplets are intercepted in the network, the user is compromised till the time all the triplets are used up. After this a new set of triplets will be requested. | If SSD is intercepted in the network, the user can be compromised for long periods of time. Call history count (COUNT) parameter is used to provide protection against this. |

use of them when needed. Table 6.8 details the authentication procedures as used in the North American protocol IS-41 Rev. C and GSM.

Figure 6.16 describes the ciphering process in GSM. GSM supports ciphering on the air interface between the mobile station and the BTS. The random number (RAND) used for authentication is used in conjunction with $K_i$ as input to algorithm A8. The output $K_C$ is used along with the frame number (which differs from transmission to transmission) as input to yet another algorithm A5. The output is known as the ciphering sequence and is used to encrypt the data. The management of the ciphering keys is similar to that of the authentication keys. On the network side the BTS decrypts each transmission using the key and sends the data to the BSC.

## 6.6    GSM Radio Channel

### 6.6.1    Frequency analysis

As mentioned before, GSM is an FDMA-TDMA system. We will now look at the FDMA aspect of the GSM radio channel. FDMA stands for frequency-division multiple access. It is a kind of access scheme whereby the entire spectrum allocated for use by the system is broken down into narrow individual radio channels. Each radio channel is specified by the frequency it is operating at and how wide the radio channel is. In GSM there are two frequency bands allocated for use. They are both 25-MHz frequency bands in the 900-MHz range. The mobile station transmits in the 890–915-MHz frequency range and the BTS transmits in the 935–960-MHz range. The mobile station transmit frequency band is separated from the BTS transmit band by 45 MHz. This separation facilitates the design of the receiver and transmitter. System performance is also improved as adjacent channel interference is reduced. In wireless terminology, the mobile station to BTS transmission direction is called an uplink and the BTS to mobile station transmission direction is called a downlink. The 25-MHz frequency bands are divided into 125 RF channels with a channel width of 200 kHz. Each RF channel is specified by an uplink and a downlink frequency pair. Thus, by specifying a particular channel number both the mobile station and the BTS transmit frequencies are known. For a particular channel number, if the mobile station transmit frequency is known, the BTS transmit frequency can be determined simply by adding 45 MHz to the mobile station transmit frequency. This scheme where separate frequencies are used for uplink and downlink is known as frequency-division duplex (FDD). Within the GSM system, an absolute radio frequency channel number (ARFCN) is used. Channels range from 1 to 124. Channel 0 is a guard band, that is, it is there to avoid interference to other systems from GSM radio channels; it is not used. The mobile stations have to be capable of transmitting within their entire frequency band, since the BTS can allocate any radio channel to a mobile station. Recently there have been additional frequency band assignments to GSM, specifically 10 MHz of additional spectrum has been allocated in the uplink and downlink bands. This makes the frequency band 880–915 for uplink and 925–960 for downlink transmission. These new frequency assignments are known as extended-GSM.

As previously mentioned, there is an upbanded version of GSM known as DCS 1800, also known as PCN. This system is primarily for microcell environment. It has a different frequency assignment from that of GSM. It uses two 75-MHz frequency bands separated by 95 MHz. The ranges in the uplink direction are 1710–1785 MHz and 1805–1880 in the downlink. Thus, there are 374 channels each of 200-KHz width. The channels are numbered from 512 to 885 to distinguish them from the channels in the primary and extended-GSM frequency bands.

In the United States there is PCS 1900, which is similar to DCS 1800 with the only difference being the frequency bands. The frequency bands allocated for PCS in the United States range from 1850–1910 in the uplink direction and 1930–1990 in the downlink. The frequency separation is 80 MHz. The entire band is not available for the same operators. The frequency bands are broken up and distributed among different operators who may use different technologies. The largest frequency band is 15 MHz.

## 6.6.2  Time analysis

The TDMA aspect of GSM will be explained in this section. So far we have seen how a given frequency spectrum is broken down into individual radio channels. GSM provides for enhanced capacity by enabling the use of a single radio channel by multiple users in the same cell while all of them are involved in a call. This is achieved by subdividing each radio channel into eight different time slots. Time slots are specified by a frequency and time metric. Each time slot can support a single user. Thus, one radio channel can now support up to eight users. Each user is allocated a time slot on the radio channel and can only transmit during that slot. As soon as that user is finished, the next user starts transmitting and it goes on in this round-robin fashion. The time slots are numbered from 0 to 7. For instance, if a user is assigned a time slot of 0, then the user transmits during time slot 0 on that radio frequency and then turns the transmitter off for the duration of time slots 1 to 7. The user again transmits on time slot 0 and so on. A set of eight time slots on the same radio channel is referred to as a *TDMA frame*. The user's transmission on a time slot is known as a *burst*. The length of a time slot or burst duration is 577 μs and within a frame is 4.615 ms (8 times 577 μs). GSM provides for staggered uplink and downlink transmission time slots. A user assigned time slot 0 will transmit on time slot 0 in the uplink direction and receive on time slot 0 in the downlink direction. But the uplink and downlink time slots do not occur at the same time. The uplink transmission of time slot 0 occurs three time slots after the reception of downlink time slot 0. Having this type of configuration is advantageous because there is no need for a duplexer in the mobile station. A *duplexer* is a bulky device and requires a lot of power to operate. It allows for simultaneous transmission and reception. Eliminating this device from the mobile station makes GSM handsets lightweight, increases their battery life, and makes them cheaper to manufacture.

### 6.6.3 Speech coding

One of the most important teleservices offered to GSM subscribers by the GSM network is voice telephony (i.e., speech transmission). The requirement is to transmit speech at a satisfactory level of quality. As we have seen GSM is a digital transmission system utilizing a TDMA scheme, where mobile stations transmit data only during specific time slots (i.e., transmission takes place in pulses). Since speech is continuous and not discrete, this means that the continuous speech signal has to be broken down into blocks, compressed so that it can be transmitted in a time slot, transmitted over the air interface, and expanded at the receiver to enable it to be representative of the original continuous speech signal. The device which does this is known as a *speech codec* or *vocoder* (voice codec). A speech codec transforms human speech into a digital stream of data suitable for transmission over the radio interface and alternatively regenerates audible speech from the received data. Every mobile station capable of voice telephony has a speech codec. On the network side the speech codecs form the TRAU. We have seen that the TRAU can exist anywhere in the BSS. GSM supports several kinds of speech codecs. Phase 1 of the GSM specifications supports the full-rate speech codec, full rate indicating that this codec uses the entire time slot of every frame for speech transmission. This codec uses RPE-LTP, regular pulse excitation with long-term prediction, which is a form of encoding speech. There is a version of this used in the United States for PCS 1900 that is called as enhanced full-rate speech codec (EFR). The speech quality of this codec is superior to that of the full-rate codec. Phase 2 of the GSM specifications support half-rate codecs. Half-rate codecs use each TDMA time slot every other frame.

Since the radio channel has restricted transmission capacity, every attempt is made to use minimal capacity while maximizing voice quality. The full-rate codec output is 13 kbps. EFR also has the same output but a superior voice quality. Half-rate codecs output is 6.5 kbps, which is why a GSM network implemented with only half-rate codecs will support twice as many users as a network with only full-rate or EFR codecs. On the other hand the voice quality of the half-rate codecs is inferior to the other two. Operators will have to struggle with the dilemma of excess capacity versus better quality until the half-rate codecs are improved.

### 6.6.3.1 Speech path and its transformation.

We shall now see how the human speech is converted and transformed as it travels along the transmission path from user to user. Figure 6.17 shows the transformation of the voice signal across the different GSM interfaces.

Human voice is converted to an electrical signal by the microphone. It is then passed through a filter so that it contains only signals within the voice band (300–3.4 KHz). This filtered signal is then sampled at 125 μs and quantized by a 13-bit word. A 13-bit word results in 8192 quantization levels. Every 20 ms the speech codec takes 160 sampled 13-bit words and analyzes them. Analysis of the samples produces filter coefficients that are sorted into four blocks of 40 samples each. The speech codec then selects the sample sequence with the most energy and compares this sequence with the previously generated sequences. It determines the sequence that is most similar to its current sequence and determines a differential value between the two. This differential value along with the compared-to-sequence identity is transmitted on the air interface. The output of the speech codec is a block of 260 bits every 20 ms, this translates to 13 kbps. The BTS receives this information and sends it, unmodified, across to the TRAU. The TRAU does the reverse process of transcoding, that is, transforming this 13-kbps digital stream into the traditional 64-kbps PCM that is used in the PSTN. The speech information, digitally encoded into 64 kbps, is received by the ISDN phone that converts it into human speech.

An additional feature of the transcoder is that it is able to detect silence and indicate this to the radio interface. This allows the MS to turn off the transmitter and save power. The network simply repeats "comfort noise" until it is updated. *Comfort noise* is a representation of the background noise of the user.

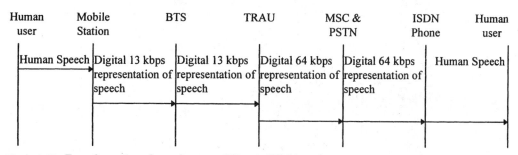

**Figure 6.17**  Transformation of speech across different GSM interfaces.

### 6.6.4   Channel coding

Figure 6.18 shows the channel coding and interleaving process. The RF environment is very destructive and adversely affects the data sent on the RF channel. There are several ways of combating these destructive affects. One of the most popular ways is channel coding. Channel coding protects the data against errors by adding redundancy bits. Adding redundancy increases the amount of data that has to be transmitted and thus can only be useful to a certain extent. Prior to channel coding the 260-bit block from the speech codec is divided into three categories according to their importance and function. The purpose is to provide the greatest protection to the bits that are most important and less protection to the bits that play a less critical role. There are three categories of bits, Ia, Ib, and II. Category Ia is the most important and consists of 50 bits out of the 260. These bits describe the filter coefficients and other codec parameters. These 50 bits receive the maximum protection from the channel coding process. They are first protected by being block encoded, which provides error detection. This process adds 3 bits. These block encoded bits are then half-rate convolution coded to provided error correction and detection capability. Half-rate convolution doubles the number of bits. Category Ib is next in importance to receive some level of protection. These 132 out of 260 bits are convolution encoded similar to category Ia bits. Least important are the category II bits (78 out of 260) which receive no protection at all. Channel coding increases the data rate from 13 kbps, as delivered by the speech codec, to 22.8. Most of the increase in the data rate is due to the addition of redundancy bits that assist in error correction and detection.

### 6.6.5   Interleaving

Once the speech data has been encoded as described in the previous section, the next stage is to build it into a burst in order for it to be transmitted on the air interface in a TDMA slot. This process is known as *interleaving*. It is used as a countermeasure against the unreliable transmission path on the air interface, specifically to combat Rayleigh fading. By the process of interleaving, the data is spread over several time slots on the radio path, thus reducing the probability of total corruption of a speech frame. Interleaving depths of 4 (most control channels), 8 (full-rate speech), and 19 (data channels) bursts are used on the air interface.

Blocks of full-rate speech are interleaved on 8 bursts, that is, the 456 bits output from the speech encoder are split into 8 blocks of 57 bits each and each block is sent on a different burst. The delay added as a result of interleaving is 65 burst periods or 37.5 ms.

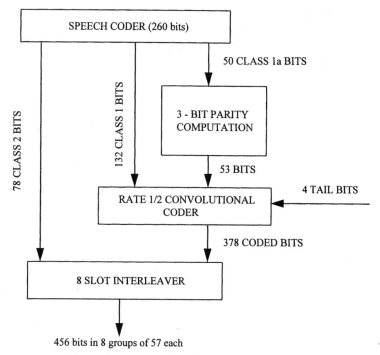

**Figure 6.18** Forward error correction for speech codec (full-rate traffic channel).

### 6.6.6  Modulation

The modulation scheme used in GSM is GMSK (Gaussian minimum shift keying). This scheme has an efficiency of 1 bit per symbol. Due to a more narrow power spectra for this kind of modulation GSM, unlike IS-54, does not require linear power amplifiers.

### 6.6.7  Channel organization

There are several different types of information that are exchanged between the mobile station and the BTS. There is user information, signaling information, channel configuration information, access information, and so on. To manage all these different kinds of information GSM organizes them into logical information. All information dealing specifically with a particular function is grouped as logical information and sent on an associated logical channel. GSM specifies 10 logical channels each with a specific function.

**6.6.7.1  Control channels.**  Control channels carry information that is necessary for the system to operate correctly. The mobile station and the BTS make use of these channels to ensure that user information is transferred correctly, to inform each other of events, to set up calls, to manage mobility and access, and so on. In addition to signaling information, control channels may also be used to carry packet-switched data, including those related to short message services. Several kinds of control channels are discussed in the following section.

### 6.6.7.2 Broadcast control channel (BCCH).

As the name implies this channel is broadcast by the BTS in its cell. It is a unidirectional downlink-only channel that carries information used by all mobile stations in its cell. The information broadcast on this channel is used by mobile stations to synchronize with the network. Based on this information the mobile station is able to decide whether and how it may gain access to the system via the current cell. Information carried on this channel also enables a mobile station to identify the network, to gain access to the network, etc. The information broadcast may be grouped into the following classes:

- *Information giving unique identification of the network and neighboring cell.* Cell identity, mobile network code (forms part of the IMSI), location area identification (LAI), and information about the frequencies of the broadcast control channels of neighboring cells, constitute this class.

- *Information describing the current control channel structure.* Control channel configuration used in the cell, periodic location updating timer, and other such information constitute this class.

- *Information defining the options supported within the cell.* Discontinuous transmission (DTX) allowed or not, cell reselect hysteresis, maximum transmit power level a mobile can use when accessing on a control channel, the minimum received signal level at the mobile station for which it is permitted to access the system, if half-rate codecs are supported, or if extended GSM frequencies are supported form part of this class.

- *Information controlling access.* Maximum number of retries, average delay between retries, whether the cell is barred for access, if call reestablishment is allowed, if emergency calls are allowed, constitute this access control class.

### 6.6.7.3 Frequency correction channel (FCCH).

The FCCH provides the mobile station with the frequency reference of the system. A mobile station uses the FCCH to correct the frequency of its internal time base in order to ease the decoding of other channel bursts. This channel also provides an indication to the mobile station about when the synchronization channel (SCH) will occur since a SCH always follows a FCCH by 8 time slots on the same frequency.

**6.6.7.4 Synchronization channel (SCH).** This channel provides the mobile station with the training sequence necessary for the mobile station to decode other channel bursts. Since the training sequence is known to both the mobile station and the BTS beforehand, the mobile station can adjust its internal timing mechanism and decode the bits correctly. Additionally, this channel provides information about the base-station information code which has the training sequence being used (there are several kinds of training sequences that can be used and they are used in a base-station color-code fashion), national color code, and the TDMA frame number.

**6.6.7.5 Common control channels (CCCH).** These channels support the establishment of a dedicated communication path (dedicated channel) between the mobile station and the BTS. There are three types of CCCH—the random access channel (RACH), paging channel (PCH), and the access grant channel (AGCH).

**Random access channel (RACH).** This channel is used by the mobile station to request a dedicated channel from the network for call setup. It is a unidirectional uplink-only channel which is shared by any mobile attempting to access the network. There is only one kind of message that is sent on the RACH, the channel request message. It is 8 bits long and has an establishment cause field and a random reference. The establishment cause provides an indication to the network of the reason for the access attempt. It allows the network to reserve resources and to prioritize the attempts based on the cause. Some of the establishment causes supported are emergency calls, answer to page, location updating, originating speech call, and originating data call. The random reference is a number chosen randomly by the mobile station to correlate the responses from the BTS with its own request.

**Paging channel (PCH).** This channel is used by the BTS to page particular mobiles in the cell. It is a unidirectional downlink-only channel shared by all mobiles in the cell. GSM allows for up to four mobiles to be paged by a single, page message. Mobiles can be paged either by their TMSIs or IMSIs. GSM also supports something known as discontinuous reception (DRX). To conserve battery life, it is essential to minimize the information that the mobile station has to decode while it is in the idle mode (i.e., awaiting a page message). One method of minimizing is to allow the mobile station to monitor only a portion of the PCH for its pages instead of the entire PCH. GSM supports this by allowing paging subchannels. Pages for a particular mobile station are scheduled only on its paging subchannel. This enables the mobile station to decode only pages sent on its paging subchannel instead of the entire PCH, thus conserving battery life. Mobile subscribers are assigned to paging subchannels in a predetermined way by taking into account the last three digits of the subscriber IMSI.

**Access grant channel (AGCH).**  Responses to mobile channel requests on the RACH are sent on the AGCH. This is a unidirectional downlink-only channel shared by all the mobiles in the cell. Successful responses include information about the allocated dedicated channel number, the timing information the mobile station needs in order to ensure that its messages do not overlap at the BTS, and the random reference sent by the mobile station in its channel request message.

**6.6.7.6  Traffic channels (TCH).**  These are channels that transport user information such as speech or data. These are bidirectional dedicated channels used between a single mobile station and the BTS. They are primarily in two flavors, traffic channel/full-rate (TCH/F) and traffic channel/half-rate (TCH/H) where full rate refers to the full-rate speech codec and the half rate to the half-rate speech codec. Thus, the information rate of a TCH/F is 13 kbps and that of a TCH/H is 6.5 kbps.

**6.6.7.7  Dedicated control channels.**  These channels are used for transferring nonuser information between the network and the mobile station. Information on channel maintenance, mobility management, and radio resource management is sent on these channels. Typical information sent includes: the details of an additional dedicated channel allocated by the network if requested by the mobile, start and stop indications of ciphering, mobile station information queries, handover messages, and the like. The three kinds of dedicated control channels are: stand-alone dedicated control channel (SDCCH), slow associated control channel (SACCH), and fast associated control channel (FACCH).

**Stand-alone dedicated control channel (SDCCH).**  This channel is used for transfer of signaling information between a mobile station and the BTS. This is a bidirectional, dedicated channel. It is typically used for location updating prior to use of a traffic channel in case of speech and data calls.

**Slow associated control channel (SACCH).**  An SACCH is always allocated in conjunction with a TCH or a SDCCH. It is a bidirectional, dedicated channel. It carries control and measurement parameters along with routine data necessary to maintain a radio link between the mobile station and the BTS. Information such as radio signal levels of current and neighboring BCCHs is sent on this channel along with timing advances/retards.

**Fast associated control channel (FACCH).**  FACCH is a channel which appears on demand. The information that can be carried on this channel is similar to SDCCH with the only difference being that an SDCCH is allocated and fixed for a duration of time till the network or user releases it. An FACCH on the other hand is used over a TCH where it steals time slots from a TCH. It is used when time-critical messages have to be exchanged between the mobile station and the network. An example of a time-critical message is a handover messages. Table 6.9 shows the different channels supported by GSM along with their characteristics.

The different transmission bursts used on these different channels are shown in Fig. 6.19. There are eight different kinds of training sequences in GSM which are used to differentiate transmissions from mobile stations on the same frequencies in neighboring cells.

### 6.6.8   Discontinuous transmission and voice activity detection

An important aspect of human speech communication is that it is not continuous. When two humans are involved in a conversation at any

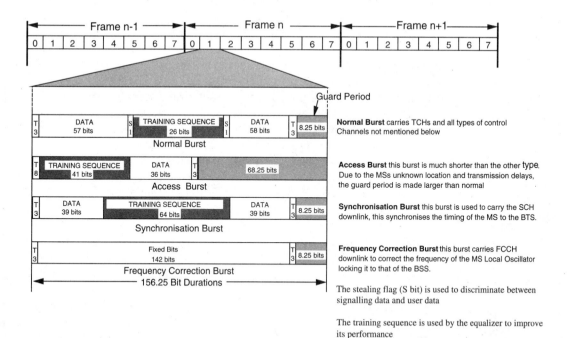

**Figure 6.19**   Different burst types used in GSM.

**TABLE 6.9    GSM Logical Channels and Their Characteristics**

| Logical channel | Uplink-only | Downlink-only | Both uplink & downlink | Point-to-point | Broadcast | Dedicated | Shared |
|---|---|---|---|---|---|---|---|
| BCCH |  | ✓ |  |  | ✓ |  | ✓ |
| FCCH |  | ✓ |  |  | ✓ |  | ✓ |
| SCH |  | ✓ |  |  | ✓ |  | ✓ |
| RACH | ✓ |  |  | ✓ |  |  | ✓ |
| PCH |  | ✓ |  | ✓ |  |  | ✓ |
| AGCH |  | ✓ |  | ✓ |  |  | ✓ |
| SDCCH |  |  | ✓ | ✓ |  | ✓ |  |
| SACCH |  |  | ✓ | ✓ |  | ✓ |  |
| FACCH |  |  | ✓ | ✓ |  | ✓ |  |
| TCH |  |  | ✓ | ✓ |  | ✓ |  |

given time, typically one person is talking and the other is listening. If both the persons start talking at the same time, we get confused and the information transfer level falls. GSM takes advantage of this natural human tendency of only one person speaking at a given time and the other being silent. When a mobile station user is involved in a conversation and is silent, there is no speech that needs to be transferred to the other user (i.e., there is incoming speech and no outgoing speech). The GSM speech codec, when it detects pauses in speech, discontinues radio transmissions for the duration of the pause. This is known as discontinuous transmission (DTX) in GSM. If DTX is enabled then speech is encoded at 13 kbps and during a pause it is encoded at 500 bps. The question may arise as to why there is a need to transmit even 500 bps when the user is not speaking. The reason is that it is very disconcerting to the listener to hear speech and then complete silence followed by speech again. This on-off effect tends to make the listener believe that the connection is broken. To avoid this GSM provides for the transmission of what is called *comfort noise*. This is basically the encoding of the background noise of the speaker when there is a pause. The background noise differs from situation to situation. It could be encoding of the car noise if the speaker is traveling in an automobile or encoding of the background office noise if the speaker is in an office. Since the same background noise is also partially encoded when the speaker is actually speaking, encoding it when there is a pause provides a sense of continuity to the listener. To enable DTX to function there has to be voice activity detection (VAD), the purpose of which is to indicate whenever there is a pause in the speech. Use of DTX tends to reduce the interference level in the system and thus increases system efficiency. Also, since by using DTX the total time the transmitter is on is reduced, the power consumption is lower and this leads to longer battery life in mobile stations.

### 6.6.9    Timing advance and power control

Both timing advance and power control are necessitated due to the varying distances of the mobile stations from the BTS. In any cell there will be mobile stations that are far away from the BTS, near the periphery of the cell, and then there will be others that are very close to the BTS, and still others that are in the middle of the cell. If, at any instance all the mobile stations transmit at the same time, these messages will arrive at the BTS at slightly different times due to the varying distances. Thus it is possible that even though two mobile stations—one close to the BTS and the other far away—transmit on adjacent time slots on the same TDMA frame, their bursts may overlap at the BTS on reception due to different propagation delays. The same holds true for transmit power. If all mobiles transmit at the same power level then mobiles closer to the BTS will be received more strongly than the ones on the periphery of the cell. Timing advance and power control are specified in GSM to allow for control of these parameters so that all bursts are time and power synchronized on reception at the BTS.

**6.6.9.1    Timing advance.**    GSM is a very tightly time synchronized system. One way to deal with varying propagation delays is to take into account the maximum propagation delay that is possible and have enough guard time at the end of each burst to compensate for it. Mobile stations would not be transmitting user data during the guard period. Even if two bursts overlap they would only overlap during the guard periods. Since no user information is transmitted during that period, no data is lost. This scheme, though having the advantage of simplicity and minimal signaling requirements, decreases the spectral efficiency of the system. The greater the nonuser information sent in each burst, the lower the efficiency of the system. The approach chosen by GSM rejects this scheme in favor of smaller guard periods and dynamic control of the timing of each burst. Timing advance allows for independent control of the uplink transmitting time of each time slot. Mobiles that are farther away from the BTS are instructed to transmit their bursts earlier than those closer to the BTS. On establishment of a dedicated channel, the BTS continuously measures the time offset between its own burst schedule and the reception schedule of the mobile station bursts. Based on these measurements, it is able to provide the mobile station with the required timing advance and does that on the SACCH at the rate of twice per second. The maximum timing advance allowed in GSM restricts the cell size to about 35 kms.

**6.6.9.2    Power control.**    To compensate for the attenuation over different distances within the cell, the BTS can instruct the mobile station to change its transmit power so that the power arriving at the BTS's receiver is the same for each time slot. This reduces the overall system interference level and increases spectral efficiency. The BTS can independently control the power level of each uplink and downlink time slot. The power level control scheme is controlled by the BSC, which calculates the power level increase or decrease and communicates it to the BTS. The power level control algorithm is very closely coupled with the handover algorithm. A BSC will try to increase the power level of a mobile station as it moves farther away from the BTS. After it makes a determination that the quality of the communication link can no longer be improved just by an increase in the transmit power level of the mobile station, it starts the handover process. Power level can be changed in steps of 2 dB increments every 60 ms.

### 6.6.10    Mobile access

The first step for a mobile getting service from the GSM network is for it to access the system. It is also the most important. Access is a means of having a mobile station indicate to the network that it wants to avail itself of the services provided by the network. The reasons for access can vary: an attempt to make a call, a response to a network request such as a page, an automatic determination by the mobile station that it needs to access the system for periodic location updating. Whatever the reason for access, the access procedure is similar. Access in GSM is on a cell-by-cell basis. Before we discuss in detail how the access mechanism works in GSM let us first try to understand the mobile environment in which the access attempt takes place. Mobile stations will be rapidly moving in and out of individual cells. At any given time there may be a random number of users in a cell, the number of which varies with time. For instance, a cell on a highway connecting the downtown area of a city to the suburbs will have a lot of users during the morning and evening rush hours, moderate traffic in the afternoon, and very light traffic at night. Moreover, a random number of these users may be involved in a call, attempting to make a call, or idle. Due to the randomness of the number of users and their movements, it is not practical to coordinate the users in their attempts to access the system. In GSM there is a single channel on which all the mobile stations in that particular cell send their access requests. Thus, the issue of access in GSM falls into the famous ALOHA problem domain. The ALOHA system is one in which uncoordinated users are competing for the use of a single, shared channel.

More specifically due to the time slot arrangement in GSM, the model is a slotted-ALOHA model. The basic idea of an ALOHA system is simple: let users transmit whenever they have an access request to send. If two users transmit at the same time there will be a collision and both of the requests may be destroyed. (It is possible that the receiver may decode one access request even though two were sent at the same time. This is possible due to the capture effect of FM transmissions.) If the transmitting agent does not receive an acknowledgment from the receiver then it waits a random amount of time and retransmits the message. The waiting time must be random or the same messages will collide again and again, in lockstep. Since GSM uses time slots, collisions can occur only during time slots; this is slotted-ALOHA. The efficiency of slotted-ALOHA schemes is shown to be close to 37 percent of the channel utilization. Though it might seem to be comparatively small, the simplicity of the scheme more than makes up for any lack of efficiency. Moreover, the capacity of the channel can always be increased. There are two ways in which the message throughput can be increased on a channel. The first and most obvious way is for the channel capacity to be increased. GSM supports this method and the access channel capacity can be engineered according to the traffic. The second method is by decreasing the message length. For a given channel capacity more messages can be sent if they are shorter in length, than if they are longer. GSM is highly efficient in this regard and the initial access request is only 8 bits in length.

GSM's backoff mechanism for mobile stations, when collisions do occur, can also be engineered on a per cell basis. Every time a collision occurs a mobile station attempts to retry to send the access message. GSM supports the control of the frequency and the number of retries by the network. Moreover, GSM supports it on a per cell basis. Every cell transmits parameters on the BCCH which indicate the number of time slots a mobile station has to wait before attempting a retry and the maximum number of retries that are allowed before the mobile station has to stop making any further attempts. Engineering of the access channel plays a critical role in the system performance. The behavior of the entire system is reflected in how mobile stations can access and be granted service. Having excess capacity on the access channel results in several users getting access, but the system would be unable to support them due to lack of traffic channels. This places an undue burden on the system because of all the signaling that has taken place prior to being denied service. If service is to be denied, it is much more efficient to have it denied as early as possible in the call-setup phase, as this avoids the use of unnecessary system resources. On the other hand, having too little capacity on the access channel denies service to users who might have gotten service from the system. This results in underutilization of system resources, such as traffic channels.

Even a well-engineered access channel will have to deal with occasional bursts of peak traffic, which cannot be controlled by the measures we have described. To enable the system to maintain a satisfactory level of performance under these brief peak traffic conditions GSM has described explicit procedures to handle overload conditions. The system can explicitly send a message to an individual mobile station to forbid it from attempting to access the system for a specified amount of time. For more severe overload conditions the concept of access classes is supported. It consists of forbidding entire classes of mobile subscribers from attempting to access the system. This is done by sending a message on the BCCH which indicates the authorized mobile station access classes that are allowed to access the channel. The entire subscriber population of a GSM operator is randomly broken down into 10 classes. The access class of a subscriber is stored in the SIM. During normal operation all access classes are allowed to access the system. Under overload conditions the system can deny services to selected access classes. This is a very efficient way of handling overload traffic, since it requires minimal signaling load on part of the network. To enable service to be provided to emergency and other government agencies during an overload condition there have been 5 additional mobile classes. Class 12 is for security personnel, 13 for public utilities, 14 for emergency personnel, 15 for GSM network personnel, and 11 is left open for the specific use of the network operator.

## 6.7    GSM Call Scenarios

Till now we have seen the different building blocks that are part of GSM. Spectrum allocations, system architecture, channel structure, transmission structure, and others have been discussed. We have seen each building block as an isolated entity with emphasis on its function and how it achieves its task. In this section, we shall integrate the different pieces and explain how individual elements are combined to achieve a single task. We will look at several network procedures, build on the previous one, introduce elements, and detail how it fits into the GSM scheme. This will illustrate and simplify the complexities inherent in the GSM system.

### 6.7.1    Mobile behavior on start-up: getting in sync

When a mobile station is turned on, it has to orient itself within the GSM network. It has no knowledge of its location, cell configuration, network options, access conditions, and so on. It has to gain all this information so that it can inform the network of its presence if need be, respond to pages, or attempt call establishment on request. To gain the information necessary for the mobile station to operate it has to first determine the BCCH frequency so that it can start decoding some of the system parameters essential for operation. In GSM 900, there are 124 radio frequencies and in DCS 1800 close to 375. It would take a mobile station quite some time to go through and decode all these frequencies to determine the BCCH. To assist the mobile station in its task GSM allows for the storage of a list of frequencies in the SIM. These frequencies are the BCCH frequency of the last cell camped on, along with the neighboring cell frequencies broadcast on that BCCH. The mobile station on power-up starts scanning these frequencies. It is mandatory in GSM that the BCCH frequency be fully powered at all times (i.e., there is no power control done on the BCCH frequency). It is also mandatory that in all idle time slots of this frequency the BTS has to transmit specially defined idle bursts. These two conditions ensure that the BCCH frequency has a higher power density than other frequencies in that cell. A mobile station simply scans the radio frequencies looking for a frequency that has a higher power density than the others. On finding a promising radio frequency, the mobile station next tries to determine the FCCH. A similar principle is used here since the power density of the FCCH is greater than that of the BCCH frequency. On finding a FCCH, the mobile station decodes it and is able to get itself synchronized with the master frequency of the system. The FCCH typically carries a fixed sequence of zeroes. Once the mobile station has determined the FCCH and is frequency synchronized, it has to correctly determine the boundaries of the time slots and frames. For that, it has to get time synchronized. The mobile station knows that a SCH will follow a FCCH on the same frequency after 8 time slots. It simply waits for 8 time slots, decodes the SCH and gets time synchronized. Now the mobile station is ready to decode the rest of the data being sent on the BCCH.

## 6.7.2   Cell selection

Prior to selecting a cell on which the mobile station will park, it has to determine if it can be provided service in that network. The BCCH carries information regarding the GSM network identity that particular cell is part of. GSM mobile stations can either manually (with intervention of user) or automatically (preferred list of GSM networks stored in SIM) determine the valid GSM networks for which there is subscription. If the current cell is not part of any authorized GSM network the mobile station simply looks for a different BCCH.

Once a valid GSM network has been chosen, the mobile station then has to choose a cell to camp on. The mobile station has a cell selection algorithm that it uses to determine the best available cell. There are several factors that go into the decision of selecting a cell. Some of the critical ones are received signal strength by the mobile station, mobile power class, and location area. Received signal strength is an indication to the mobile station of how well it is receiving the BTS in that cell. Obviously if this indicator is too low, then the mobile station knows that any communication link it establishes in future will not have very good quality. Keep in mind that the BCCH frequency is being transmitted at the maximum allowed power for that cell. Any dedicated channel established by the mobile station will be at or below the power level of the BCCH frequency due to power control. In fact one of the parameters transmitted on the BCCH is the minimum received signal level at the mobile station for which it is permitted to access the system. If the received signal level is below this value the mobile station is forbidden to park on that cell.

Location area at first seems like an odd criterion to be using for cell selection. Its purpose is not to improve the quality of the radio link from the mobile station to the BTS, but to decrease signaling traffic between the mobile station and the network. Every new location area update from a mobile station uses a lot of signaling bandwidth and the attempt of every network operator is to minimize this traffic without affecting performance. Every time a mobile station is powered down, the serving location area identity is stored in the SIM. Upon power-up and during cell selection, this stored location area identity is compared with the broadcast location area identity. If they do not match, that particular cell is weighted down by a factor. Thus, if there are two identical cells in different location areas, with equally good radio link characteristics as determined by the mobile station, it will choose the cell where it does not have to do a location update.

The third criterion used in the cell selection process is the mobile power class, specifically, the maximum transmit power of the mobile station. Even if the mobile station is able to receive the BTS perfectly well, there is no guarantee that the BTS will be able to receive the mobile station. This can happen, for instance, if a cell has been designed for a power class 1 GSM 900 mobile station whose maximum transmit power level is 20 W. A power class 5 GSM 900 mobile station is only capable of a maximum transmit power level of 0.8 W. This mobile station, if it is on the periphery of that cell, cannot ensure that its transmission will be received by the BTS at an acceptable power level. This highlights a very important point; namely, that the cell boundaries are not fixed and vary from mobile station to mobile station depending on their power class.

Using the cell selection criteria and algorithm, the mobile station scans the BCCHs of the cells in its vicinity and makes a list of all the cells that have passed its selection criteria. The cell with the highest value is chosen from this list and the mobile station decides to park on this cell. If the chosen cell is in a new location area, then a location update procedure is undertaken. Until now all the activity has been undertaken by the mobile station. The network is not even aware of the existence of the mobile station in that particular cell. All that the BTS has done is provide information in a passive manner, broadcasting the necessary system parameters for the benefit of any mobile station that happens to listen to its BCCH. From now on the network plays a more active role—exchanging messages with the mobile station.

### 6.7.3  Location updating

Once a mobile station has selected a cell to park on, its next step is to determine if this cell is part of the previously registered location area. It decodes the location area information from the BCCH and compares it to the previous location area stored in its SIM. If the location areas are the same, the mobile station then enters the idle mode and awaits either a user request for a call origination or a page from the network.

If the location areas are different, then this indicates that the network does not have the correct location of the mobile in its databases. During this time any call termination toward the mobile will be unsuccessful, since the network will try to terminate the call by paging in the mobile station's previous location area. Since the mobile is not in that location area, the call will be unsuccessful. Thus, it is imperative that the mobile station inform the network of its new location area as soon as possible, so that the network can update its database records and successfully terminate future calls toward the mobile station. On determining that a location update is necessary, the mobile station immediately attempts the location update procedure described later on in this chapter.

### 6.7.4  Establishing communication link

Before the mobile station can actually initiate the location updating procedure, it has to establish a communication link with the network over which the location update messages can be exchanged. The communication link establishment procedure is initiated by the mobile station tuning to the RACH and transmitting a channel request message. As mentioned previously, this message and the establishment cause is coded as location updating. It then tunes to the AGCH and awaits a response from the network. The channel request is received by the BTS and is forwarded to the BSC with some additional information about the transmission delay. As discussed before, the BSC is able to estimate the delay by comparing it with its own burst schedule. The channel request message sent by the mobile station has a large guard period since it is the first message sent by the mobile station and both the network and the mobile station do not have any idea about the expected delay. Having a large guard period ensures that even if this message overlaps with the message in the next time slot, on reception at the BTS, the information content is not lost. The BSC chooses an available channel (typically a SDCCH), calculates the estimated timing delay, and informs the BTS to activate the channel. It then sends a channel assignment message to the BTS to be sent on the AGCH. This message carries a reference number so that the mobile station, on reception, can correlate this response to its channel request. It also has information on the allocated channel (type, frequency, time slot, etc.), timing advance to use on that channel, and the initial transmission power level to be used by the mobile station.

### 6.7.5  Initial message procedure

The mobile station, on reception of the channel assignment message, tunes to the allocated channel on the indicated frequency with the adjusted timing advance and power level. It then transmits a message known as a service request (sent on the SDCCH). This message indicates that the mobile station is requesting a service from the network. In the case of location updating, the request is a location updating request. This message has more detailed information about the identity of the mobile. This is the first instance when the network is made aware of the mobile identity. Until now the network has granted service in good faith (i.e. without the knowledge of whether the subscriber is authorized to receive service from the network or not). The location updating request has information as to the specific type of location updating being initiated (normal, periodic, etc.), information about mobile identity (typically, the TMSI), power class, frequency capability, ciphering algorithm supported by the mobile station (e.g., A5), and more. This message is passed on by the BSC to the MSC for further processing via an A-interface message. The MSC in turn passes the message on to the VLR via a MAP-B message.

### 6.7.6  Authentication

Once the current VLR has successfully received the initial message with the appropriate cause (location updating, call setup, etc.) it initiates the authentication and encipherment procedures. The purpose of the authentication procedure is twofold, first to permit the network to check whether the identity provided by the mobile station is acceptable or not, and second to provide parameters enabling the mobile station to calculate a new ciphering key. The authentication procedure is always initiated by the network. As discussed before, the authentication algorithms reside in the AuC in the network side and in the SIM on the mobile station side. The AuC, for every subscriber, chooses a random number and inputs it into the A3/A8 algorithms along with the subscriber's unique key. The output is what is called in GSM jargon as a triplet (i.e., RAND, SRES, and $K_c$). As explained previously, this calculation is very CPU intensive, and to save time the AuC calculates several triplets at a time and provides them to the HLR. The VLR requests these triplets in a send parameters message and the HLR provides them in the response.

Whenever the VLR has to perform authentication, it sends an authentication request to the mobile station (via the SDCCH). This message contains the random challenge (i.e., the RAND). This value of RAND is taken from one of the several triplets that the VLR has stored against this particular subscriber. In essence the VLR is challenging the mobile station to prove it is what it claims to be. The mobile station with the help of the SIM generates the response, SRES, and sends it to the VLR. It also generates a new ciphering key, $K_c$, which it stores. The VLR on receiving the SRES compares it with its internally stored value and if they match the subscriber is flagged as being legitimate.

### 6.7.7    Location update procedures

Figure 6.20 shows the location update call flow. At this time there are several possibilities that can occur. If the location area at which the mobile station is currently parked is controlled by the VLR which receives this message (current VLR), then that means the VLR has already gotten all the information it needs about the subscriber and it can proceed with completing the location update procedure.

Another scenario occurs if the VLR has no prior record of this subscriber. In this case the current VLR has to request the subscriber information from the subscriber's HLR. The VLR then sends a MAP-D location update message to the HLR. This message has the mobile identity, along with the VLR address which enables the HLR to query the VLR in case of mobile terminated calls. The HLR looks up the subscriber's subscription in its internal database records and determines if the subscriber should be provided service in the current VLR. This decision is made depending on the subscriber's subscription. If the subscriber is entitled service in that VLR area, then the HLR returns a successful result to the current VLR. If the subscriber is not entitled service it returns a failure result. But the HLR is not done yet. If a snapshot of the network database was taken at this moment, we would find (1) that the mobile station was registered in the current VLR (since the HLR has returned a successful result), (2) there is information in the HLR about the current VLR address, and (3) the mobile station is also registered in the previous VLR where a successful location

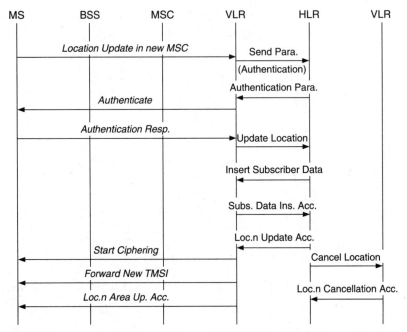

**Figure 6.20**  Location update call flow: only MS, VLR, and HLR interactions are shown.

update was done prior to the current location update. The network has to erase the record in the previous VLR to have consistent data throughout the network. To achieve this the HLR sends a cancel location message to the previous VLR, which cancels the subscriber record in that VLR. The HLR then sends the subscriber data to the current VLR via an insert-subscriber data message that provides all the necessary information to the VLR providing service to the subscriber.

### 6.7.8  Ciphering

The VLR then initiates the ciphering procedure. It informs the MSC, which in turn sends a message to the BSC along with ciphering key to be used. The BSC informs the mobile station via the BTS to start encrypting all future transmissions. Prior to that, the BTS is also informed to expect encrypted messages and is given the key so that it can decipher the messages. The BTS decrypts the messages and sends them to the BSC. An indication is sent to the VLR that the ciphering procedure has been initiated and that all future communications will be encrypted.

### 6.7.9  TMSI reallocation

The VLR now sends a successful result to the mobile station's location update request. It still has one more piece of unfinished business to take care of—the mobile's TMSI. Remember that the TMSI used for initial communication was from the previous location update request. Since the location area has changed, a new TMSI has to be assigned to the mobile station. This new TMSI is assigned by the VLR and piggy-backed on the successful location update indication to the mobile station. The mobile station, on receipt of the TMSI, overwrites the previous value and stores it in the SIM. This value will be used on all subsequent location updates.

The location updating procedure described in the previous section applies for all types of location updates, whether normal location updates or periodic updates.

### 6.7.10  Release of communication link

Once the location update procedure has been terminated successfully, the communication links between the mobile station, BTS, BSC, and MSC are terminated. The mobile station returns to an idle mode awaiting either a user-initiated event or a network page.

### 6.7.11  Mobile origination

Figure 6.21 shows the mobile origination call flow. The mobile origination process is similar to the location update procedure. The mobile origination call-setup procedure is preceded by the establishment of a communications-link procedure, initial-message procedure (cause is originating call setup), authentication, and ciphering procedures. The operation of these procedures is almost identical to those described in the previous section.

Once these procedures have been completed successfully, the mobile station initiates origination-specific procedures. These procedures are derived from ISDN and are for the most part identical. The mobile station sends a setup message on the established link (SDCCH). This message has the called party number, bearer capability, and other fields which are used by the network for establishing the connection to the PSTN. Bearer capability would indicate if the call was going to be a speech or data call. Details such as circuit or packet call, synchronous or asynchronous, and user-data rate (can vary from 300–9600 bps) will be provided. The MSC uses this information to ensure that the required capabilities are supported and provisioned. The MSC also queries the VLR via a MAP-B message to determine if there are any supplementary service restrictions. The called party number sent by the mobile station could be a CUG number, in which case it requires translation. The VLR checks the subscription restrictions to ensure that the call can be allowed. If the user has any kind of call barring restrictions the VLR ensures that this call does not violate them. Once the call has been validated and allowed to proceed the MSC sends a call-proceeding message to the mobile station, which basically informs it that the setup message has been received and processed and that the network is attempting to terminate the call.

If we take a snapshot of the network at this stage, we realize that the radio channel allocated to the mobile station is an SDCCH. If the user is requesting a speech connection, that means that we have to allocate a traffic channel (TCH) since SDCCHs cannot be used to carry speech. To achieve this, the BSC allocates a new radio channel on command from the MSC. This channel would be a traffic channel (full-rate or half-rate depending on if the mobile station and network support half-rate). The BSC informs the BTS of the new channel and the BTS activates the new channel. The BSC then allocates TRAU resources for speech decoding. After all the resources on the network side have been allocated for handling traffic channels, the BSC sends a assignment-command message to the mobile station, informing it of the new channel to use in future transmissions. The mobile station tunes to the new radio channel and starts transmitting on it. It sends a assignment-complete message on the new channel, indicating that it has successfully tuned to the new channel. The BSC then releases the old channel.

Meanwhile, the MSC initiates the call-setup procedure through the network. For instance, if the connection to the PSTN switch is via ISUP (ISDN user part) then an IAM (initial-address message) is sent to the

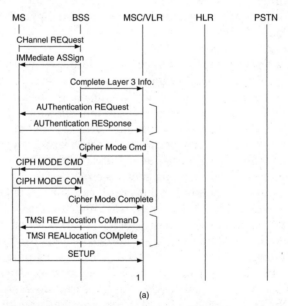

(a)

**Figure 6.21** (*a*) Mobile-originated call, mobile to land call flow (part 1 of 2).

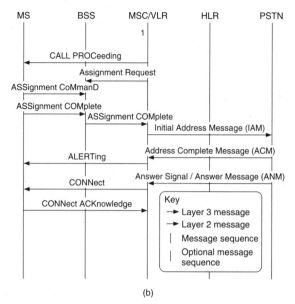

**Figure 6.21** (*b*) Mobile-originated call, mobile to land call flow (part 2 of 2).

PSTN. An ACM (address-complete message) is sent backward from the terminating switch to the MSC, indicating that the called party is being alerted. When the MSC receives the ACM, it sends an alerting message to the mobile station. On receipt of this message the mobile station generates an alerting tone to inform the user that the called party has been contacted and that the phone is being alerted. When the called party answers the phone, the ANM (answer message) is sent across the network to the MSC, which informs the mobile station of the event by a connect message. Now the two parties are connected in a call and can exchange information.

### 6.7.12  Mobile termination

Figure 6.22 shows the call termination call flow. A mobile termination procedure is different since it involves gateway MSC functionality, location updating, and paging. A call from a wire-line subscriber is placed to a GSM subscriber by dialing the mobile subscriber's MSISDN. The call is routed by the PSTN to the mobile subscriber's home GSM network gateway MSC. This will essentially be an IAM message if ISUP is being used in the PSTN. The called party number in the IAM will be the MSISDN number. This speech call is parked at the gateway MSC while it tries to determine the location of the subscriber. Using MAP-C signaling procedures it asks for routing information from the subscriber's HLR. The gateway MSC is able to determine the subscriber's HLR since it has translation tables which indicate the HLR associated with an MSISDN. The HLR, on receiving the request, maps the MSISDN provided into an IMSI number with the help of its internal tables. It then looks up the subscriber profile associated with that IMSI number and, depending on what subscriber features are activated, may do one of several things.

If the subscriber has CFU, then the HLR returns the forwarded-to number to the gateway MSC which reroutes the parked speech call to the destination switch, handling the forwarded-to number. If the subscriber has BAIC, the HLR denies service.

In the normal case, where there are no restrictions on call termination, the HLR determines the address of the VLR where the mobile subscriber is currently registered (the VLR address is stored as part of the subscriber's profile information at the HLR), and using MAP-D procedures queries the VLR for a routing number—specifically, the mobile station roaming number (MSRN). Even though the HLR has the VLR address, it is not a PSTN routable number. The MSRN is essentially a PSTN routable number. It has a similar format to that of MSISDN. The MSRN when provided to a PSTN will allow a speech call to be routed to the destination MSC controlling the radio resources in the location area in which the mobile subscriber has registered. The VLR allocates a MSRN from a pool of available MSRNs and returns it to the HLR. It also marks the subscriber record with the allocated MSRN. The HLR returns the MSRN to the gateway MSC which then uses it to route the parked speech call to the destination MSC. This portion of the call is again routed via an IAM message. The called party number in this case would be modified to be the MSRN number in place of MSISDN. The destination MSC informs the associated VLR about the incoming call and provides the MSRN. The VLR looks up the subscriber's record, based on the MSRN, and determines the current location area. It then sends out a message to all the BSCs controlling the cells in that location area. This message contains the mobile identity (TMSI or IMSI), IMSI (this is sent on the air interface, but used by the BSC to determine the paging subchannel in support of DRX), and a list of cells in which to page. The BSC, in turn, sends out a paging command to the BTSs instructing them to send out pages for the subscriber over the paging channel.

The mobile station, on receiving its page, initiates the establishment of the communications-link procedure, the initial-message procedure, authentication, and ciphering procedures. Wherever necessary, the cause for establishment is marked as responding to a page. The MSC is the network entity on which the incoming speech call is parked. Therefore, it is responsible for coordinating the page responses with the parked incoming speech calls. Once it determines that a valid page response has been

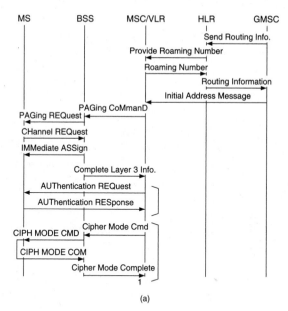

**Figure 6.22** (*a*) Mobile-terminated call, land to mobile call flow (part 1 of 2).

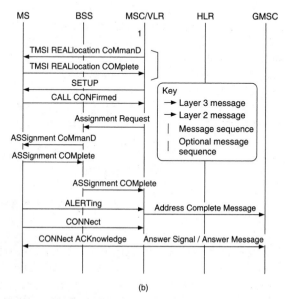

**Figure 6.22** (*b*) Mobile-terminated call, land to mobile call flow (part 2 of 2).

received, it sends a setup message over the established communications link. On receipt of the message, the mobile station sends a call-confirmed message to the network, which informs it that the setup message has been received. The network initiates the channel-assignment procedure allocating a traffic channel to the mobile station. On successful completion of this procedure, the mobile station starts providing an audible alert tone to the user. It simultaneously sends an alerting message to the MSC, which in turn sends an ACM message to the PSTN user. This message indicates to the PSTN user that the mobile subscriber is available and is being alerted. The mobile station may, optionally, while it is providing the alerting tone, also display the number of the calling party. When the mobile station user answers the phone, the mobile station sends a connect message to the MSC and the MSC, in turn, sends an ANS to the PTSN user, whereas the call is connected in both directions and the user communication proceeds.

### 6.7.13   Handover

Handover is the process of continuing a call when the mobile station changes cells. Handover is a GSM-specific term; the same process is known as handoff in North American cellular systems. Handover becomes necessary to avoid losing the call when a mobile station changes cells. As is evident, all the radio resources are allocated on a per BTS level. It becomes obvious that if a BTS is changed then the existing radio resources cannot be used. Moreover, due to radio engineering, if a mobile station is about to change a cell it has to be on the outer periphery of the cell, which indicates that the radio signal level is not very good. This is another reason for handover.

To understand the handover process in GSM we will break it up into three phases, prehandover processing, handover execution, and posthandover processing. In the prehandover processing phase the network collects all the data necessary for making the handover decision and if a handover is deemed to be necessary, information about which cell is a suitable candidate for handover is also collected. In the handover execution phase the actual handover procedure is executed and the mobile is connected to the new BTS. In the posthandover phase all the network resources which are not required anymore are released and the system returns to a stable phase.

**6.7.13.1   Prehandover processing.**   The handover process in GSM is a mobile assisted handover (MAHO). The mobile station plays an active role in the handover process and provides information that forms input for the handover algorithm. The decision to perform a handover and the identification of the most suitable new BTS are both based on several different measurements that are performed by the MS and the BTS. Parameters describing the capabilities of the MS and the BTS also form part of the input to the handoff algorithm. Some of these parameters and measurements are maximum transmit power of MS, serving BTS, neighboring BTSs, cell capacity, load, uplink transmission quality and reception level, downlink quality and reception level, and downlink reception levels from neighbor cells.

### 6.7.13.2    Mobile measurements.

To assist in the handover process, the mobile station takes quality and received signal strength measurements for the currently serving cell and received signal strength measurements for the neighboring cells and reports them to the serving BTS. Quality measurements are the raw-bit error rates of the currently serving downlink channel, converted to a value between zero and seven. The received signal strength of the serving BTS is converted into a 6-bit number. Measurements of the received signal strengths of neighboring cells is also made. This measurement is done between the uplink transmission and downlink reception time slots. During this time period the mobile station tunes to the neighboring cell BCCH and measures the downlink received signal strength. These measurements along with the details about the BCCH frequency are sent to the serving BTS. A mobile station may report on up to six neighboring cells in addition to the measurements of the serving cell. The BCCH carrier frequencies of the neighboring cells on which a mobile station must report are contained in the information transmitted on the BCCH and the SACCH.

The measurements taken by the mobile station are reported to the BSC via the BTS. The SACCH is used to carry this information. One SACCH frame is transmitted every 120 ms, but due to interleaving (used to mitigate the effects of Rayleigh fading by spreading the transmission across several bursts) the complete frame is received at the BTS after a delay of 480 ms. Also to overcome the effects of short term, transient radio link degradation the measurements are averaged at the mobile station, BTS, and the BSC.

The uplink is measured by the serving BTS. It does both quality and received signal strength measurements and sends this information to the BSC, along with the received mobile station's measurements.

### 6.7.13.3    Handover execution.

Figure 6.23 shows the handover call flow. Once the decision has been taken to initiate a handover and the most suitable new cell has been identified, the mobile station and the network enter the handover-execution phase, in which the connection with the currently serving BTS is terminated and a connection with a new BTS in the new cell is made. The handover-execution process along with the signaling involved is dependent on the choice of the new cell. If the new cell is controlled by the BSC which also controls the old BTS then the handover is termed as an intra-BSC handover and the signaling is limited to the BSC and does not involve the MSC. If the new BTS is controlled by a different BSC from the one controlling the old BTS, and if both the old and new BSCs are controlled by the same MSC then it is an intra-MSC handover, if the BSCs are controlled by different MSCs then it is inter-MSC handover. Each of these handovers requires signaling that is a little different. The inter-MSC handover execution utilizes MAP-E messages. The first step of the handover phase is for the new BSC to be informed by the old BSC that a handover is required. Except for the case where both the old and new BTSs are controlled by the same BSC, this information is passed via the MSC(s) to the new BSC. A communication path is established between the new BSC and the new MSC. In the case where the old and new MSCs are the same, the communication path established takes the form of resources on the A-interface. If the old and new MSCs are different, it also includes resources between the two MSCs. Once the new BSC has been informed of the handover, it allocates a channel on the new BTS, and once it is successful, sends the information about the newly allocated channel to the old BSC (if the new BSC is not the same as the old BSC). A handover message is generated and sent to the mobile station via the old BTS. This message has information about the

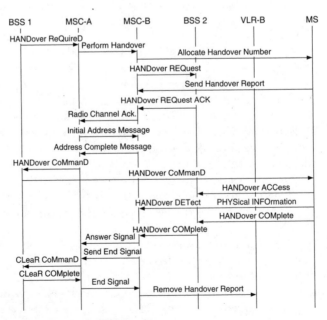

**Figure 6.23**   Inter-MSC handover.

new channel, handoff number, time synchronization information and so on. The mobile station retunes to the new BTS at the specified frequency and commences normal transmission.

**6.7.13.4  Posthandoff processing.** Once the mobile station has synchronized itself with the new network, it sends a handover-complete message to the new BTS. This message is sent via the network to the old BSC. This BSC then releases the old radio resources along with all the old terrestrial resources allocated for the mobile station on the A-bis and A-interface.

## 6.8  Summary

GSM is the most popular digital technology in use today. It has been accepted in a large number of countries all over the world, used in different frequency bands (900, 1800, 1900 MHz), and a number of operators are part of the GSM MoU. All of this is an indication of the popularity of GSM technology. This is due in part to GSM's layered architecture and standardized interfaces between network entities. It enables the operator the option of choosing between different network equipment vendors and also allows the infrastructure manufacturers to build only specific components, in place of building the entire system. For instance, a manufacturer can choose to make only BSS equipment or HLRs or network management systems. This allows innovative approaches to be adopted by manufacturers and lets them specialize in certain network components, enabling general availability of better, lower-cost equipment. GSM equipment vendors are also rolling out their second- and third-generation equipment which have a higher degree of component integration, quality, and lower cost due to economies of scale. This is in contrast to other competing digital technologies which are developing their first generation systems.

GSM as a technology is several years old. This means that there is a lot of experience and knowledge in the wireless industry as relates to GSM. Knowledge about operating GSM networks, network and systems engineering, RF planning, and service deployment is prevalent and easily available. New operators can quickly roll out service by tapping this existing GSM knowledge base. One of the drawbacks of being a mature technology is that there will be newer and enhanced technologies available. GSM, by allowing for rollout of services in different phases, can incorporate these new technologies and remain competitive. For instance, the North American PCS market wanted a better quality GSM vocoder. The GSM equipment vendors together with the operators developed a new full-rate GSM vocoder known as enhanced full rate

(EFR) which complies with the vocoder bit rate and other technical requirements of GSM. This new vocoder will be used for PCS service in the United States and is also being standardized to enable it to be offered as an option to subscribers all over the world. GSM's phase 2 and other future phases will incorporate the features and services requested by sophisticated operators and subscribers. Some of the new services that will be part of GSM are higher user-bit rates allowing for the use of more than 1 time slot in a frame, support of large diameter cells allowing for transmissions in every other time slot (analogous to increasing the guard time), capacity on demand, and video. As ISDN evolves, most of the new ISDN services will also be supported by GSM. By constantly evolving, GSM will remain a viable wireless technology for today and for years to come.

# 7

# D-AMPS 1900

## 7.1 Introduction

D-AMPS 1900 is a North American dual-band/dual-mode digital technology for personal communications systems. D-AMPS stands for digital advanced mobile phone service and is so named due to compatibility and similarity with AMPS. Dual-band implies that D-AMPS 1900 technology can be used in the two wireless bands in the North America, the cellular band at 800 MHz and the PCS band at 1900 MHz (1850–1990 MHz). Dual-mode implies that D-AMPS 1900 phones can operate either in the digital or analog mode. D-AMPS 1900 uses time-division multiple access (TDMA) technology in the digital mode and AMPS in the analog mode. In North America, TDMA digital cellular systems built according to the D-AMPS standard have been in operation in the cellular band since 1992. D-AMPS was the first digital technology to be in commercial operation in North America and is currently deployed in 13 countries around the world. As of June 1995, there were over 1.5 million subscribers worldwide using D-AMPS technology, and forecasts show that by the end of 1996 there will be approximately 4 million D-AMPS subscribers worldwide on the 800-MHz cellular band. D-AMPS uses the North American IS-41 standard for interoperability between networks. Major manufacturers of D-AMPS network equipment are Alcatel, AT&T, Ericsson, Hughes, and Northern Telecom.

D-AMPS systems currently in commercial service are based on the IS-54 Rev. B TIA standard. It is a dual-mode standard that allows both analog AMPS and digital D-AMPS services to be offered in the same network. D-AMPS phones are also dual-mode, meaning they can access services either from the analog or digital network. Dual-mode operation is achieved by having the same channel bandwidth for AMPS and D-AMPS radio channels. IS-54 Rev. B also uses a common analog control channel for both digital and analog operations. This allows for common access techniques between analog and D-AMPS phones. Thus, current IS-54 Rev. B-based phones share the control channel with analog phones and use digital TDMA technology for traffic channels. The D-AMPS standard applicable for PCS networks, designated as IS-136 for 800 MHz, introduces the digital control channel for support of new applications and teleservices. D-AMPS 1900 is the upbanded version of the D-AMPS system that operates at 800 MHz. It allows for dual-band operation between cellular and PCS frequency bands and feature transparency. In this chapter we will refer to D-AMPS in the PCS band as D-AMPS 1900, and the technical specification supporting it by its popular 800 MHz equivalent, IS-136.

We will first study the IS-54 Rev. B-based D-AMPS technology and see how it differs from AMPS. Later we will look at IS-136-based modifications to IS-54 Rev. B and learn how dual-band operation is supported along with several of the enhanced services that IS-136 provides.

## 7.2    Digital Radio

What exactly do we mean by digital? How is digital radio different from analog radio? Digital radio consists of two processes: voice coding and modulation. (See Fig. 7.1.) *Voice coding* is the process of converting a continuous analog speech signal into a discrete digital form (binary form). *Modulation* is the process of impressing this discrete digital information onto a radio signal, effected by varying some key parameter (amplitude, frequency, phase) of the signal in a controlled manner.

### 7.2.1    Digital voice coding

The voice-coding process occurs in three stages: sampling, quantization, and coding. The process of voice coding takes a continuous waveform as input and produces a discrete output that contains all the information of the input signal. The input signal is first sampled. *Sampling* is the process of taking instantaneous measurements of a continuously changing parameter, for instance, amplitude of a waveform. The sampling rate is critical to capture all the information in the continu-

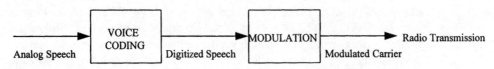

**Figure 7.1**   Digital radio.

ously varying waveform. According to a well-known communications equation, if the sampling rate is twice that of the highest component frequency in the waveform, then all the information is captured. Thus, for voice at 4 kHz the sampling rate should be 8000 times a second in order to capture all the information of the original signal.

Sampling is followed by quantization. Sampling results in a number of discrete values. For instance, sampling of a voice signal at 4 kHz results in 8000 discrete values per second. *Quantization* is the process of reducing these collected samples to a limited number of known values. A measurement scale is used which has a minimum value and a maximum value and the values in-between are graded according to a particular scheme. The grading can be linear, logarithmic, or so on. For instance, a common scale used is one of 256 values ranging from 0 to 255. The collected samples are graded against this scale and the closest approximation from the scale is chosen to represent that sample value. Obviously this results in some loss of information, since the exact sampled value is not used. Using a more granular scale reduces this error known as quantization noise. Hypothetically speaking, if we are using a 0–255 linear scale, then a sampled value of 121.8 would be quantized to 122 according to the 256 level scale. Since 256 values can be represented in binary form by 8 bits, then this forms the basic block for each speech sample. Thus the sampled value of 121.8 would be represented by the binary number 01111010 (122 in binary). This scheme is very common in telecommunication systems and is known as pulse code modulation (PCM). It uses a logarithmic quantizing scale. In PCM each 4-kHz voice signal is converted into 64 kbps (8000 samples × 8 bits per sample). There is a trade-off involved between the granularity of the quantization scale and the number of bits to be transmitted. The finer the granularity, the more the number of levels and hence the larger the number of bits required to represent the quantized value. A larger number of bits results in more data to be transmitted.

Digital radio voice coders have several objectives that are related to the limited bandwidth availability of the radio channel, the mobile environment, and the speech signal generation environment. The primary objective of the voice coder (vocoder) is to reduce the number of bits that need to be transmitted, while maintaining wire-line voice quality. This objective is directly related to the limited RF spectrum. Since the spectrum is limited, fewer bits can be transmitted, hence the requirement of efficient vocoders. The 64-kbps PCM coder is a balanced waveform coder, that is, it assumes nothing about the input speech signal and codes all values as being equally probable. But studies have shown that there is very strong correlation and redundancy in human speech, especially over small periods of time, such as 20–100 ms. This means that a particular speech sample is very similar to the previous sample, differing from it by only a few degrees. Vocoders have been designed that make use of this fact. They essentially code the change of the current sample from the previous sample and transmit this change. On the other end, the receiving vocoder applies this change to the previous sample and obtains a copy of the actual sample. Since changes between samples are transmitted instead of the actual sample itself, fewer bits are needed. Another objective of the vocoder is to withstand the high error rate of the radio channel. The vocoders also have to be designed to operate efficiently in a noisy speaker environment, such as in a car or airport where there are other noise sources.

### 7.2.2  Modulation

We have now converted the analog signal into digital information in the form of binary bits. This information has to be transmitted to the receiver. Transmission can be done in a number of ways. It depends on the transmission media. For instance, if the transmission media is a wire or a cable the binary information can be transmitted directly by electrical energy pulses. A 1 may indicate the presence of energy, and a 0 the absence of it. In a fiber-optic cable the energy could be light energy. There are more complex schemes used to transmit the digital information, depending on the requirements of the media. If the transmission environment is very noisy, then the encoding has to be very robust so that the receiver can accurately estimate the information transmitted. In radio transmission, it is preferable to have a continuous transmission and to vary some characteristic of the transmitted radio wave. These variations can be coded to carry the binary information generated by the analog to digital (A/D) converter. This process is known as modulation. In digital modulation systems the changes to the carrier are discrete compared to analog systems wherein the changes to the carrier are continuous. Modulation expands the bandwidth that is needed to transmit the signal. Since bandwidth is a limited commodity in wireless systems, multilevel modulation schemes are used in which each modulation symbol carries more than one information bit. A quaternary phase shift keying (QPSK) can transmit two bits of information in each symbol (see Fig. 7.2). In Fig. 7.2 the distance of the symbols from the origin of the circle indicates amplitude. Since all the symbols are equidistant from the origin, this implies that only phase is varied and amplitude is constant. The trade-off of using two bits/symbol modulation is the added complexity in the receiver since it has to detect phase changes with finer granularity.

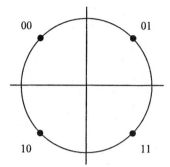

**Figure 7.2** Phase diagram for quarternary phase shift-keying modulation.

### 7.2.3    Digital measures against multipath

Digital systems differ from analog systems in that they have a wider range of countermeasures to use against the deleterious effects of multipath. We have already seen two of them: voice coding and modulation. Though the primary objective of these schemes is not to combat multipath, measures against it can be incorporated into them to provide some sort of protection against its effects. *Adaptive equalization* and *forward error correction* (FEC) are digital methods whose primary objective is to combat the effects of the radio environment. Equalization is used in the demodulation process to correctly interpret the transmitted, modulated symbols. It uses a known initial training sequence and a feedback algorithm in demodulating the bits. The known initial sequence is used by the equalizer to compare it with the received symbols. Since the equalizer knows what the received pattern should be, it is able to estimate the effects of the radio channel and use that knowledge in subsequent demodulation. It also uses a feedback algorithm wherein it compares a demodulated sequence with the received information and adjusts its demodulation process. Equalization requires fast digital signal processing to be effective.

Forward error correction (FEC) techniques work on the principle that if certain bit sequences are coded with redundant information the receiver is able to detect and correct the errors. The redundant information is added in such a way that it is easy for the receiver to detect certain bit errors and possibly correct them. FEC provides a measure of protection against the effects of the radio environment. The trade-off is that since redundant bits are added, it increases the bandwidth and thus can only be done to a certain extent.

## 7.3   Analog Control Channel and Digital Traffic Channel (D-AMPS)

### 7.3.1   Network architecture

The D-AMPS network architecture is shown in Fig. 7.3. It is identical to the AMPS architecture. Most of the digital-specific functions are isolated to the base station which is responsible for modulation/demodulation and digital voice coding/decoding, along with RF management.

### 7.3.2   TDMA

D-AMPS uses TDMA for voice channels. Dual-mode indicates that by complying with this standard this mobile station can operate in two modes: analog and digital. TDMA stands for time-division multiple access and is another way of accessing the network as opposed to FDMA (frequency-division multiple access). TDMA schemes divide the entire available radio spectrum into individual channels just as FDMA schemes do. TDMA goes further and divides each individual radio

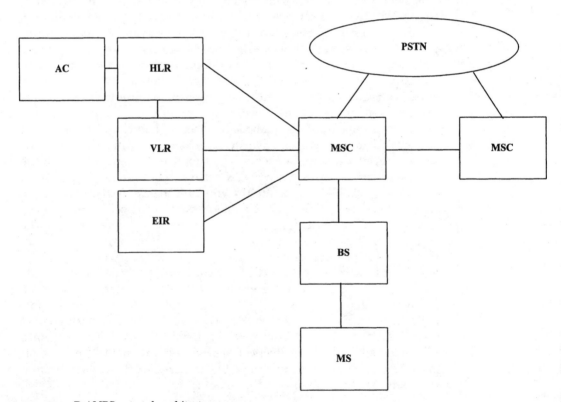

**Figure 7.3**   D-AMPS network architecture.

channel into time slots (see Fig. 7.4). Users can transmit and receive only during their particular time slot. Others can use the same radio channel as long as they are on a different time slot for their communication. In IS-54 each radio channel has the same bandwidth as an AMPS channel (i.e., 30 kHz).

A TDMA standard was found necessary because of the limitation of the existing AMPS radio interface. Specifically, AMPS did not have the capacity to support some of the very dense markets. Additionally, analog technology could not support some of the new services and features being requested by consumers.

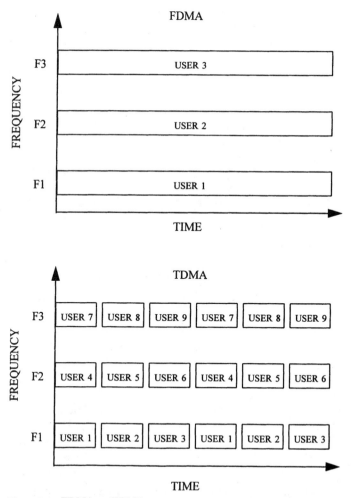

**Figure 7.4**   FDMA vs. TDMA.

### 7.3.3  Dual-mode operation

Dual-mode operation in IS-54 systems is achieved by sharing a common analog control channel for AMPS and IS-54 mobiles. One of the key criteria for achieving dual-mode operation was that the existing installed base of AMPS base stations and mobile stations should not be affected. This is achieved in IS-54 by having analog control channels, analog voice channels, and digital voice channels (henceforth known as digital traffic channels). The channel bandwidth and frequency assignments for cellular band of all the channels are identical to AMPS. A dual-mode base station has analog control channel, analog voice channel, and digital traffic channel radios. A dual-mode mobile station can operate in the digital mode wherein it accesses the system using the analog overhead control channel. When it sends an access request, the dual-mode mobile station indicates to the base station that it is capable of operating on a digital traffic channel. The base station allocates a digital traffic channel to the mobile station if one is available. The allocation procedure will indicate to the mobile station the channel number along with the time slot on that particular radio channel the mobile station has to use to communicate with the base station. The mobile station tunes to that particular radio channel and time slot and starts exchanging data with the base station. If a digital traffic channel is unavailable at the base station, it can assign an analog voice channel to the mobile station. The mobile station, on receiving indication of the assigned channel, will tune to the analog voice channel and start operating in the analog mode. This operation is identical to the operation of an AMPS mobile station.

The dual-mode operation of the IS-54 system is very attractive to operators. Operators wishing to roll out digital service in their markets can do so incrementally, due to the coexistence of AMPS and IS-54 mobiles in the same band. Typically, the operator would replace some of the analog radios with either digital-only radios or dual-mode analog/digital radios in the base station. Using digital-only radios will result in loss of some analog capacity in the cell, whereas dual-mode analog/digital radios will not result in loss of any analog capacity in the cell. For instance, if a base station has 20 analog radios, then at any given time it can support a maximum of 20 simultaneous analog voice calls. If an operator replaces one of the analog radios with a digital-only radio, then there are 19 analog radios and 1 digital radio available. The maximum number of simultaneous analog voice calls that can now be supported in the cell is reduced to 19, while the number of voice calls supported increases to 22 (19 analog and 3 digital). If the operator replaces an analog radio with a dual-mode analog/digital radio, one that is capable of operating in the analog or digital mode, then the maximum number of simultaneous analog voice calls that can now be supported in the cell is 20, provided there is no digital traffic channel in use at that time. As the IS-54 penetration level increases in the operator's network, more and more analog radios can be replaced by digital radios, providing for a smooth migration from analog to digital.

### 7.3.4  Power class and frequency assignments

The AMPS specification allows the minimum mobile station power to be −22 dBW (7 mW). The IS-54 specification allows the minimum mobile station power to be −34 dBW (400 μW). The corresponding power class is class IV. The AMPS minimum cell size is limited to a ½-mile radius. With a lower power output IS-54 mobiles can operate in cells with a much smaller radius.

Dual-band D-AMPS 1900 mobiles can gain service in the 800-MHz AMPS band and the 1900-MHz PCS band.

### 7.3.5  Digital traffic channel

As stated previously, IS-54 specification specifies three kinds of channels—analog control channel, analog voice channel, and digital traffic channel. The analog control and voice channels are identical to the AMPS (EIA-553) specification and will not be shown here. Chapter 7 contains a description of the analog channels.

Digital traffic channels are time-division multiplexed into time slots. A sequence of time slots makes up a frame. Frames are 40 ms long and consist of six time slots. There are two types of user channels: full-rate and half-rate. Full-rate and half-rate are indicative of the output of the vocoder. In the uplink direction the full-rate channel uses two out of the six time slots for transmit. In the downlink direction a full-rate channel uses two out of the six time slots for reception. To eliminate the need for a duplexer wherein the mobile station has to receive and transmit at the same time (this is done in AMPS), the time slots for the uplink and downlink are staggered by 8.5 ms (see Fig. 7.5). This means that a mobile station assigned time slot one will transmit on the uplink in time slot one and receive in the downlink in time slot one which occurs 8.5 ms later. Note that even though the duplexer is not needed for digital operation it may be required in the dual-mode mobile station for AMPS operation. The duplexer is bulky, costly, and very power inefficient. Presently half-rate channels are not in use, but when they are put in service the mobile station will transmit in one time slot in the uplink and receive on one time slot on the downlink. This will increase capacity since half-rate systems can then support six users per radio channel as compared to three users per radio channel for full-rate systems. The digital traffic channel gross bit rate is 48.6 kbps. Accounting for control overhead and number of users per channel results in a gross bit rate of 13 kbps for a full-rate channel and 6.5 kbps for a half-rate channel.

Time slots are arranged in a TDMA block which consists of three consecutive TDMA slots (1–3 or 4–6). Two consecutive TDMA blocks make a TDMA frame. A TDMA frame is 40 ms in duration (25 frames per second). A frame transmits 1944 bits and each full-rate TDMA user transmits twice in each frame. A full-rate traffic channel is made up of a pair of time slots, either one and four, two and five, or three and six. This means that a full-rate traffic channel user assigned to time slot one will transmit on time slots one and four.

### 7.3.6 Full-rate vocoder

The speech coding algorithm used in IS-54 is a member of a class of speech coders known as code excited linear predictive coding (CELP). These techniques use codebooks to vector quantize the excitation signal. The speech coding algorithm is a variation of CELP called vector-sum excited linear predictive coding (VSELP). VSELP uses a codebook which has a predefined structure such that the computations required

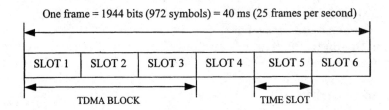

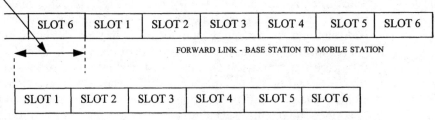

**Figure 7.5**   TDMA frame structure.

for the codebook search process can be significantly reduced. Before we see exactly how this vocoder works let us understand some terms. *Vector quantization* is a process which works by the help of codebooks. Codebooks are constructed in advance and are predefined and stored in read-only memory for access by the processor. They contain bit sequences known as vectors, against which an input speech block is compared. The vector matching the closest speech block is transmitted in place of the speech. Since vectors have substantially fewer bits than the original speech sequence there is a lot of data compression that can be achieved. This vocoder models the human vocal tract. It models the speech signal as the output of a vocal-tract filter whose input is an appropriate excitation signal. An *excitation signal* is the source of the sound energy. It has pitch and loudness associated with it. The vocal-tract filter's frequency response is specified by 8–10 parameters. The excitation signal can be digitally encoded at a data rate much lower than the original speech signal. The vocal tract and excitation information is updated every 20 ms. The vocoder works on 20-ms speech samples and compresses them into 159 bits of speech. The digital vocoder reduces the 64 kbps input to 7950 bps.

### 7.3.7  Channel coding

Refer to Fig. 7.6 for a description of the channel coding process. The channel error control for the speech coder data employs three mechanisms for the mitigation of channel errors. The first is to use a rate one-half convolutional code to protect the more vulnerable bits of the speech coder data stream. The second technique interleaves the transmitted data for each speech coder frame over two time slots to mitigate the effects of Rayleigh fading. The third technique employs the use of a cyclic redundancy check over some of the most perceptually significant bits of the speech coder. After the error correction is applied at the receiver, these cyclic redundancy bits are checked to see if the most perceptually significant bits were received properly.

The first step in the error correction process is the separation of the 159-bit speech coder frame's information into class 1 and class 2 bits. There are 77 class 1 bits and 82 class 2 bits in the 159-bit speech coder frame. The separation of bits into classes is done on the level of protection required for each class of bits. Class 1 bits are categorized as needing the most protection, and are protected by a half-rate convolutional encoder which basically replaces 1 bit of speech with 2 bits of encoded data. The class 2 bits are not protected at all. A 7-bit CRC is used for error detection purposes and is computed over the 12 most perceptually significant bits of the class 1 bits for each frame.

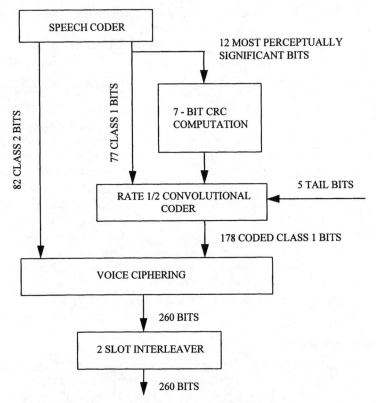

**Figure 7.6**   Forward error correction for speech codec.

Interleaving is the next process in the channel coding process. Before transmission, the encoded speech data is interleaved over two time slots with the speech data from adjacent speech frames. Stated another way, each time slot contains information from two speech coder frames. Interleaving is useful against Rayleigh fading because it causes errors in sequential bits. By spreading the data over two time slots, the probability that sequential data is lost is reduced. Loss of nonsequential data is easier to correct by using FEC methods.

### 7.3.8   Modulation

The modulation method used in IS-54 is known as the $\pi/4$ shifted, differentially encoded quadrature phase shift keying (DQPSK). As explained before, this results in spectral efficiency since each symbol carries information about 2 bits. The two conflicting objectives in choosing a modulation technique are spectral efficiency and narrow power spectra. Spectral efficiency is important since it allows for transmission of more bits for a given channel bandwidth. Narrow power spectra is necessary to avoid intersymbol interference. The broader the power spectrum the more will be the intersymbol interference due to delay spread. Unfortunately, narrow spectrum modulation techniques are not spectrally efficient and vice versa. DQPSK is a spectrally efficient technique that belongs to the family of linear modulation techniques. The drawback of these techniques is that they require linear power amplifiers that are expensive. European TDMA standard GSM uses GMSK (Gaussian minimum shift keying) which does not require linear power amplifiers, thus reducing the cost of the TDMA equipment.

### 7.3.9   Slot formats

There are three kinds of slot structures that exist in the IS-54 specification for digital traffic channel (DTC)—uplink speech slot, downlink speech slot, and shortened burst slot (refer to Fig. 7.7). Each slot type has dedicated fields, and slots are composed of 324 bits (162 symbols).

Synchronization word is used for time alignment and provides the known bit sequence to the equalizer. The slow associated control channel (SACCH) allows for transfer of signaling information while the mobile station is involved in a call. The data bits are the vocoder bits which are error protected and interleaved. The coded digital traffic color code (CDVCC) is essentially the digital voice color code (DVCC) appended with 4 parity bits to form the 12-bit CDVCC. The function of this color code is to provide cochannel identification similar to what is done in AMPS by SAT. Guard time is allocated to account for the different distances the mobile stations would be from the base station. Note that the guard time is present only for uplink time slots. It prevents the overlapping of received bursts due to radio signal transit time. The ramp time allows gradual rising and falling of the RF energy within the time slot.

### 7.3.10   Time alignment and shortened burst transmissions

*Time alignment* is the process of controlling the time of TDMA time slot burst transmissions from the mobile by advancing or retarding the mobile transmit burst. It is done so that the burst arrives at the base-station receiver in the proper time relationship to other time slot burst transmissions. An error in time alignment is caused by the arrival of transmissions from two different mobile transmitters simultaneously at the base-station receiver. This in turn causes errors in both signals. This overlap will occur at the head or tail of a time slot. Upon detection of an overlap condition, the base station must send an appropriate time alignment message to the mobile station using the appropriate downlink signaling channel.

Slot Format  MS - Base Station

| Guard Time | Ramp Time | Data | Sync | Data | SACCH | CDVCC | Data |
|---|---|---|---|---|---|---|---|
| 6 | 6 | 16 | 28 | 122 | 12 | 12 | 122 |

Slot Format  Base Station - MS

| Sync | SACCH | Data | CDVCC | Data | RSVD | CDL |
|---|---|---|---|---|---|---|
| 28 | 12 | 130 | 12 | 130 | 1 | 11 |

Shortened burst slot format

| G1 | R S D S D V S D W S D X S D Y S | G2 |
|---|---|---|

The shortened burst contains:
G1:  3 symbol length guard time
R:    3 symbol length ramp time
S:    14 symbol length sync word; the mobile station uses its assigned sync word
D:    6 symbol length CDVCC
G2:  22 symbol length guard time
V:    0000
W:    00000000
X:    000000000000
Y:    0000000000000000

**Figure 7.7**   Digital traffic channel slot formats.

Shortened bursts are sent when a mobile station begins operating in a large diameter cell because the propagation time between the mobile and base is unknown. The mobile station transmits shortened burst slots until the base station can calculate the required time offset. The default delay between the receive and transmit slots in the mobile is 44 symbols. The difference can be reduced in 30 increments each of a half-symbol duration. This results in a maximum distance at which a mobile station can operate in a cell, which is 72 miles for an IS-54 cell.

Time alignment is also necessary during handoff, when a mobile station hands off to a digital traffic channel. Handoff orders contain estimated time alignment information used by the mobile station when handing off from one digital traffic channel to another. For smaller diameter cells, this estimated time alignment information will be used to adjust the mobile station transmit timing so that there will be no burst collisions at the base station. For systems with sector to sector handoff, the estimated time alignment information will also be used to adjust the mobile station transmit timing so that there will be no burst collisions at the base station. For large diameter cells, however, this estimated time alignment information may not be accurate enough to avoid burst collisions at the base station. In such cases the mobile station will have to transmit shortened burst messages prior to getting timing alignment from the base station.

### 7.3.11   Digital traffic channel signaling

Digital traffic channel uses two channels to support control and supervision of the mobile station during a call—fast associated control channel (FACCH) and slow associated control channel (SACCH).

**7.3.11.1   FACCH.** FACCH is a signaling channel for transmission of control and supervision messages between the system and the mobile. The FACCH block replaces the user information block whenever it is to be transmitted (i.e., FACCH messages replace speech data with control and supervision messages). There is no limit on the number of speech frames that may be replaced and the vocoder quality degradation with lost frames is not linear. FACCH messages, replacing speech frames (260 bits), are interleaved and a message is composed over two slots exactly as is done for speech. The data is protected by a quarter-rate convolutional encoder for extra protection; this reduces the net transmission rate to 3250 bps from 13 kbps. There are no fields within a standard slot which identify it as speech data or an FACCH message. To determine if an FACCH message has been sent, the mobile must attempt to decode the slot as speech. If it decodes in error, it must then attempt to decode the slot as an FACCH slot. If the CRC then calculates correctly, it is an FACCH message.

**7.3.11.2   SACCH.** SACCH is a signaling channel for transmission of control and supervision messages between the digital mobile and the system. SACCH uses 12 coded bits per TDMA burst. It transmits speech in parallel without replacing speech information. A SACCH channel is made possible by the dedication of 12 bits per slot, so information may be sent while speech information is processed uninterrupted. Due to the small number of bits allocated to the SACCH channel per time slot (12 bits), it takes up to 22 slots for a single SACCH message to be transmitted. The data is half-convolutionally encoded and the net data rate is 300 bps. Tables 7.1 and 7.2 detail some of the messages on the digital traffic channel, along with their function.

**TABLE 7.1    Messages on the Downlink Digital Traffic Channel**

| Message | Number of transmissions | Channel | Function |
|---|---|---|---|
| Alert with info | Multiple | FACCH | Causes audible or visual signaling at the mobile relating to the initiation of a mobile-terminated call. Conveys info about call, such as calling party ID. |
| Measurement order | Multiple | FACCH | Informs the MS that it will begin the channel quality measurements and reporting. It contains an ordered list of RF channels (up to 12) for the MS to scan and report on. |
| Stop measurement order | Multiple | FACCH/ SACCH | Informs the MS that it will terminate the channel quality measurements and reporting. |
| Handoff | Multiple | FACCH | To order the MS from one traffic channel to another. Contains info about the target channel, rate, time slot, color code, SAT/DVCC, VMAC, time alignment, etc. |
| Physical layer control | Multiple | FACCH/ SACCH | To order the MS to set up or change all relevant parameters, such as output power level, time-alignment value, DTX allowed or not, power change. |
| Release | Multiple | FACCH | Informs MS that current call is terminated. |
| Base station ack. | Single | FACCH | This message acknowledges messages CONNECT, RELEASE, STATUS. |
| Maintenance | Multiple | FACCH | Sent in order to check the operation of an MS. |
| Audit | Multiple | SACCH | Sent in order to check whether a mobile is active in a system. |
| Local control | Multiple | SACCH | Sent in order to customize the operation of an MS. |
| Flash with info ack. | Single | FACCH | Ack for flash with info message. |
| Send burst DTMF | Multiple | FACCH | Used to request generation of DTMF tones. |
| Send burst DTMF ack. | Single | FACCH | Used to ack. send burst DTMF request. |
| Send continuous DTMF | Multiple | FACCH | Used to request sending of a single in-band DTMF digit. |
| Send continuous DTMF ack. | Single | FACCH | Ack. to send continuous DTMF request. |
| Flash with info | Multiple | FACCH | Used by both BS and MS to convey appropriate data to the other station. BS will use to indicate msg. waiting, connected number, etc. |
| Parameter update | Multiple | FACCH | Used to cause the MS to update its internal call history parameter that is used in the authentication process. |
| Status request | Multiple | FACCH | Used to query a mobile for its status (serial number, call mode, etc.). |

**TABLE 7.2    Messages on the Uplink Digital Traffic Channel**

| Message | Number of transmissions | Channel | Function |
|---|---|---|---|
| Connect | 3 | FACCH | This message is sent to indicate a call answer by the mobile subscriber. |
| Measurement order ack. | 1 | FACCH | Message acknowledges the start of the channel quality measurement in the MS. |
| Channel quality msg. 1 | 1 | FACCH/ SACCH | Contains the measurement report for the 1st RF channel to the 6th RF channel from the MS. The information carried in the information element "RSSI of 1st RF channel" will correspond to values measured on the RF channel designated as "1st RF channel" in the measurement order and so on. |
| Channel quality msg. 2 | 1 | FACCH/ SACCH | Contains the measurement report for the 7th RF channel to the 12th RF channel from the MS. The information carried in the information element "RSSI of 7th RF channel" shall correspond to values measured on the RF channel designated as "7th RF channel" in the measurement order and so on. |
| Release | 3 | FACCH | Informs BS that a call is terminated. |
| Mobile ack. | 1 | FACCH/ SACCH | Acknowledges alert with info, stops measurement order, release, maintenance, audit, local control and handoff. |
| Flash with info | 3 | FACCH | Message is sent to indicate that a user desires to invoke a special service. |
| Flash with info ack. | 1 | FACCH | Acknowledgment for flash with order from BS. |
| Send burst DTMF | 3 | FACCH | Requests sending of DTMF tones on the land line. |
| Send continuous DTMF | 3 | FACCH | Acknowledgment for send burst DTMF request. |
| Send burst DTMF ack. | 1 | FACCH | Requests the sending of continuous DTMF tones on the landline. |
| Send continuous DTMF ack. | 1 | FACCH | Acknowledges for sending of continuous DTMF tones on the landline. |
| Physical layer control ack. | 1 | FACCH | Sent to acknowledge the physical layer control message. |
| Status | 3/1 | FACCH | This message is used either as a reply to the STATUS request message or as a spontaneous message from the MS to inform the BS of a change in status. |
| Parameter update ack. | 1 | FACCH | This message acknowledges a mobile station's update of its internal call history parameter. |

**7.3.11.3   Transmission and acknowledgment of messages on the DTC.**   A mobile station transmits a message on either the FACCH or the SACCH. If the retransmission timer expires and the message can be retransmitted, the mobile retransmits the message on the same associated control channel (FACCH/SACCH). After transmitting a message which requires an acknowledgment from the base station, the mobile waits for the corresponding acknowledgment from the base station before transmitting another message. However, the channel quality message can be transmitted in the absence of an acknowledgment of the previous message.

The time-out interval for acknowledgment response to a message requiring an one begins after the last bit of the message is transmitted. The time-out interval is determined by the channel on which the message requiring an acknowledgment was transmitted, regardless of which channel may be used to respond with the acknowledgment. The time-out interval values are 200 ms for FACCH and 1200 ms for SACCH on the full-rate DTC.

### 7.3.12   Mobile assisted handoff (MAHO)

MAHO is a technique used by digital mobiles to aid the network in making handoff decisions. It also substantially speeds up the handoff process. The mobile station measures the RSSI and quality of its existing forward traffic channel and other forward RF channels of other base stations in the vicinity. It does this during the idle periods when it is neither transmitting nor receiving (see Fig. 7.8). These measurements are reported back to the base station which uses them as input to the handoff decision algorithm. Since the mobile is reporting the RSSI measurements, the network does not have to request neighboring base stations to measure the signal strength of the mobile station. The reporting also reduces the need for a scanning receiver at a base station. IS-54

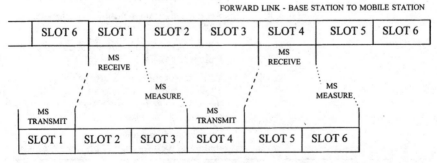

**Figure 7.8**   Mobile behavior on the digital traffic channel.

supports analog-to-analog, analog-to-digital, digital-to-analog, and digital-to-digital handoff. In some of these scenarios the mobile station may have to start transmitting shortened bursts on the new digital traffic channel so that the base station can time-align its transmissions.

The MAHO operation is initiated by the base station sending a start-measurements message to the mobile station. This message identifies those forward RF channels which the base station requires the mobile station to measure. Upon receipt of this message, the mobile station begins measurements on the current traffic channel and all forward RF channels identified in the measurement message. The mobile station reports the measurements to the base station in the order in which they were received in the start-measurements message. The measurement results are sent by the mobile station on the SACCH and the base station does not acknowledge the reception of the message.

Most of the system selection, cell selection, access, call processing, maintenance, and termination procedures are functionally identical to that of AMPS, the only difference being that some of the messages will have digital specific information that can be deciphered only by a digital mobile. An analog-only mobile, on receiving that information, will not be able to decipher it and it will be discarded.

## 7.4    IS-136—Digital Control Channel

Recently, the wireless industry has specified a new specification IS-136 which calls for a digital control channel (DCCH) for use in conjunction with IS-54. IS-136 is meant to replace the analog control channel which was part of IS-54. With the specification of DCCH, a digital mobile will be able to operate entirely in the digital domain, that is, use the digital control channel for system and cell selection as well as for access and the digital traffic channel for traffic. DCCH has several logical channels defined which have different purposes.

### 7.4.1    Logical channels

DCCH comprises the logical channels shown in Fig. 7.9. The uplink DCCH consists of a RACH. The downlink DCCH consists of SPACH, BCCH, SCF, and reserved channels.

**7.4.1.1    Random access channel (RACH).** RACH is used to request access to the system. It is a unidirectional (uplink only), shared (by all mobiles in a cell attempting access), point-to-point (MS to BS) channel. Access messages such as origination, registration, page response, audit confirmation, serial number, and message confirmation are sent on this channel. It also carries messages that provide information on authentication, security parameters update, and short message service (SMS)

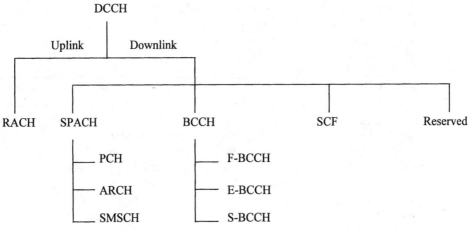

Figure 7.9   DCCH logical channels.

point-to-point messages. RACH can operate in two modes using contention resolution, as is done in analog control channel, and it can also operate in a reservation mode for replying to a base-station command.

**7.4.1.2   SMS point-to-point, paging, and access response channel (SPACH).** SPACH is used to transmit information to specific mobile stations. Information transmitted is in the form of SMS point-to-point messages (SMSCH), paging messages (PCH), or access response messages (ARCH). Each of these types of messages is sent on its individual, logical subchannel. The SPACH channel is unidirectional (downlink only), shared (shared by all mobiles in the cell), and point-to-point (BS to individual mobiles). It carries messages related to a single mobile or a small group of mobiles and allows for breakdown of large messages into small blocks for transmission. SPACH can be considered to be broken down into three logical channels.

**Paging channel (PCH).**   PCH is a logical channel subchannel of the SPACH, dedicated to delivering pages and orders. Paging messages, message-waiting messages, and user-alerting messages are sent on this channel. A single-page message may carry up to five mobile identifiers. Page messages are always repeated once. Messages such as call history count updates and shared secret data update, used for the authentication and encryption process, are sent on this channel.

**Access response channel (ARCH).**   ARCH is a logical subchannel of the SPACH. A mobile station autonomously moves to an ARCH upon successful completion of contention- or reservation-based access on a RACH. ARCH may be used to convey assignments to another resource or other responses to the mobile station access attempt. Messages assigning a mobile to an analog voice channel, a digital traffic channel, or redirecting the mobile to a different cell are sent on this channel. Registration access (accept, reject, or release) messages are also sent on this channel.

**SMS channel (SMSCH).** SMSCH is used to deliver short messages to a specific mobile station, in the context of short message services (SMSs). Point-to-point short messages are carried on this channel. Each message can be up to a maximum of 200 characters of text. Mobile-originated SMS is also supported. Broadcast short message service, wherein a base station can broadcast a short message for several mobile stations, is not supported in IS-136.

**7.4.1.3   Broadcast control channel (BCCH).**  BCCH is an acronym used to refer to the F-BCCH, E-BCCH, and S-BCCH logical subchannels. These channels are used to carry generic, system-related information. This channel is a unidirectional (downlink only), shared (shared between all mobile stations in the cell), and point-to-multipoint channel (BS to multiple MSs).

**Fast broadcast control channel (F-BCCH).** F-BCCH is used to broadcast DCCH structure parameters and those that are essential for accessing the system. DCCH structure parameters include information about the number of F-BCCH, E-BCCH, and S-BCCH slots in the DCCH. The mobile station uses this information when it first accesses the system to determine the beginning and ending of each logical channel in the DCCH. F-BCCH also includes information about the hyperframe count; access parameters that include information necessary for authentication and encryption; information for mobile access attempt such as number of access retries, access burst size, initial access power level; and indication if the cell is barred or not. Information about different types of registration, registration periods, and system identification information, such as network type, mobile country code, and protocol revision, is also provided. There are also additional optional messages that can be sent on this channel. They have to do with the optional services provided and supported by a particular base station.

**Extended broadcast control channel (E-BCCH).** E-BCCH carries broadcast information that is less time-critical than F-BCCH for the mobile stations. This channel carries information about the neighbor cell analog and TDMA cells. Optional messages include emergency information, time and date message information, and the services supported by neighboring cells.

**SMS broadcast control channel (S-BCCH).** This logical channel is used for SMS service. Individual short messages for a mobile station are sent on this channel.

**7.4.1.4   Shared channel feedback (SCF).**   SCF is used to support random access channel operation. It provides information about which time slots the mobile station can use for access attempts and also if a mobile station's previous RACH transmission was successful.

## 7.4.2   Digital control channel structure

**7.4.2.1   Superframe (SF).**   The full-rate digital control channel (DCCH) has a bit rate of 13 kbps, the same as a full-rate digital traffic channel. A DCCH is similar to a digital traffic channel in the use of time slots. DCCH information is sent in time slots which are grouped as TDMA blocks and TDMA frames. A full-rate DCCH on a particular frequency will use two time slots from the TDMA frame and a half-rate DCCH will use one time slot from each TDMA frame. For the purposes of organizing the data sent on the FDCCH, TDMA frames are grouped into superframes (SFs). Each superframe consists of 16 TDMA frames and is 640 ms long. Each superframe on the forward digital control channel (FDCCH) is comprised of an ordered sequence of logical channels as shown in Fig. 7.10. The superframe phase (SFP) increments every TDMA block. The first slot in a superframe (SFP = 0) is always a F-BCCH slot. The F-BCCH channel is followed by E-BCCH, S-BCCH, reserved, and SPACH logical channels. Assuming that a full-rate F-DCCH is being used and that the slots allocated to the F-DCCH are slots one and four, Fig. 7.10 shows how the logical channels of a superframe are mapped into individual time slots. Thus, a full-rate F-DCCH uses 32 out of the possible 96 time slots that can be transmitted during a superframe. The time slots not being used by the F-DCCH can be used for other F-DCCHs on the same frequency or as traffic channels. Table 7.3 lists the slot allocations for the different logical channels for a full- and half-rate F-DCCH.

Due to the critical nature of the F-BCCH information, some of which is used by the mobile in the cell selection decision, the F-BCCH information cannot be carried over to another superframe, that is, all the F-BCCH information has to be transmitted within a single superframe. The minimum number of slots designated for F-BCCH are three and the maximum 10.

All the mandatory F-BCCH messages such as DCCH structure, access parameters, registration parameters, system identity, and control channel selection parameters are sent during the time slots allocated to the F-BCCH channel, along with any optional messages. Once all the F-BCCH information has been transmitted, mandatory E-BCCH messages such as neighbor cell and regulatory configuration messages are sent, followed by optional E-BCCH messages. E-BCCH messages, unlike F-BCCH messages, do not all have to be sent during a single superframe. These messages can spill over into the next super-

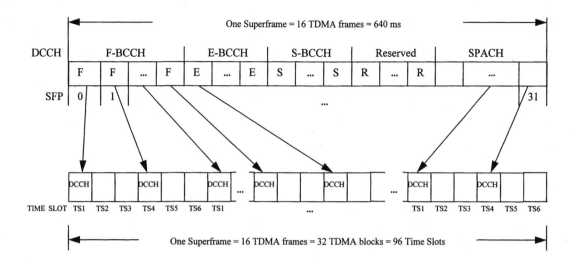

One Superframe = 16 TDMA frames = 640 ms

**Figure 7.10** DCCH superframe structure.

frame. Currently there are no S-BCCH messages defined and the S-BCCH logical channel is present as a place marker for future use. SPACH messages follow and can be PCH, ARCH, or SMSCH messages. SPACH messages carry mobile-specific information and may span several F-DCCH slots which may be part of different superframes.

**7.4.2.2 Hyperframe.** A hyperframe consists of two superframes: one primary and the other secondary. Figure 7.11 shows the structure of a hyperframe and the organization of the logical channels in a hyperframe. F-BCCH information in each superframe of a hyperframe is identical. E-BCCH information may be different from SF to SF, as the information repeats after a certain period. SPACH information, if it is a PCH, has to be identical in both the primary and secondary superframes of a hyperframe. The rest of the SPACH information may be different from SF to SF. There are 32 hyperframes, after which the hyperframe counter is reset to 0. The duration of a hyperframe is 1.28 seconds.

**7.4.2.3 Paging frame (PF).** A paging frame is defined by a paging frame class (PFC). A paging frame class defines the hyperframe in which a particular mobile looks for its pages. Since a hyperframe consists of a primary superframe and a secondary superframe, both of

**TABLE 7.3    Slot Allocations for Logical Channels on the F-DCCH**

| | Full-rate F-DCCH | | Half-rate F-DCCH | |
|---|---|---|---|---|
| | Min. (time slots) | Max. (time slots) | Min. (time slots) | Max. (time slots) |
| F-BCCH (F) | 3 | 10 | 3 | 10 |
| E-BCCH (E) | 1 | 8 | 1 | 8 |
| S-BCCH (S) | 0 | 15 | 0 | 11 |
| Reserved (R) | 0 | 7 | 0 | 7 |
| SPACH | 1 | 32–(F+E+S+R) | 1 | 16–(F+E+S+R) |

which should carry the same paging channel information, a mobile looks for its pages in either the primary or secondary superframe of a hyperframe. Paging frame classes are used for discontinuous reception (DRX) or mobile sleep mode. This is a technique whereby the battery life of a mobile is extended by allowing it to power down for certain periods of time. During these periods the mobile does not scan the control channels for information and can only originate calls. DRX is achieved in IS-136 by assigning different PFCs to mobile stations. For instance, mobiles assigned to $PFC_1$ can expect a page in every hyperframe, mobiles assigned to $PFC_2$ can expect a page in every other hyperframe (i.e., they can sleep for two hyperframes), and so on. The lowest PFC supported is $PFC_1$, and it is also the default PFC for all mobiles prior to registration. A mobile can be assigned to a different PFC from the default one by a command from the network. All SPACH messages (directed either to a specific mobile or a group of mobiles) have a field in their header known as paging frame modifier (PFM), which either increments or decrements the PFC of the mobile. The highest PFC supported is $PFC_8$, and it indicates that the mobile stations assigned to it can expect a single page over 96 hyperframes. Thus,

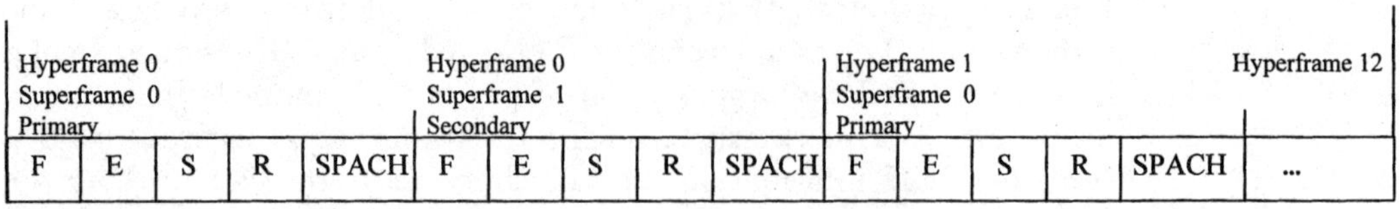

| Hyperframe 0 Superframe 0 Primary | | | | | Hyperframe 0 Superframe 1 Secondary | | | | | Hyperframe 1 Superframe 0 Primary | | | | | Hyperframe 12 |
|---|---|---|---|---|---|---|---|---|---|---|---|---|---|---|---|
| F | E | S | R | SPACH | F | E | S | R | SPACH | F | E | S | R | SPACH | ... |

| | | |
|---|---|---|
| F | F-BCCH | Information carried on this logical channel is the same for every superframe for a given DCCH |
| E | E-BCCH | Information carried on this logical channel can differ from superframe to superframe |
| S | S-BCCH | Information carried on this logical channel can differ from superframe to superframe |
| R | RESERVED | |
| SPACH | | Information carried on this logical channel can differ from superframe to superframe except when it is a PCH. In this case the information carried in the primary and secondary superframes must be the same for a given hyperframe |

**Figure 7.11**    DCCH hyperframe structure.

the sleep periods supported by IS-136 are from a few milliseconds ($PFC_1$) to over two minutes ($PFC_8$).

A mobile station is allocated a specific paging subchannel within its paging frame to monitor for pages. A mobile station only monitors this paging subchannel, instead of monitoring the entire paging frame, for pages. The mobile's paging subchannel is calculated by using its mobile identity. In case of multiple DCCHs on the same frequency, a mobile station calculates the DCCH to camp on by using its mobile identity.

### 7.4.3 DCCH burst formats

There are three kinds of burst formats used on the DCCH. The normal and abbreviated slot formats are used on the RACH in the mobile to base-station direction. The slot format is used in the base station to mobile station direction.

The abbreviated slot format is used by the mobile station when the exact timing advance is not known. This usually occurs when the mobile station is accessing the system for the first time. Since the timing advance is not known, this burst format has a larger guard period

Normal Slot Format MS - Base Station on RDCCH

| Guard Time | Ramp Time | Preamble | Sync | Data | Sync+ | Data |
|---|---|---|---|---|---|---|
| 6 | 6 | 16 | 28 | 122 | 24 | 122 |

Abbreviated Slot Format MS - Base Station on RDCCH

| Guard Time | Ramp Time | Preamble | Sync | Data | Sync+ | Data | Ramp Time | Abb. Guard Time |
|---|---|---|---|---|---|---|---|---|
| 6 | 6 | 16 | 28 | 122 | 24 | 78 | 6 | 38 |

Slot Format Base Station - MS on FDCCH

| Sync | SCF | Data | CSFP | Data | SCF | RSVD |
|---|---|---|---|---|---|---|
| 28 | 12 | 130 | 12 | 130 | 10 | 2 |

**Figure 7.12**  Burst formats on the DCCH.

known as the abbreviated guard field. This field is 38 bits long and mitigates burst collisions at the base station. If collisions do occur, they will occur while the base station is decoding the abbreviated guard field; since there is no useful data in this field, no data is lost. The shared channel feedback field (SCF) is used by the base station to control accesses on the RACH. It indicates to the mobile station if the RACH access slot is busy, idle, or reserved by another mobile station. It also indicates to the mobile station if its previous RACH transmission was received by the base station. User data bits (coded information bits) are mapped onto the data field for transmission. In the forward direction, the field is 260 bits in length. In the reverse direction, the length of the data field is 244 bits for the normal slot format and 200 bits for the abbreviated slot format.

CSFP is used to convey information regarding the superframe phase (SFP) so that the mobile station can find the start of the superframe. The content in this field may also be used to discriminate between DCCH and digital traffic channel (DTC) in that the CSFP of a DCCH and CDVCC of a DTC have no common code words.

The synchronization field acts as a time slot identifier. It is a 28-bit field which is used for slot synchronization, equalizer training, and time slot identification. A mobile station is able to uniquely determine a time slot number with the help of the SYNC field and the CSFP. The CSFP assists the mobile station in determining the start of the superframe, and the SYNC assists the mobile station in determining a time slot in a particular TDMA block (SFP). The SYNC+ field is used to provide additional synchronization information to improve base-station receiver performance. Note that the SYNC+ field exists only in the mobile to base-station direction and is sandwiched between two DATA fields. Its primary purpose is to provide additional information to the equalizer and improve its performance. Since the equalizer has foreknowledge of the bit pattern of the SYNC+ field, it adjusts itself based on the received bit pattern.

### 7.4.4 Channel encoding

The logical channels BCCH, SPACH, and RACH (normal and abbreviated) use a half-rate convolutional encoding. The encoding scheme is similar to that a full-rate digital traffic channel with the exception that the user's data is not broken down into different classes. The entire user information, along with any layer two headers, 16-bits CRC, and 5-tail bits, are sent to the encoder. The output is 260 bits of encoded data for SPACH and BCCH, 244 bits for RACH (normal length), and 200 bits for RACH (abbreviated length).

### 7.4.5  Interleaving

For all channel types and burst lengths, all bits are sent within one burst (i.e., only intraburst interleaving is performed). This differs from interleaving on a digital traffic channel where two time slot interleaving is done.

### 7.4.6  Message mapping

IS-136 uses a structured, layered protocol for message generation, unlike IS-54 which does not use a layered approach. User information is put in a layer 3 message with a protocol discriminator and a message type. The maximum length of a layer 3 message is 255 octets, with the exception of the F-BCCH which is further limited by the number of F-BCCH slots allocated per superframe (3–10 time slots). A layer 3 message is sent to layer 2 which adds a layer 2 header, tail bits, and CRC bits. The length of a layer 3 message is determined by a layer 3 length indicator which is carried as part of layer 2 header. The length of an layer 2 frame is fixed, determined by the specific logical channel. Some of the information included in the layer 2 header of an RACH message is the type of burst (that is, if it is the first burst or a continuation of the previous burst), mobile station identity, number of layer 3 messages, and layer 3 length indicator. Layer 2 header information for F-BCCH and E-BCCH channels includes flags to indicate if F-BCCH or E-BCCH information has changed since the last transmission. This is very useful since a mobile station, if it has already read a full cycle of F-DCCH information, simply looks at these bits to determine if it should update its internally stored F-DCCH information. Layer 2 header information for SPACH channel (which is a channel directed to specific mobiles) has information about change in paging frame class, number of mobiles at which this message is targeted and their mobile identities, type of receiving mobile station's access attempt (contention/reservation) in response to this message, DCCH barring, and more. This layer 2 message is then channel coded, interleaved, and transmitted in a time slot. The length of the DCCH time slots are fixed, but the number of information bits sent differ. Figure 7.13 shows an example of how one layer 3 message is mapped into several layer 2 frames, an example of a layer 2 frame mapping onto a time slot, and an example of time slot mapping onto a DCCH channel.

Layer 1 functions on the base station include power setting, modulation/demodulation, timing offset control, synchronization detection, random access subchannel number mapping, forward error correction (FEC/channel coding), interleaving/deinterleaving, and SPACH message continuation. Layer 2 functions include header formatting, SCF control, layer 3 message concatenation, CRC generation/verification,

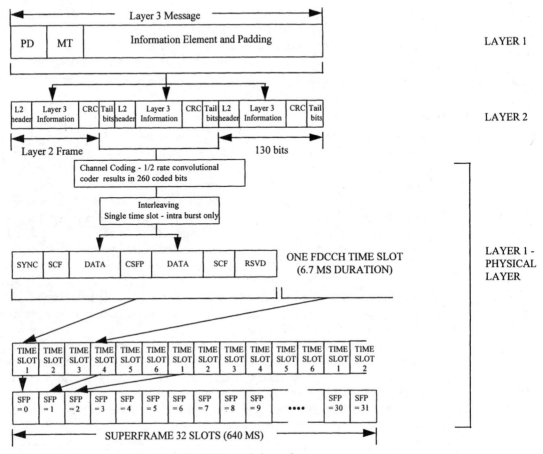

**Figure 7.13**   Message mapping for a FDCCH logical channel.

padding, mobile identity management, BCCH change notification, layer 2 frame segmentation and reassembly, SCF setting, and burst usage. For most parts the functions of the different layers on the mobile station side are also identical.

### 7.4.7  Digital control channel frequency assignments

Digital control channels can be placed anywhere in the entire frequency band, unlike analog control channels which are isolated to channels 313–333 for 800-MHz A-band and 334–354 for 800-MHz B-band. Moreover, a dual-mode/dual-band mobile station would potentially also scan for DCCHs in the PCS bands. Although a mobile station can scan all channels for a DCCH, it will take considerable time. Several strategies are defined in IS-136 to speed up the DCCH selection process in the mobile station. The mobile station may store DCCH allocation information to assist in the location of control channels. For example, a mobile station may store the DCCH allocation of its home system or its last visited system.

A more generic strategy suggested in IS-136 is to divide the available radio spectrum into one or more frequency bands reflecting the number of operators supported in a geographic area. To aid the mobile station in searching for a DCCH, any given frequency band is comprised of 16 probability blocks. Probability blocks are assigned a relative order of probability regarding their potential for DCCH support. The ranking of probability blocks is only a relative measure of probability and shall not be interpreted as an absolute probability for finding a DCCH. Within a probability block a mobile station searches for a DCCH in ascending order of channel number. For instance, an 800-MHz dual-mode TDMA mobile station that prefers system A and follows the recommended probability blocks begins searching for a DCCH according to Table 7.4.

Note that the DCCHs are recommended to be placed on the opposite end of the frequency band (channels 1–26) from the analog control channels (channels 313–333).

A 1900-MHz TDMA mobile station that follows the recommended probability blocks shall begin searching for a DCCH according to Table 7.5.

**TABLE 7.4    Recommended 800-MHz A-band DCCH Allocation**

| Block number | Channel number | Band | Number of channels | Relative probability |
|:---:|:---:|:---:|:---:|:---:|
| 1 | 1–26 | A | 26 | 4 |
| 2 | 27–52 | A | 26 | 5 |
| 3 | 53–78 | A | 26 | 6 |
| 4 | 79–104 | A | 26 | 7 |
| 5 | 105–130 | A | 26 | 8 |
| 6 | 131–156 | A | 26 | 9 |
| 7 | 157–182 | A | 26 | 10 |
| 8 | 183–208 | A | 26 | 11 |
| 9 | 209–234 | A | 26 | 12 |
| 10 | 235–260 | A | 26 | 13 |
| 11 | 261–286 | A | 26 | 14 |
| 12 | 287–312 | A | 26 | 15 |
| 13 | 313–333 | A | 26 | 16 (lowest) |
| 14 | 667–691 | A′ | 26 | 3 |
| 15 | 692–716 | A′ | 26 | 2 |
| 16 | 991–1023 | A″ | 26 | 1 (highest) |

TABLE 7.5    Recommended 1900-MHz A-band DCCH Allocation

| Block number | Channel number | Band | Number of channels | Relative probability |
|---|---|---|---|---|
| 1 | 1–31 | A | 31 | 16 (lowest) |
| 2 | 32–62 | A | 31 | 15 |
| 3 | 63–93 | A | 31 | 14 |
| 4 | 94–124 | A | 31 | 13 |
| 5 | 125–155 | A | 31 | 12 |
| 6 | 156–186 | A | 31 | 11 |
| 7 | 187–217 | A | 31 | 10 |
| 8 | 218–248 | A | 31 | 9 |
| 9 | 249–279 | A | 31 | 8 |
| 10 | 280–310 | A | 31 | 7 |
| 11 | 311–341 | A | 31 | 6 |
| 12 | 342–372 | A | 31 | 5 |
| 13 | 373–403 | A | 31 | 4 |
| 14 | 404–434 | A | 31 | 3 |
| 15 | 435–465 | A | 31 | 2 |
| 16 | 466–499 | A | 31 | 1 (highest) |

### 7.4.8    DCCH identification

Since a DCCH can be located anywhere across the entire frequency spectrum and since this spectrum can also be occupied by digital traffic channels which have a slot format similar to DCCH, it could be a problem for a mobile station to correctly identify a DCCH. A mobile station makes use of the following information to distinguish a DCCH from a DTC:

- *CSFP and CDVCC fields.*    The CSFP and CDVCC fields used in the DCCH and DTC slot formats always have 4 bits out of 12 which are different in every pair of CDVCC and CSFP code words regardless of which code word is transmitted by a base station. The CDVCC field content is fixed from slot to slot on a DTC, whereas the content of the CSFP changes in a predictable fashion from slot to slot on a DCCH.

- *DATA fields.*    Channel coding and interleaving employed on the DTC is different from the one employed on a DCCH. A DTC employs two-slot interleaving, whereas the DCCH employs only intraburst interleaving.

- *SACCH and RESERVED fields.*    SACCH and RESERVED fields have different functionality on a DCCH, which can be used to distinguish a DCCH from a DTC.

### 7.4.9    Control channel selection

Once a mobile station has scanned the frequency band and located a DCCH, it has to determine if it is eligible to park on that particular

DCCH. IS-136 mobiles make use of two criteria to determine which DCCH to park on, the signal strength and the services aspect criteria.

**7.4.9.1 Signal strength criteria.** The signal strength criteria is a complex equation the mobile station uses to determine the eligibility of the current DCCH to serve itself. Some of the key parameters of this equation are the signal strength of the current DCCH as received by the mobile station (RSS, received signal strength), the minimum received signal level required to access the cell (RSS_ACC_MIN, received signal strength for access minimum), the maximum nominal output power that the mobile station can use when initially accessing the network, and the mobile station power class.

The received signal strength is taken into account; if the mobile station does not have good reception of the control channel (which is typically broadcast at the maximum power level the cell is capable of), then it is useless to park on that cell since subsequent signaling and traffic messages will also not be received at an acceptable level. To account for the multipath effects of the radio channel, which could adversely affect a particular RSS measurement, the RSS is averaged over the last five measurements, the minimum time between any two consecutive measurements being 20 ms. The RSS_ACC_MIN is a parameter broadcast on the F-BCCH. It defines the minimum received signal level required to access the cell. The mobile factors this into the equation.

The mobile station power class also plays a role in the signal strength criteria. Due to the existence of classes of mobile stations based on their output power capacities, it is possible that a class IV mobile station on the edge of a large cell may receive the DCCH at an acceptable level, but the base station may not receive the mobile station's transmission due to low output power of a class IV mobile. Thus, the power class of the mobile station has to be taken into account. This also means that the cell size is not fixed, but differs from mobile station to mobile station based on their power class. For instance, in the previous example a class II mobile may have been able to access the cell and receive acceptable service.

With the help of these parameters the cell can be optimized for mobile access. It also enables hierarchical cell structures wherein smaller cells are contained within larger macrocells. The smaller cells form a microcell underlay to the macrocell. This type of service is used in downtown areas where the microcells provide capacity and the macrocells coverage.

**7.4.9.2 Service aspects criteria.** Once the current DCCH has fulfilled the signal strength criteria, the mobile station then has to determine if it is allowed to park on this cell. The mobile may not be able to park on this cell due to the following service reasons:

- *Cell barred.*   The base station broadcasts the status of the cell in the BCCH. The cell status can be barred, in which case all mobile stations are prohibited from parking on this cell. There can also be a time specified as to when this cell will be barred. A base station may bar a cell for a variety of reasons, such as overload conditions and network outage.

- *Invalid network type.*   The base station transmits its network type on the BCCH. This network type identifies if the cell is part of a public, private, or residential cellular system. The mobile station has to ensure that it subscribes to the broadcast network type before parking on the cell.

Once a DCCH has fulfilled all the signal strength and service aspect criteria, the mobile station can park on that cell.

### 7.4.10   Mobile access

Prior to making any kind of access attempt the mobile station has to read a full cycle of F-BCCH and E-BCCH information broadcast on the BCCH. This is so that the mobile station has knowledge of all the parameters needed for access and is aware of the cell control channel structure. IS-136 supports an access attempt either on a contention or a reservation basis.

**7.4.10.1   Contention-based access.**   The access mechanism is an ALOHA type of access. In an ALOHA access several uncoordinated users attempt to use a shared resource. Essentially, each user is allowed to begin transmission at any time, independent of other users. When two users start their transmissions at the same instant, their transmissions will collide and ALOHA specifies graceful back-off mechanisms, wherein each user waits for a random amount of time before reattempting access. IS-136 uses a slotted ALOHA-based protocol specifically CSMA/CD (carrier sense multiple access with collision detection). Since IS-136 makes use of TDMA, the access burst cannot be sent at any time, just at the beginning of a time slot, hence slotted-ALOHA. The mobile, prior to attempting transmission, reads the busy-idle-reserved status of the channel (broadcast by the base station on the DCCH in the SCF). If the channel is busy, it awaits a random amount of time (between zero to six TDMA blocks) before reattempting access. This ensures that once a channel is captured by a mobile it will be able to complete its transmission fully.

**7.4.10.2   RACH subchanneling.**   RACH subchanneling is a process by which the available RACH bandwidth is broken down into different paths to allow for simultaneous mobile access attempts. It is primarily designed to allow for sufficient time at both the mobile station and the base station to process an access attempt. There is a finite amount of time required between a mobile station transmission and reception of it by the base station. Furthermore, the base station has to process the message and then, if valid, set a flag in the FDCCH to indicate to the mobile station that its transmission was received. RACH subchanneling accounts for all of these delays.

As described previously, the SCF field is broadcast by the base station on the FDDCH. The SCF information is part of the slot format on the FDCCH and is added as part of the layer 2 function. Therefore, it is present in every FDCCH transmission independent of the logical channel (SPACH or BCCH). It serves two purposes: (1) it indicates to the mobile station the availability status (i.e., busy, idle, reserved) of the corresponding RACH slot (the RACH slot occurring after 64.8 ms), and (2) it indicates the reception status of the previously transmitted RACH burst. Thus, the SCF provides information about the availability of the upcoming RACH slot and reception status of previously transmitted RACH burst simultaneously. If a full-rate DCCH exists, then its RACH and FDCCH slots are multiplexed so as to create six distinct access paths (three distinct access paths for a half-rate DCCH) as shown in Fig. 7.14. Figure 7.14 illustrates RACH subchanneling. Assuming that path 1 (P1) in the FDCCH indicates that the next P1 burst on the RACH is available (i.e., idle) and is selected for an access attempt, a mobile station will begin sending the first burst of its access at that time (64.8 ms after receiving the full P1 slot in the FDCCH). The mobile station then begins reading the SCF flags in the next P1 FDCCH slot (41.8 ms after completing transmission of its access burst) to determine if its previous transmission was received correctly by the base station.

Figure 7.15 explains in detail the different fields transmitted by the base station and its use by the mobile station for an access attempt. Since a full-rate DCCH transmission occupies only two time slots in a TDMA frame, the slot numbers are compressed in the figure to show only DCCH transmission bursts (assuming slots one and four are used for DCCH). The corresponding SCF information is also displayed. This is followed by the RACH subchanneling path. The arrows show the order of events associated with an access attempt. Thus, following the arrows from left to right on RACH subchannel P1, the first SCF field indicates the availability of the next P1 burst on the RACH. A mobile station reads that and transmits a burst in the corresponding RACH slot (indicated in the figure by a Z). The mobile station then reads the

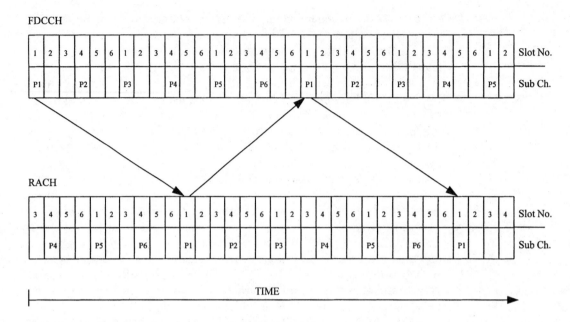

P   Different paths available for RACH messages. For a full-rate DCCH there will always be 6 paths for RACH messages, i.e., a single DCCH can support 6 simultaneous user access attempts.

**Figure 7.14**   RACH subchannels.

second field of the SCF information in the next P1 FDCCH slot to determine the reception status of its transmitted burst. If the SCF information indicates that it was received, then the mobile station sends any additional bursts remaining, beginning in the next P1 burst in the RACH. On successfully transmitting all the information, the mobile station begins looking for an access response from the base station (an ARCH message). The base station then responds with a message sent over two FDCCH slots (marked with a Y in the figure) within the expected time frame.

**7.4.10.3   Reservation-based access.**   IS-136 supports reservation-based access for certain SPACH messages. A reservation-based access is controlled by the base station and the base station orders a mobile station to make a reservation-based access attempt. Figure 7.15 shows the reservation-based access attempt. The base station marks the SCF information fields with a reserved status and the mobile station makes an access attempt on the corresponding RACH slot.

FDCCH

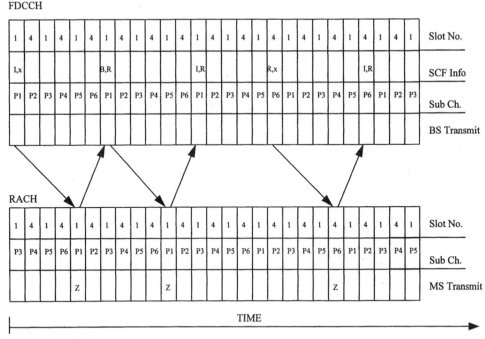

"x" indicates "do not care"
"Z" indicates "transmit slots for mobile station"
"Y" indicates "base station response on the ARCH"
The first field of the SCF info represents the busy/idle/reserved status of the corresponding RACH slot.
The second field of the SCF info represents if the MS transmit in the previous RACH slot was received.

**Figure 7.15**   Mobile access using SCF and RACH subchannels.

### 7.4.11   Registration

IS-136 supports the following kinds of registration.

**7.4.11.1   Power-down registration.**   Power-down registration is done by the mobile station when its power is switched off. The base station broadcasts a bit indicating if power-down registration is supported in the cell or not. If power-down registration is supported, a mobile, prior to powering down, sends a registration message with a power-down indication to the base station. The base station will forward this message to the network, which will put a flag in the subscriber's user profile to indicate that the subscriber has powered down. Power-down registration is used to cut down on network paging and signaling load. If a call is placed to a user whose mobile station had previously sent a power-down registration, the network does not have to page this mobile station, since it knows that the mobile station will not respond to the page.

**7.4.11.2 Power-up registration.** Power-up registration is the opposite of power-down registration. When a mobile station's power is switched on and if the network supports power up registration, the mobile station will send a registration message with a power-up indication to the base station. This message is sent to the network, which will remove the flag from the subscriber's profile so that any future incoming calls will result in a page for the subscriber.

**7.4.11.3 Deregistration.** Deregistration occurs whenever a mobile station decides to acquire control channel service on a different type of network (public, private, or residential). If the network supports deregistration the mobile station sends a registration message to the base station with a deregistration indication. It can optionally also include information about the new network on which it is going to park.

**7.4.11.4 New system/location area registration.** If the broadcasted system ID (SID) value does not match the SID value stored in the mobile station or if the mobile station desires service on a new private SID or residential SID, it sends a registration message to the base station with a new system indication. Similarly, when the location area of the mobile station has changed, it sends in a registration message.

**7.4.11.5 Periodic registration.** A mobile station may be instructed to periodically register with the network. The registration interval is in multiples of 94 superframes and can range from 94 to 48,128 super-frames (approximately 1 minute to 8.5 hours). Whenever the registration interval expires, the mobile station sends a registration message to the base station with a periodic registration indication.

**7.4.11.6 Forced registration.** A network may, under certain circumstances, force all mobile stations to register. This is done by broadcasting a bit on the BCCH which informs all mobile stations to register. The mobile stations then send a registration message to the base station with an indication that this is a forced registration.

## 7.5    IS-41

Interim standard 41 (IS-41) documents the intersystem cellular operations and intersystem signaling standards as specified by the Telecommunications Industry Association (TIA). It deals specifically with defining operations between two cellular systems that can be from different equipment vendors, between different operators, or using different radio technologies (AMPS, IS-54, etc.). IS-41 has evolved over a period of ten years. It has gone through four revisions up to now. IS-41 Rev. 0 was published in February 1988, IS-41 Rev. A was published in January 1991, IS-41 Rev. B in December 1991, and IS-41 Rev. C in December 1994. It is under continuous revision to address the new services that require standardized intersystem procedures. IS-41 Rev. 0 dealt with handoff-forward and handoff-back procedures for analog cellular stations. Intersystem data communications were on point-to-point links between systems using SS7 (signaling system number 7) or X.25 (packet protocol). Voice traffic was on direct links. The transaction capabilities application part (TCAP) specified by CCITT was used for association control and remote operations. IS-41 Rev. A changes included replacement of CCITT TCAP with ANSI TCAP, data communications over X.25 or SS7 networks (restriction of Rev. 0 of dedicated links was removed), automatic roaming, call delivery, and remote feature activation/deactivation procedures. IS-41 Rev. B added path-minimization procedures to optimize the intersystem voice trunks involved in a handoff call. Support for IS-54 mobiles was included. The global title translation (GTT) capability of the SS7 SSCP was added.

In the following sections we focus on the latest revision of IS-41, Rev. C, since this includes all the aspects of the previous revisions, as well as some new services.

### 7.5.1    IS-41 Rev. C

IS-41 Rev. C supports the EIA-553 AMPS, IS-54/IS-136 DAMPS/DCCH, and IS-95 CDMA air interfaces. We have already discussed all of these air interfaces except IS-95 which will be described in Chap. 8.

IS-41 Rev. C differs from IS-41 Rev. B in its support of the following functionalities:

- Intersystem authentication and encryption
- Intersystem operations for dual-mode CDMA terminals
- Border cell problem resolution
- Expanded feature support

IS-41 Rev. C is organized into the following sections:

- Functional entities
- Intersystem handoff procedures
- Automatic roaming procedures
- Authentication and encryption
- IS-41 Rev. C features

IS-41 Rev. C specifies a single set of application-specific services called the mobile application part (MAP). These MAP services reside on top of the ANSI TCAP. ANSI TCAP can be used on top of SS7 layers (SCCP, MTP3, MTP2, MTP1) or X.25.

IS-41 Rev. C also defines a network reference model as shown in Fig. 7.16. This figure presents the functional entities and the associated interface reference points that may logically comprise a cellular network.

For the purposes of IS-41 procedures described in this chapter, the reference points B, C, D, and E are relevant. Only these functional entities relevant for our understanding of the cellular intersystem operations will be discussed.

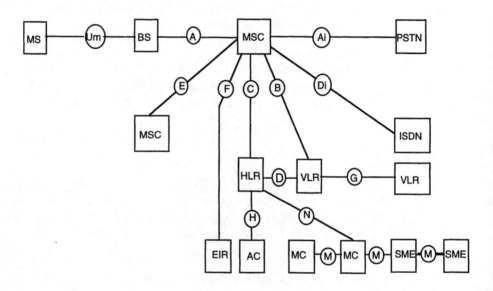

| AC | Authentication Center |
| BS | Base Station |
| EIR | Equipment Identity Register |
| HLR | Home Location Register |
| ISDN | Integrated Services Digital Network |
| MC | Message Center |
| MS | Mobile Station |
| MSC | Mobile Switching Center |
| PSTN | Public Switched Telephone Network |
| SME | Short Message Entity |
| VLR | Visitor Location Register |

**Figure 7.16**   IS-41 Rev. C network reference model.

### 7.5.1.1  Functional entities

**Authentication center (AC).** The AC is an entity that manages the authentication information related to the mobile station. The AC may or may not be located within or be indistinguishable from an HLR. An AC may serve more than one HLR.

**Base station (BS).** The base station is the common name for all the radio equipment located at one and the same place used for one or several cells.

**Home location register (HLR).** The HLR is the database to which a user identity is assigned for record purposes such as subscriber information (e.g., electronic serial number [ESN], profile information, or current location). The HLR may or may not be located within and be indistinguishable from an MSC. One HLR may serve more than one MSC.

**Mobile station (MS).** MS is the interface equipment used to terminate the radio path at the user side. It provides the capabilities to access network services by the user.

**Mobile switching center (MSC).** MSC is an automatic system that constitutes the interface for user traffic between the cellular network and other PSTNs or MSCs in the same or other cellular networks.

**Visitor location register (VLR).** VLR is the location register other than the HLR used by an MSC to retrieve information for handling of calls to or from a visiting subscriber. The VLR may or may not be located within and be indistinguishable from an MSC. The VLR may serve more than one MSC.

## 7.5.2  Intersystem handoff

Intersystem handoff refers to the general provisions by which a call in progress on a radio channel under the control of a current serving MSC may be automatically transferred to a different radio channel under the control of another MSC without interruption to the ongoing communication. Figure 7.17 shows the intersystem handoff process.

There is a call involving the served MS X. The call is between a wireline subscriber and the MS X. The figure at the top shows the voice connections prior to the intersystem handoff. As the mobile station's signal strength drops (either due to RSSI measurements at the base station in AMPS, or MAHO in IS-54), the serving MSC elects, based on its internal algorithm, to determine if a handoff to an adjacent candidate MSC is appropriate. The serving MSC sends a handoff measurement request (HANDMREQ) to the candidate MSC (the serving MSC may send several handoff measurement requests to different candidate MSCs). The handoff measurement request basically instructs the

## Initial Connections

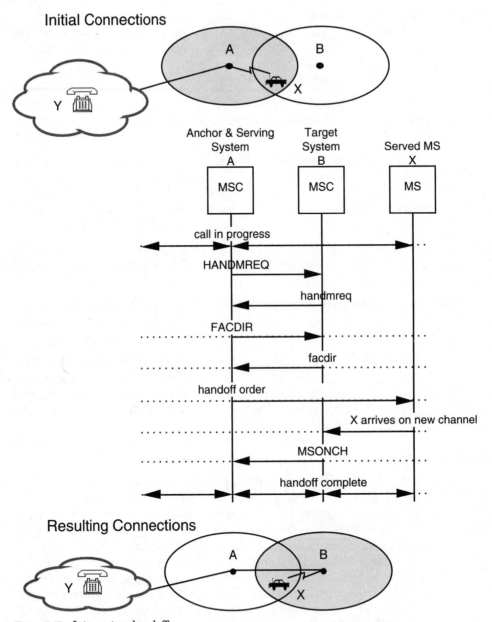

**Figure 7.17**   Intersystem handoff.

receiving MSC to perform signal strength measurements on a particular channel and report back the results. The candidate MSC performs the measurements and returns the results to the serving MSC in a response. The response includes a list of cells in which the mobile could be heard, along with the corresponding received signal strength values. The serving MSC may receive several such reports from different candidate MSCs to whom it sent out the measurement requests. The serving MSC determines that the call should be handed off to the candidate (now target) MSC and that the target MSC is not already on the call path. It sends a facilities directive (FACDIR) to the target MSC. This message essentially informs the target MSC to allocate resources such as voice channel and voice trunks between the serving and target MSC. Once this is done by the target MSC, it responds back with the new voice channel number MS X should use on the new cell. The serving MSC on receiving the response sends a handoff order to the MS X informing it to tune to new voice channel and establish communications. MS X tunes to the new voice channel controlled by target MSC and establishes communications. Once MS X has successfully transferred over to the new voice channel, target MSC completes the voice path between the voice channel and the inter-MSC trunk and sends a mobile station on channel (MSONCH) to the serving MSC. The serving MSC on receiving this message connects the call path on the inter-MSC trunk and the wire-line subscriber, completing the handoff process.

### 7.5.3  Automatic roaming

Automatic roaming refers to the general provisions for automatically providing cellular services to the MS that are operating outside their home service area but within the aggregate service area of all participating MSCs. In the most general implementation these include:

- Timely indentification of the current serving MSC
- Automatic service qualification of the roaming MS, including feature privileges, credit validation, and feature control
- Automatic call delivery (incoming call termination) to the roaming MS

**7.5.3.1  Registration.**  Registration (shown in Fig. 7.18) is the process by which an MS not currently in its home network can inform the visited network of its presence and can get credit validation and access to subscriber's feature set as provisioned at the home network. The visited system detects the presence of the roamer via autonomous registration. After determining that a roaming MS is now within its service

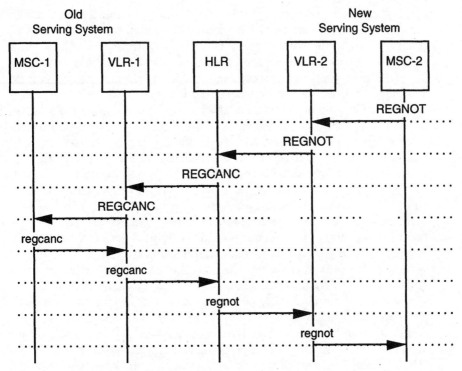

**Figure 7.18**  Autonomous registration.

area, the new serving MSC (MSC-2) sends a registration notification (REGNOT) to its VLR (VLR-2). MSC-2 may detect the MS's presence through autonomous registration, call origination, call termination, and so on. Assuming that the VLR does not have any record associated with the roamer, it forwards the REGNOT to the HLR associated with the MS. The HLR makes a record of the new location of the MS. If the MS was previously registered elsewhere, the HLR sends a registration cancellation (REGCANC) to the previously visited VLR (VLR-1). VLR-1 erases the MS's record from its internal database and sends the REGCANC to the MSC-1 to do the same. On a successful response from MSC-1, VLR-1 sends a REGCANC to the HLR, which sends a response to the VLR-2. This response will have all the subscriber information, active feature indications, and so on. VLR-2 forwards the response to the MSC-1.

**7.5.3.2  Call delivery.**  Figure 7.19 shows the call flow for a call delivery to a mobile station. A call origination and the dialed MS address digits are received by the originating MSC. The MSC determines that it has

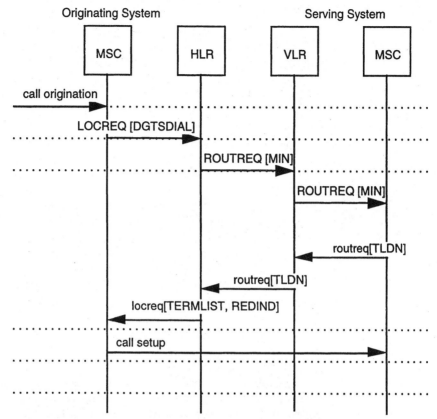

**Figure 7.19** Call delivery.

no record of the particular MS. It sends a location request (LOCREQ) to the HLR requesting it to provide the location of the mobile subscriber. The HLR (from the last registration) knows the address of the VLR where the MS is currently registered. It sends a route request (ROUTREQ) to the VLR, requesting a number to route the call to. The VLR forwards this message to the associated MSC. The serving MSC allocates a temporary local directory number (TLDN) and returns this information to the VLR in the response. This TLDN is a directory number controlled by the MSC and as such when dialed, routes the call to the MSC. It is transient in nature and once the call is completed, is released. The VLR forwards the response to the HLR. The HLR creates a LOCREQ response by populating it with the TLDN, MIN, and ESN of the mobile. It sends this to the originating MSC. The originating MSC establishes a voice path to the serving MSC via the PSTN using the TLDN specified in the response.

### 7.5.4  Authentication and encryption

Authentication is the process of verifying the identity of the mobile phone. Encryption is the process of protecting a user's data. In IS-41 the term *authentication* refers to the process during which information is exchanged between a mobile station and the base station for the purposes of enabling the base station to confirm the identity of the mobile station. A successful outcome of the authentication process occurs only when it can be demonstrated that the mobile station and the base station possess identical sets of shared secret data (SSD). To support authentication and encryption, IS-41 Rev. C introduces a new network element known as the authentication center (AC) which is connected to the HLR. Only the HLR can communicate with the AC. The purpose of the AC is to store all authentication and encryption parameters necessary for the validation of each mobile station.

Shared secret data is a 128-bit pattern stored in the mobile station and available to the base station. SSD is partitioned into two subsets: SSD-A, used to support the authentication process, and SSD-B, used to support the encryption process. The mobile station and the AC share a common authentication and encryption algorithm known as the CAVE algorithm. Due to the unique characteristics of this algorithm, even if the algorithm is known, it is very difficult to predict the output of this algorithm without having knowledge of all the inputs. The input to this algorithm is 152 bits.

The mobile station also stores a random value known as RAND which is used as input to the CAVE algorithm along with the SSD. This RAND value is broadcast by the base station on the broadcast control channel. Thus, this RAND value will be common to all the mobile stations in the cell. The mobile station also has an A key which is 64 bits long and is used for the updating of the SSD. This A key is known only to the mobile station and the AC. There is another parameter known as the call history count (COUNT) which is in the mobile and the AC. It is used by the network to prevent use of cloned phones (i.e., phones that are essentially duplicates of an authorized phone). To recap, the following information resides in the mobile station: the CAVE algorithm which does not change, SSD which can be changed on request from the network, RAND which is common to all the mobile stations in the cell and changes frequently, the A key which does not change, and the COUNT which is typically updated on each successful access attempt. On the network side the CAVE algorithm resides at the AC, SSD resides at the AC and the VLR, RAND at the MSC, the A key at the AC, and the COUNT at the AC and the VLR.

Now that we have looked at the security parameters used for the authentication and encryption process, we will see how these parameters are used as part of the authentication process.

**7.5.4.1  Authentication process.**  The primary authentication mechanism in IS-41 Rev. C is what is known as global (broadcast) challenge as opposed to unique challenge as used in GSM. IS-41 Rev. C supports unique challenge also, but is used only when the primary mechanism fails. The authentication process is initiated with a registration access attempt.

**7.5.4.2  Initial registration with authentication.**  The MSC transmits on the cell broadcast channel information about whether authentication is necessary for access attempts and, if so, the value of the RAND parameter. When a mobile station attempts registration, it reads the RAND value from the broadcast control channel and uses it as input, along with the electronic serial number (ESN), MIN1, and SSD-A to the CAVE algorithm. Figure 7.20 shows the input to the CAVE algorithm for registrations.

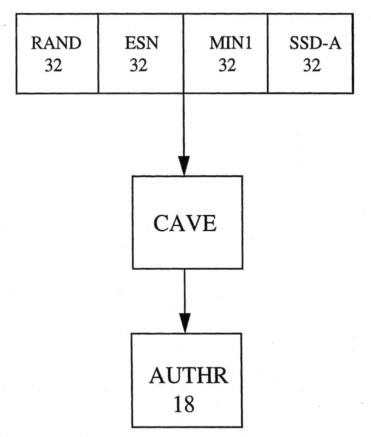

**Figure 7.20**  Computation of AUTHR for authentication of mobile station at registrations.

MIN1 is derived from the MIN. AUTHR is set to the 18 bits of the CAVE algorithm output. The mobile station sends the AUTHR, COUNT, and RANDC (derived from RAND) in the access message to the network. The mobile station also provides its MIN and ESN in the message. It will depend on the air interface being used whether the message is IS-136, IS-95, or so on. The MSC-2 verifies that the RANDC supplied by the MS is the same as that broadcast in the cell. It then sends the information sent by the mobile to the new VLR, replacing RANDC with RAND. The VLR forwards this message to the HLR, which in turn forwards it to the AC. The AC verifies the MIN and ESN reported by the mobile station. On verification the AC now has all the information it needs to generate its own AUTHR. It uses the RAND provided by the MSC, ESN, and MIN1 provided by the mobile station and its own stored SSD-A as the input to the CAVE algorithm. It compares its own internally generated AUTHR with the AUTHR sent by the mobile station, and if they match the mobile station is validated. The AC then verifies that the COUNT received from the MS is consistent with the value stored by it. The AC then replies by sending a response indicating success. This gets forwarded to the MSC-2 which then forwards the REGNOT to the HLR via the VLR. If the MS was previously registered in another system, the HLR sends a REGCANC to the old serving VLR (VLR-1). Figure 7.21 shows the registration with authentication call flow.

As can be seen, the authentication process essentially verifies that the versions of SSD-A residing at the mobile station and the AC are identical. Since SSD-A is a 64-bit parameter, it will be very difficult for a fraudulent user to correctly guess its value. Note that the rest of the parameters used for authentication such as RAND, MIN1, and ESN can be captured by eavesdropping on the access channel. Thus, the strength of the IS-41 authentication process is dependent on the large SSD-A value along with the CAVE algorithm. IS-41 Rev. C provides for updating of the SSD parameter in the mobile station on command from the AC. This ensures that the SSD parameter value changes with time and makes it more difficult for any fraudulent user to impersonate a legitimate user. It is possible that the mobile phones may be duplicated (cloned). To detect cloned phones, IS-41 Rev. C uses the call history count (COUNT) parameter. This parameter is updated in the mobile station and the network typically on successful access attempts. If there are two mobile phones in the network with the same identity, then at some point in time the counts at the mobile station and the network will be out of sync. The network detects this and takes measures for invalidating the cloned mobile.

This authentication process is more or less similar for other access attempts such as origination and termination. The inputs to the CAVE

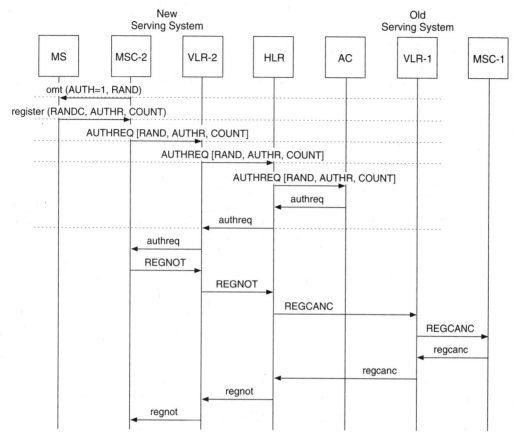

**Figure 7.21**  Registration with authentication.

algorithm are different for each access and the authentication response from the AC for originations and terminations will include encryption parameters. These parameters will be used by the MSC to encrypt the user's communication on the voice channel if so desired.

### 7.5.5  IS-41 Rev. C features

In this section we will discuss some of the new services and features that are present in IS-41 Rev. C.

**7.5.5.1  Call delivery.**  This feature has been enhanced in IS-41 Rev. C to handle location uncertainties in border cells. *Border cells* are those which are on the border of two systems. It supports intersystem paging, that is, paging in cells in both the current system and the neighboring system if the location of the mobile is in a bordering cell. It also allows for call routing to the system where the page response originates. Unsolicited page response is also supported. Unsolicited page response occurs in border cells when the mobile is paged in one system and the page response is received in the other system. The MSC receiving the unsolicited page response will be able to complete the call with the help of IS-41 Rev. C messages, even though it did not receive a routing request (ROUTREQ).

**7.5.5.2  Calling number identification presentation (CNIP).**  This feature enables the number of the calling party to be presented to the called party. Presentation is done during call setup when the called user is alerted. IS-41 Rev. C supports CNIP during call termination at alerting time, call-waiting time at call-waiting indication, and call forwarding unconditional. The last one requires special mention. IS-41 Rev. C has enhanced the capabilities of call forwarding unconditional (CFU). CFU, when activated, allows for forwarding of all calls to the mobile subscriber without first providing an opportunity to the mobile user to receive the call. It allows for an alert to be provided to the called party whenever a call is forwarded. This allows a subscriber to deactivate CFU, in case the CFU is left activated by mistake. CNIP can be provided as part of this alert to mobile user.

**7.5.5.3  Calling number identification restriction (CNIR).**  This feature enables the calling party to restrict the presentation of its number to the called party. IS-41 Rev. C allows for activation/deactivation of CNIR on a per call basis.

**7.5.5.4  Flexible alerting (FA).**  This is a work group kind of service. For instance, there may be a help center which consists of three people who are on call. Each person can be located at a different place. One of them may be a mobile home subscriber, another located at a wire-line phone, and still another roaming in another network. The help center has one number which a user dials to reach any one these three people. IS-41 Rev. C supports this kind of service with the help of the FA feature. When the user dials the FA pilot number (the number used to reach the help center, not the individual persons manning the help center), the call is routed to the MSC. The MSC queries the HLR with the pilot number. The HLR recognizes that this number is a pilot number and retrieves the database records of the individual members of the group. The HLR provides the list of individual phone numbers (PSTN number, MIN, roaming number, etc.) to the MSC. The MSC originates calls to all the work group members simultaneously and connects the call to the work group member who first answers. Subsequently it drops the remaining calls. Several different flavors of this basic feature are supported (single work group member busy, all members busy, etc.).

**7.5.5.5 Message waiting notification.** This feature allows the mobile subscribers to be informed of any messages waiting for them in their voice mail. There are several ways of informing the subscriber, via a pip tone, pip tone during call origination, pip tone during call termination, via an indication on the phone, and more. There are also schemes wherein the number of messages in the voice mail can also be indicated.

**7.5.5.6 Mobile access hunting.** This feature is very similar to flexible alerting (FA) and used for similar kinds of work groups. In FA, simultaneous alerting of all the work group members is done. In mobile access hunting (MAH), work group members are alerted sequentially. The MSC originates a call to the first work group member on the list, and if the call is not answered in a specified amount of time the call is terminated and the MSC tries an origination to the second work group member on the list, and so on till the call is answered. On answering, the work group member is connected to the calling party.

**7.5.5.7 Password call acceptance (PCA).** This feature allows for restricting call originations to only those subscribers who are privy to a password. The HLR is queried on all call originations. The HLR determines from the user's profile that PCA is activated. It then requests the MSC to collect the password digits from the subscriber. The subscriber is given a prompt to enter the password. The password, once it is collected at the MSC, is sent to the HLR which verifies it against the stored value. On successful verification, the call is allowed to proceed and normal call origination takes place. This feature is handy when the same phone is used by several users, some of whom can only receive but not originate calls, while others can originate calls. The users authorized to originate calls are given the password and the others are not.

**7.5.5.8 Selective call acceptance (SCA).** This feature allows for calls originating from a select group of numbers to terminate to a particular user. On receiving an incoming call, the MSC queries the HLR. The MSC also provides the calling party number to the HLR. The HLR determines from the subscriber profile that SCA is active. It compares the calling party number received against the screening list of the subscriber. If the calling party number is present in the screening list, the call is allowed to proceed, otherwise it is denied.

**7.5.5.9 Short message service (SMS).** IS-41 Rev. C supports short message service. SMS is a service which allows for sending of short alphanumeric messages to mobile stations.

## 7.6  Summary

D-AMPS 1900 is a technology which has been designed for coexistence with AMPS and D-AMPS 800 networks. As a technology it is on par with other existing digital technologies such as GSM/PCS 1900 in terms of variety of services provided, voice quality, level of privacy, and confidentiality provided. The primary advantage for a D-AMPS 1900 operator is the dual-band/dual-mode aspect of D-AMPS 1900 technology. The dual-band system gives operators an important, competitive edge when it comes to market expansion. It enables them to fully utilize the functionality of both the 800- and 1900-MHz bands simultaneously in order to grow and offer a variety of services. Since roaming and handoff between AMPS, D-AMPS, and D-AMPS 1900 is supported, the PCS operators can provide full coverage from day one through cooperation with cellular operators in the same geographical area. Extended coverage through cooperation with other cellular operators in different geographical areas is also possible. From an initial capital outlay, existing cellular D-AMPS network infrastructures (MSCs, VLRs, HLRs, trunks, etc.) can be used for PCS service, greatly reducing the cost of the network. This also enables rapid deployment and rollout of service. D-AMPS 1900's support of IS-41 signaling means that intersystem services can easily be provisioned.

Future subscriber growth can be accommodated by use of half-rate traffic channels which should double the capacity of full-rate D-AMPS 1900 systems. D-AMPS support of private and residential networks allows operators to provide custom innovative services, such as tiered billing. Customers' mobile phone call rates are billed according to location; rates are lower at home and higher elsewhere. D-AMPS 1900 as a technology has several unique features which should enable it to gain wide customer and operator acceptance.

# Code-Division Multiple Access (CDMA)

## 8.1  Introduction

CDMA is one of the technologies being used in the United States for the 800-MHz cellular bands and the 1900-MHz PCS bands. It differs from analog cellular and D-AMPS/GSM in the respect that users are differentiated from each other by a unique code rather than a frequency assignment as in AMPS, or frequency and time slot assignment as in GSM and D-AMPS. CDMA falls under the class of spread spectrum systems. There are several kinds of spread spectrum systems in the world. Some of them are described in the next section.

## 8.2  Spread Spectrum

FDMA and TDMA access schemes were discussed in the previous chapters. FDMA is the access scheme used in AMPS, and TDMA is the scheme used in GSM and D-AMPS. Common to both is the channelization approach to frequency management. The entire, available radio spectrum is divided into narrowband radio channels to be used for one-way communication between the mobile station and base station. On the transmission side, efforts are then taken to ensure that transmitted information is restricted to the assigned narrowband channels. On the receiver end, efforts are concentrated on receiving information only from the specified narrowband channel and rejecting information from outside the channel. Due to frequency reuse, the same radio channel is reused a short distance from a particular cell. Interference from co-channels is reduced by controlling how far their emissions can be transmitted. Due to the limited availability of bandwidth, most of the efforts in narrowband schemes are concentrated on transmitting acceptable quality speech on narrower bandwidth channels. Limited spectrum indirectly results in more narrowband channels. This allows more users for a given radio spectrum.

Spread spectrum systems do not follow the channelization principle of traditional radio communication systems. In these systems, channels and communications are combined into the same channel. In place of using narrow channels for each communication, information is transmitted (spread) over a very wide channel with several simultaneous users using the channel. Interference is built into the system to aid communications, so there is no limit on the number of subscribers that are supported. As more and more subscribers are added to the system, there is a graceful degradation of the quality of communication. Spread spectrum communication systems have been around for several decades (used primarily by the military). It is a very useful system for military applications since it prevents eavesdropping and jamming by the enemy and also makes the location determination of the transmitter very difficult. It has only been lately that spread spectrum communication systems have been introduced into the commercial arena.

Two of the most popular spread spectrum systems are frequency-hopping and direct-sequence. These techniques are characterized by transmission of information over a bandwidth much larger than that normally used in narrowband radio systems, coding of the transmission by a pseudorandom sequence shared by the transmitter and receiver, and the use of unique random sequences to differentiate between users. In code-division multiple access (CDMA), the code in frequency-hopping systems is used to generate that sequence and in direct-sequence systems it is used to generate high bit-rate pseudorandom noise.

### 8.2.1 Frequency-hopping spread spectrum

Frequency-hopping spread spectrum is a technology which was primarily used by the military to ensure antijam communications and protection from being monitored in the battlefield. Jamming is a scheme whereby the enemy transmits noise at a very high power level at a particular frequency. Due to this high interference level, there can be no communication on that frequency. The basic principle of frequency-hopping systems is to break up a message into fixed size blocks and to transmit each block in a different frequency. The frequency-hopping sequence (i.e., the order in which the frequencies are selected in order to send transmissions) is known prior to start of the message transmission by both the transmitter and the receiver. The transmitter transmits one block of the message on one particular frequency, hops to another frequency, transmits the next block of the message, and so on. As soon as it finishes receiving a single block on the other end, the receiver retunes to the next frequency where the subsequent message block is expected. This process of changing frequencies for every message block is continued until termination of the message. Every single transmitter will have a different hopping sequence so as not to interfere with other users using the same sequence. In military applications, this scheme very effectively ensured antijam communications because the enemy had to have prior knowledge of the hopping sequence to be able to jam the entire message. Not having information of the hopping sequence ensured that the enemy could, at most, jam only a few of the message blocks (those transmitted on frequencies on which the enemy happened to be jamming). To further reduce this possibility, the hopping sequence was a pseudorandom sequence having no correlation between previously transmitted frequencies. It can easily be seen that the efficiency of the system increases as the message block size is reduced and the number of frequencies in the hopping sequence is increased.

This scheme can very easily be applied to commercial cellular. The available radio spectrum is divided into radio frequencies which are a part of the hopping sequence. In each cell there will be several mobile stations all communicating to the base station. Each of the mobile stations has its own unique frequency-hopping sequence that is known to the base station. Whenever the mobile station wishes to start a communication, it uses its frequency-hopping sequence to break a message down into blocks and transmit it on the particular frequency. It then hops to the next frequency as soon as it is done. The jammer is, in effect, the multipath and Rayleigh fading which are frequency specific. Any message loss due to interference is limited to a particular message block and not the entire message. At the receiver's end, the message blocks are collated and the original message is formed. By using forward error correction (FEC) techniques, these types of loss of message blocks can easily be corrected. Care has to be taken in this system that no two users in the same cell share the same hopping sequence. Message blocks will be lost due to mutual interference from two mobiles if both of them transmit on the same frequency at the same time. Figure 8.1 shows the frequency-hopping concept. In the traditional non-frequency-hopping system, the message is sent on a single frequency. In frequency-hopping systems, the message is broken down into blocks and each block is transmitted on a different frequency.

## 8.2.2 Direct-sequence spread spectrum

In a direct-sequence spread spectrum system, a low bit-rate information signal is taken and added to a known high bit-rate pseudorandom noiselike signal. This pseudorandom signal is known to both the trans-

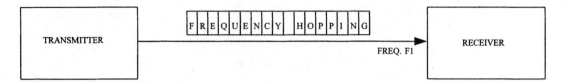

Non-Frequency-Hopping System. Frequency F1 used for T seconds

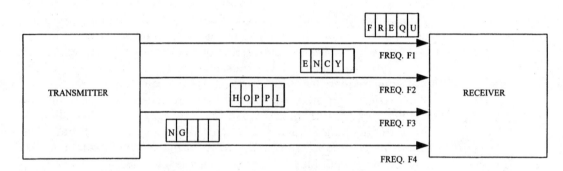

Frequency-Hopping System. Frequency F1,F2,F3,F4 used for T/4 seconds

**Figure 8.1**   Frequency-hopping system.

mitter and the receiver. This final signal has both the original information signal, which carries useful information, and the noiselike signal, which carries no useful information. Upon receiving this signal, the receiver subtracts the known pseudorandom signal and is left with the information signal. It is possible for several parallel communications to exist. All these communications will be differentiated from each other by a different pseudorandom signal.

In commercial radio systems the available radio spectrum is broken up into a few broadband radio channels which have a much higher bandwidth than that of the voice signal. The digitized voice signal is then added to the generated high-bit-rate signal and transmitted such that it occupies the entire bandwidth. The receiver receives this signal and extracts the voice signal. Adding a high-bit-rate pseudorandom noise signal to a voice signal makes it more robust and less susceptible to interference. It enables lower-power transmissions to take place, resulting in cheaper mobile stations with longer-lasting battery life. Unlike in frequency-hopping systems, communications from different mobile stations can exist at the same frequency and at the same time in direct-sequence systems. As for frequency-hopping systems in direct-sequence systems, Rayleigh fading is frequency specific and does not affect the entire bandwidth but only a small, very narrow portion of the channel.

One of the important spread spectrum system performance metrics is the processing gain. For frequency-hopping systems, the gain is approximately equal to the log of the ratio of the total channel bandwidth to the bandwidth occupied during each hop. For example, the processing gain of a frequency-hopping system with a 30-MHz broadband channel bandwidth with 1000 hops 30-kHz wide would be 30 dB. For direct-sequence systems the processing gain is the log of the ratio of the bit rate of the transmitted signal to the bit rate of the information signal. Thus, if the transmitted signal is at 30 Mbps and the information rate is 30 kbps, then the gain is 30 dB. This processing gain is the equivalent gain of the processed signal and results in good reception of even very weak signals. Mobile stations can transmit at lower-output power levels resulting in lower cochannel interference.

A common analogy used to describe the differences between FDMA, TDMA, and CDMA is the office building analogy. Imagine that there is a large office hall used by a company to do business. This hall is divided into work spaces by installing office cubes. The hall could be analogous to the frequency band and the cubes to individual frequency channels. In an FDMA environment such as AMPS, each cube has a pair of people conversing with each other. As additional users enter the hall they are assigned an empty cube to conduct their business. Each cube can accommodate only two people who can converse only with each other. If all the cubes are occupied, any new users entering the hall will have to wait till a cube becomes empty. It is easily seen that as more and more people enter the hall the total noise level is going to increase in the individual cubes. This is because over the cube walls the users in a cube can hear the conversations of other people in the hall. As the background noise increases, the users in each cube will have to speak louder and louder to have themselves heard.

In a TDMA system such as GSM, to apply the same analogy, the cubes will have to be reconfigured. The cubes are made larger, to hold up to eight pairs of people who will be communicating in pairs. Again, the analogy here is that the hall is the frequency band and the cubes are the individual frequency channels (AMPS frequency channel is 30 kHz vs. 200 kHz for GSM). In GSM, each cube holds eight people who take turns talking to each other, but never at the same time. Imagine that there is a referee in the hall who is responsible for ensuring that people speak only when it is their turn. The referee holds up flag one, which is an indication to all the first pair of users in each cube to begin their conversations. After a fixed amount of time the referee holds up flag two. The first pair stops talking and the second pair in each cube takes over. This process is repeated until the eighth pair is complete, then the referee returns to the first pair. In this way each pair gets to talk, but not at the same time. Any new users entering the hall will be assigned a pair number along with a designated cube in which they can converse. When all the allocated pairs in all the cubes are filled up, subsequent users will be denied until one becomes available.

In a CDMA system analogy, the hall would remain but without the cubes. All the cubes in the hall would be removed and the entire hall available for users to converse. To ensure intelligible conversation, each user pair would be conversing in a different language. Therefore, even if there is a lot of background noise, the users are able to converse since they are all using unique languages and the ear is able to discern the relevant information from the background noise. The analogy to CDMA is that the hall is the frequency band and the different languages are the unique hopping sequences, such as frequency-hopping spread spectrum or unique, high-bit-rate pseudorandom noiselike signal in direct-sequence systems. There are no hard limits, such as number of cubes that limit the number of users in the hall. As more and more users enter the hall, the noise level increases till all conversation has to cease. Thus, CDMA is more graceful in accommodating users into the system when compared to FDMA or TDMA

## 8.3    IS-95

IS-95 is a mobile to base station compatibility standard for dual-mode wideband spread spectrum. It is used in the United States for the 800-MHz cellular bands and has a modified version for the PCS 1900-MHz frequency bands. This standard was proposed by Qualcomm, Incorporated, of San Diego, California, in conjunction with AT&T, Motorola, and others. IS-95 is a direct-sequence CDMA scheme wherein users are differentiated on the basis of unique pseudorandom codes.

In 1988, the Cellular Telecommunications Industry Association (CTIA) released the User's Performance Requirements (UPR) document which specified the cellular carriers' requirements for the next-generation cellular technology. These requirements for digital technology included:

- Ease of transition and compatibility with existing analog system
- Early availability and reasonable costs for dual-mode radios and cells
- Substantial capacity increase over analog system capacity
- Privacy
- Long life and adequate growth of second-generation technology

- Quality improvements (voice quality, service quality in terms of dropped calls, constant voice quality level, etc.)

- Ability to introduce new features easily

IS-95 was designed to fulfill most of these requirements. Dual-mode operation is achieved by enabling CDMA (henceforth, CDMA and IS-95 will be used interchangeably) channels to exist in the AMPS frequency band. Whenever an operator deploys a CDMA system, a finite set amount of AMPS channels have to be given up to free up the necessary RF spectrum for CDMA. The CDMA channels are then deployed within the freed up RF resources.

A CDMA-AMPS dual-mode phone will operate in both the CDMA and AMPS environment. When the phone is powered up, it looks for the CDMA control channels. If it finds one, it starts operating in the CDMA mode and initiates the CDMA cell selection and access protocol. If it does not find any CDMA control channels, it begins looking for an available AMPS control channel. Upon finding an AMPS control channel, the dual-mode phone recognizes that CDMA resources do not exist in that channel band. It then initiates the AMPS cell selection and access procedures. The phone then starts operating in the AMPS mode and behaves identically to that of an AMPS phone. Spread spectrum systems are inherently more secure than traditional narrowband architecture since the eavesdropper has to be privy to the pseudorandom code.

## 8.3.1 CDMA concept

In the IS-95 version of CDMA, each signal consists of a different pseudorandom binary sequence that modulates the carrier and spreads the spectrum of the waveform. A large number of CDMA signals share the same frequency spectrum. If CDMA is viewed in either the frequency or time domain, the multiple access signals appear to be on top of each other. The signals are separated by a correlator in the receiver which can be thought of as a filter that accepts signal energy from the selected binary sequence and removes the known high-bit-rate noise signal. The other user's signals, whose codes do not match, are not despread and as a result contribute only to the noise and represent a self-interference generated by the system. This is a distinct advantage of CDMA in the fact that there is no difference between noise and the other user's transmissions. As far as the correlator is concerned, it treats both of them as noise. Figure 8.2 shows the different access schemes in AMPS, D-AMPS, and CDMA. As shown, in the AMPS, users are separated by frequency, in D-AMPS, by frequency and time, and in CDMA, by code.

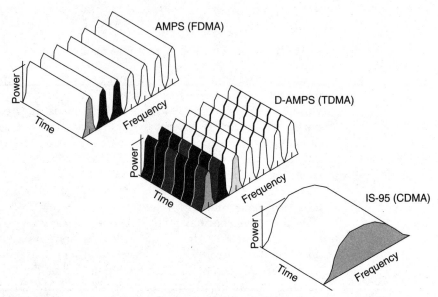

**Figure 8.2**   Access methods in AMPS, D-AMPS, and CDMA.

### 8.3.2   Frequency assignments

A single CDMA channel is 1.23 MHz wide and typically about twenty subscribers share the same channel, simultaneously. CDMA channels have been defined for the 800-MHz cellular band and the 1900-MHz PCS band. Figure 8.3 shows the frequency spacing between two consecutive CDMA channels. It also shows the cellular spectrum and the amount of spectrum in the A- and B-bands that can be used for CDMA. The first CDMA carrier in the A-band is at channel number 283, and the first CDMA carrier in the B-band is at 384. Guard bands have to be introduced between different bands to ensure that the CDMA carriers do not interfere with other users. Since the spectrum, available between channel 667 and 716, is only 1.5 MHz in the A-band, an operator will have to obtain permission from the B-band carrier to deploy a CDMA carrier in that portion of spectrum. If a CDMA carrier is being deployed next to a non-CDMA carrier, the carrier spacing has to be 1.77 MHz. There can potentially be nine CDMA carriers for the A- and B-band operator. In the PCS band, since the A and B operators have 30 MHz, there can potentially be up to 11 CDMA carriers.

### 8.3.3   CDMA capacity advantage

AMPS spectral efficiency in a cellular system is determined by the modulation spectral efficiency (i.e., the information bit rate per hertz of bandwidth and the frequency reuse factor). The AMPS channel band-

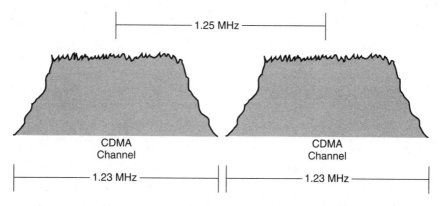

Frequency separation between two CDMA channels

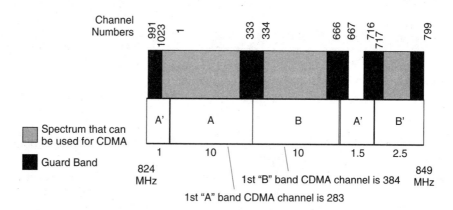

CDMA channels in cellular band

**Figure 8.3**   Frequency separation and CDMA channels in cellular band.

width is 30 kHz. It employs FM modulation, resulting in a modulation efficiency of one call per 30 kHz of spectrum. For the sake of simplicity, the information rate used for a call is assumed to be the same for AMPS, GSM, and CDMA. On a CDMA channel with a frequency bandwidth of 1.25 MHz there will be approximately 42 AMPS channels (1.25 MHz/30 kHz). Because of interference, the same AMPS frequency cannot be reused in every cell. The frequency reuse with AMPS is $\frac{7}{21}$; whereas, the numerator is used if all cells are omni and the denominator if all cells are sectored. This means that there are only six channels available in an omnicell and two in a sector assuming 120° sectors. This results in a capacity increase of two calls per sector.

In GSM, the channel is 200 kHz wide and supports eight calls. A frequency bandwidth of 1.25 MHz results in 6.25 channels. The fre-

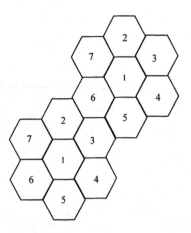 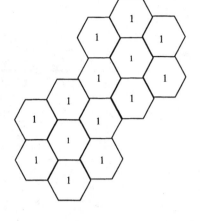

Frequency reuse in AMPS cell - Available spectrum is divided into seven frequency blocks and each block is used in individual cell. The same frequency blocks are separated by distance to avoid cochannel interference.

Frequency reuse in CDMA cells - All available spectrum can be reused in each cell/sector, i.e., frequency reuse of 1. Eliminates frequency planning.

**Figure 8.4**    CDMA frequency reuse.

quency reuse of GSM ranges from a three to nine gain in capacity. Going through the same procedure as with AMPS, the number of users/sector for GSM is 5.6 which translates to about 2.8 times capacity of AMPS.

The number of users per sector for a CDMA system is equivalent to the number of users per a 1.25-MHz CDMA channel, which is 20. In CDMA, special efforts are taken to counter the effects of interference due to the same CDMA channel being used within sectors of the same cell and within neighboring cells. This means that the frequency reuse factor for CDMA is ⅓, compared to $7/21$ for AMPS and D-AMPS, and ⅗ for GSM. Thus, the gain for CDMA compared to AMPS is close to 10 times. This is the primary source for capacity gain in CDMA systems. Reusing the same CDMA carrier in each and every sector and every cell in the system means that frequency planning, as is done in traditional narrowband cellular systems, is not required. Figure 8.4 shows the CDMA frequency reuse of 1.

### 8.3.4   AMPS to CDMA migration

CDMA channels coexist within the AMPS frequency band by having the wireless operator clear 1.25 MHz of spectrum to accommodate CDMA. A single CDMA channel takes up 41 of the 30-kHz AMPS channels. Since coexistence of frequency bands is possible between AMPS and CDMA, it is possible to reuse a CDMA channel frequency band for AMPS in a different cell. Frequency coordination has to be done to ensure that the two cells do not interfere with each other. *Frequency coordination* is the process of assigning frequency bands to neighboring or coexisting systems to minimize interference. This is traditionally done in narrowband systems by implementing a frequency reuse pattern, wherein the same frequency is used in cells sufficiently apart to minimize cochannel interference. At first glance, this kind of frequency coordination seems unnecessary in CDMA since all cells and sectors can reuse the same frequency. This may be true if there are no plans for AMPS and CDMA coexistence. But if CDMA and AMPS cells are coexisting, then the CDMA channel has to have a guard zone of approximately 900 kHz from the CDMA channel center frequency on each end. The guard zone ensures that there is no interference to CDMA cells from neighboring AMPS frequencies. Figure 8.5 shows the guard zone separation.

In order to convert an AMPS system to CDMA, an operator will have to clear 1.77-MHz AMPS spectrum, three AMPS channels per sector, for the first 1.25-MHz CDMA carrier. This has to be followed by creating a buffer zone of cells between the CDMA and AMPS cell using the same frequency. This buffer zone has to be 1.77 MHz. Studies have shown that the capacity increases quickly as more and more AMPS cells are converted to CDMA.

## 8.3.5  Multiple forms of diversity in CDMA

Diversity is used to combat multipath fading. CDMA makes use of the traditional forms of diversity as well as some nontraditional methods to combat multipath fading. There are essentially three kinds of diversity: time, frequency, and space. Traditional time diversity techniques such as convolutional coding (FEC), error detection (CRC), and interleaving of data are also used in CDMA. Downlink channels employ half-rate convolutional coding and uplink channels employ both one-third-rate, and one-half-rate convolutional coding.

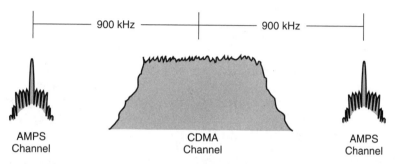

AMPS and CDMA channels have to be placed 900 kHz apart to avoid interference

**Figure 8.5**  Guard zone separation between CDMA and AMPS channels.

Both downlink and uplink also employ interleaving, which helps in mitigating loss of an entire block of data. Frequency diversity is achieved since each signal in the CDMA system covers 1.25 MHz, a relatively wide part of the spectrum. Since multipath fading usually affects only a 200–300 kHz portion of the signal bandwidth, it appears as a notch in the entire CDMA signal. Typically, delays between alternate paths at the receiver of less than 0.8 µs will cause a fade for a CDMA signal. Delays greater than 0.8 µs will cause only a power reduction of the CDMA signal. Maximum fade is caused when the delay spread is exactly 0.8 µs. Space (path) diversity is obtained in three ways by providing (1) multiple signal paths through simultaneous links from the mobile station to two or more cell sites called soft handoff (as shown in the section on handoff), (2) exploitation of the multipath environment through spread spectrum processing (rake receiver) allowing signals arriving with different propagation delays to be received separately and combined, and (3) multiple antennae at the base station (receive diversity).

### 8.3.6    Rake receiver

There will be several different paths at the receiver due to multipath. GSM attempts to minimize multipath with the help of the equalizer. With CDMA using a rake receiver, it uses the multipath to improve reception. Each rake receiver has four fingers, each of which independently tracks a received path. Each finger can adaptively cancel the delay spread, adjust the phase, and equalize the level of output of its received signal. The output from three fingers is combined in the mobile rake receiver. The fourth finger is a roving finger and is searching for the next signal to despread. Due to the presence of these four fingers, the mobile rake receiver can receive four simultaneous signals of the same CDMA carrier. This fact is taken advantage of in soft handoff as described later. The base station rake receiver does not use the fourth finger as a roving finger, thus all four finger outputs are combined.

### 8.3.7    Variable-rate vocoder

CDMA specifies a variable-rate vocoder in place of a fixed-rate vocoder as used in D-AMPS and GSM. The rate is varied by a dynamic algorithm. The rate to be used is determined through the use of adaptive thresholds. The thresholds change according to the background noise level which activates the higher vocoder rates only on the user speech. The result is a suppression of the background noise and good voice transmission even in a noisy environment. The vocoder uses a voice activity detection mechanism. In a typical, full, two-way full-duplex voice conversation, the duty cycle of each voice is approximately 50 percent. This means that any one person is talking only 50 percent of the time. CDMA takes advantage of this by reducing the transmission rate when no speech is present. Use of speech activity detection reduces interference to subscribers in the same cell in case of CDMA; whereas, the use of speech activity detection reduces interference between cells in case of D-AMPS and GSM. The use of voice activity detection also reduces the average mobile station transmit power requirements. The coding algorithm used in CDMA is known as QCELP (Qualcomm code excited linear prediction). The primary vocoder for an IS-95 CDMA system is an 8-kbps variable-rate vocoder (IS-96A) operating within a standard 9.6-kbps digital data stream. This is known as rate set 1. CDMA vocoders can vary their rate in full, ½, ¼, or ⅛ levels. Thus, the variable rates supported for rate set 1 are 9.6 kbps, 4.8 kbps, 2.4 kbps, and 1.2 kbps of digital data. CDMA also provides for an optional higher variable-rate vocoder providing better voice quality at the expense of reducing capacity and coverage loss of approximately 30 percent. This is the 13.25-kbps vocoder operating within a 14.4-kbps digital data stream and is known as rate set 2. The variable rates supported for this vocoder are 14.4 kbps, 7.2 kbps, 3.6 kbps, and 1.8 kbps of digital data.

## 8.4 CDMA Downlink Channels

CDMA downlink channels consist of broadcast channels used for control, and traffic channels used to carry user information. The pilot, sync, and paging channels constitute the broadcast channels. All of these channels are on the same 1.23-MHz CDMA carrier. The mobile station is able to differentiate between logical channels due to the unique code assigned to each channel. This code is known as a Walsh code and has orthogonal properties. Figure 8.6 shows the different downlink channels supported in CDMA. An example assignment of the code channels transmitted by a base station is shown in this figure. Out of 64 code channels available for use, the example depicts the pilot channel (always required), 1 sync channel, 7 paging channels (the maximum number allowed), and 55 traffic channels. Another possible configuration could replace all the paging channels and the sync channel one for one with traffic channels for a maximum of 1 pilot channel, 0 paging channels, 0 sync channels, and 63 traffic channels.

### 8.4.1 Downlink traffic channels

The downlink traffic channels support both rate set 1 (9.6 kbps) and rate set 2 (14.4 kbps) vocoder traffic. Figure 8.7 describes the different functions which act on the data, resulting in the creation of a downlink traffic channel. We will look in detail at each of these processes.

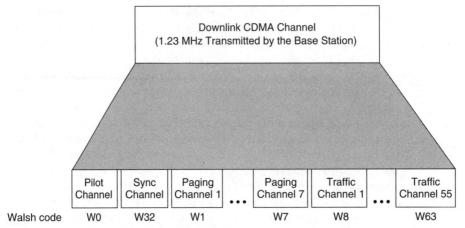

**Figure 8.6**   Structure of downlink CDMA channel.

**8.4.1.1   Vocoding.**   The CDMA vocoder is a variable-rate vocoder operating at full, ½, ¼, and ⅛ rates. There are currently two types of vocoders: the 8-kbps vocoder operating in a 9.6-kbps data stream, and the 13.3-kbps vocoder operating in a 14.4-kbps data stream, referred to as rate set 1 and rate set 2, respectively. Rate set 1 contains four elements—specifically, 9600, 4800, 2400, and 1200 bps. Rate set 2 contains four elements—specifically, 14,400, 7200, 3600, and 1800 bps. The mobile station has to mandatorily support rate set 1 while rate set 2 support is optional. The channel structure differs for rate set 1 and rate set 2. Both the vocoders support voice activity detection and utilize it to reduce interference in the system.

Figure 8.8 shows the downlink traffic channel frame structure of rate set 1 and Fig. 8.9 shows the frame structure of rate set 2. Information from the vocoder is structured in 20-ms frames. The full-rate output of the rate set 1 vocoder is 8.6 kbps, which translates to 172 bits every 20 ms. Frame quality indicators, which are essentially cycle redundancy checking (CRC) checks, are added to the vocoder bits along with an 8-bit tail (set to 0s). The frame quality indicators serve two purposes. They allow the receiver to determine if the frame is in error since the CRC is calculated on all the 172 information bits, and they assist in the determination of the data rate of the received frame. The 172 information bits from the vocoder, 12 frame quality indicator bits, and the 8-bit tail are structured in a 20-ms 9600-bps frame. The 9600-bps frame is a result of 192 bits (172 + 12 + 8) transmitted every 20 ms. A similar process is employed for the 4800-bps frame. The 2400- and 1200-bps frames, do not have frame quality indicator fields since these frames have a higher processing gain and most of the information sent in these frames is background noise.

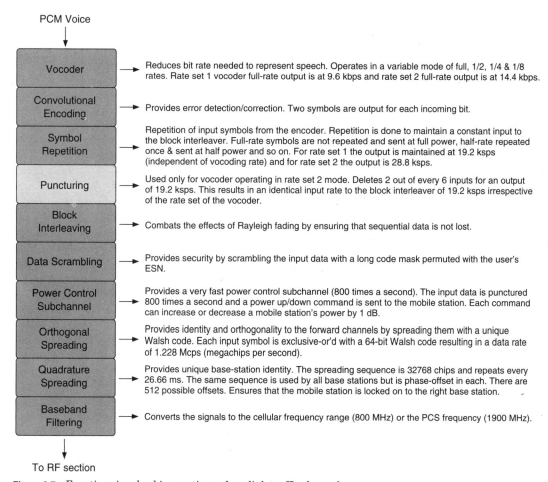

PCM Voice

| Vocoder | Reduces bit rate needed to represent speech. Operates in a variable mode of full, 1/2, 1/4 & 1/8 rates. Rate set 1 vocoder full-rate output is at 9.6 kbps and rate set 2 full-rate output is at 14.4 kbps. |

Convolutional Encoding → Provides error detection/correction. Two symbols are output for each incoming bit.

Symbol Repetition → Repetition of input symbols from the encoder. Repetition is done to maintain a constant input to the block interleaver. Full-rate symbols are not repeated and sent at full power, half-rate repeated once & sent at half power and so on. For rate set 1 the output is maintained at 19.2 ksps (independent of vocoding rate) and for rate set 2 the output is 28.8 ksps.

Puncturing → Used only for vocoder operating in rate set 2 mode. Deletes 2 out of every 6 inputs for an output of 19.2 ksps. This results in an identical input rate to the block interleaver of 19.2 ksps irrespective of the rate set of the vocoder.

Block Interleaving → Combats the effects of Rayleigh fading by ensuring that sequential data is not lost.

Data Scrambling → Provides security by scrambling the input data with a long code mask permuted with the user's ESN.

Power Control Subchannel → Provides a very fast power control subchannel (800 times a second). The input data is punctured 800 times a second and a power up/down command is sent to the mobile station. Each command can increase or decrease a mobile station's power by 1 dB.

Orthogonal Spreading → Provides identity and orthogonality to the forward channels by spreading them with a unique Walsh code. Each input symbol is exclusive-or'd with a 64-bit Walsh code resulting in a data rate of 1.228 Mcps (megachips per second).

Quadrature Spreading → Provides unique base-station identity. The spreading sequence is 32768 chips and repeats every 26.66 ms. The same sequence is used by all base stations but is phase-offset in each. There are 512 possible offsets. Ensures that the mobile station is locked on to the right base station.

Baseband Filtering → Converts the signals to the cellular frequency range (800 MHz) or the PCS frequency (1900 MHz).

To RF section

**Figure 8.7** Functions involved in creating a downlink traffic channel.

Rate set 2 frames employ a similar process as rate set 1. The differences are that all rate set 2 frames have frame quality indicators. This is due to the higher transmission rates which reduce the processing gain of these frames. There is also a 1-bit reserved field.

**8.4.1.2 Convolutional encoding.** Convolutional encoder provides protection to the information bits by providing error correction/detection capabilities. The rate set 1 and rate set 2 frames are sent to a half-rate convolutional encoder. A half-rate convolutional encoder replaces each single input bit with two symbols.

**8.4.1.3 Symbol repetition.** The symbol repeater follows the convolutional encoding which repeats the data as necessary to result in an out-

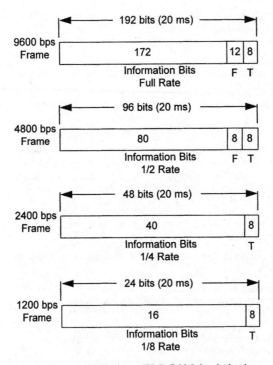

F is a Frame Quality Indicator (CRC) field. It is calculated on information bits only. Not calculated for 2400 bps and 1200 bps frames. T is encoder tail bit. Set to zeroes on all frames.

**Figure 8.8** Downlink/uplink traffic channel frame structure for rate set 1.

put of 19.2 ksps (kilosymbols per second) for rate set 1 and 28.8 ksps for rate set 2. For example, to achieve this, rate set 1 does not repeat any symbols if the input is 19.2 ksps. It repeats each symbol twice if the input is 9.6 ksps, it repeats each symbol four times if the input is 4.8 ksps, and so on. Symbol repetition provides for additional countermeasures against the deleterious effects of the radio channel. Repeated symbols have a lower power level than full-rate symbols. The aggregate power level across all symbols in a rate set is the same, but the power per symbol is reduced, since the power is distributed across several repeated symbols.

**8.4.1.4 Symbol puncturing.** The symbol puncturing process is done only for rate set 2 frames. A decision was made in IS-95 to use the same block interleaver for both rate sets. This meant that the input symbol rate for the block interleaver had to be the same. This is achieved in CDMA by puncturing the 28.8-ksps stream to result in a

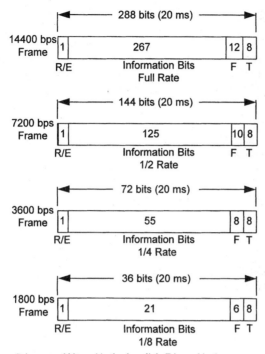

R is reserved bit used in the downlink, E is used in the reverse link to indicate bad frame reception by MS to BS. F is a Frame Quality Indicator (CRC) field. It is calculated on information bits only. Calculated for all frames. T is encoder tail bit. Set to zeroes on all frames.

**Figure 8.9**  Downlink/uplink traffic channel frame structure for rate set 2.

19.2 ksps. Puncturing is achieved by deleting every two out of six inputs.

**8.4.1.5  Block interleaving.**  Interleaving is used to combat the effects of Rayleigh fading. Rayleigh fading is a frequency selective fading which causes errors in large blocks of contiguous data. If contiguous information is sent in consecutive frames, then Rayleigh fading will make it very difficult to reconstruct the information at the receiver. Interleaving shuffles the information around so that it does not occur contiguously. Thus, even if errors occur they will not affect contiguous information. Errors which are distributed in a block of information can be more easily corrected using error-correction techniques.

The block interleaver for the downlink traffic channels accepts 384 modulation bits every 20 ms. These bits are input into a 24 × 16 array column. The interleaver shuffles them, and the output is sent to the next process.

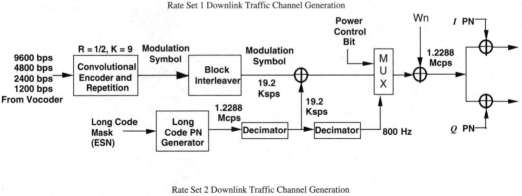

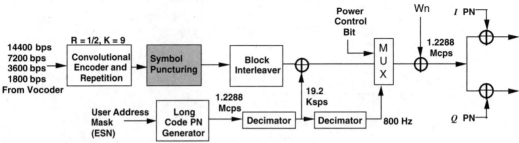

**Figure 8.10**   Rate set 1 and 2 downlink traffic channel generation.

**8.4.1.6  Data scrambling.**   Data scrambling provides security and privacy to the downlink traffic channel. A long code mask, which has a period of about 40 days, is used in conjunction with the permuted ESN of the mobile station currently using the downlink traffic channel. The base station can determine the mobile station's long code mask since the mobile station sent its ESN as part of the access message. If the encryption procedure is being used on the downlink traffic channel, then the mobile station's private long code mask is used. The long code mask provides security and repeats itself once every 40 days, making it very difficult for an eavesdropper to determine where in this long period a user's transmission is located. Alteration of the long code mask by a mobile station's permuted ESN provides additional security. Figure 8.10 shows the rate set 1 and rate set 2 downlink traffic channel generation.

**8.4.1.7  Power control subchannel.**   A power control subchannel is continuously transmitted on the downlink traffic channel. It is used to control the mobile station's output power on the uplink. The subchannel transmits at a rate of one bit (0 or 1) every 1.25 ms (i.e., 800 bps). A 0 bit will indicate to the mobile station to increase the mean output power level and a 1 bit will indicate to the mobile station to decrease the mean output power level. The amount the mobile station increases and decreases its power for every power control bit is 1 dB. This kind of fast power control is required in CDMA due to the near-far problem. This occurs when a mobile station near to the base station transmits at a higher power than the mobile station near the end of the cell. The mobile station near the base station could potentially overpower the farther mobile station's transmission. This is avoided in CDMA by use of the fast power control subchannel.

The base station downlink traffic channel receiver estimates the received signal strength of the particular mobile station to which it is assigned over a 1.25-ms period. The base station then uses the estimate to determine the value of the power control bit (0 or 1). It transmits the power control bit on the corresponding downlink traffic channel using the puncturing technique. Using the puncturing scheme, the power control bit which is two symbols long replaces two consecutive downlink traffic channel modulation symbols. This puncturing occurs at an 800-Hz rate with complete disregard for how important the sacrificed bits may be. The mobile's decoder is left with the task of separating the power control subchannel from what remains of the traffic data, and then repairing the damage to the remains of the encoded data stream. This technique is used in spite of its drawback of affecting the quality of the link due to its fast response time. The mobile station can decode the power control bit very quickly without having to decode the headers and other frame information. Once the power control subchannel is recovered, the mobile uses the data within to make fine adjustments to its RF output power.

**8.4.1.8  Orthogonal channel spreading.**  The 19.2-kbps data stream, which has been punctured with the power control bits of the power control subchannel, is modified by a Walsh code. Typically, Walsh codes W8 to W63 are used for downlink traffic channels. Walsh codes are orthogonal codes which give a unique identity to each downlink traffic channel. Codes are orthogonal if the result of exclusive-or-ing them is an output consisting of an equal number of 1s and 0s. The mobile station can differentiate between individual downlink traffic channels because of the unique Walsh codes used for each traffic channel. Figure 8.11 and Table 8.1 show the exclusive-or truth table for the description of orthogonal spreading of the downlink channels. Each bit of the 19.2-kbps input is exclusive-or'd with a 64-bit Walsh code which is unique to that channel. Figure 8.12 shows Walsh code 20 being used. The result is the output which will be transmitted. The input thus is spread (19.2 kbps × 64 bits = 1.228 Mbps) over a larger bandwidth.

**TABLE 8.1    Truth Table for Exclusive-or**

| Input $X$ | Input $Y$ | Output |
|-----------|-----------|--------|
| 1 | 1 | 0 |
| 1 | 0 | 1 |
| 0 | 0 | 0 |
| 0 | 1 | 1 |

When the transmission on the mobile station receiver is received, it is again exclusive-or'd with the same Walsh function used by the base station on this channel. (The identity of the Walsh code to be used on a particular channel is provided by the base station to the mobile station as part of the channel allocation process.) The mobile station considers 64-bit/symbol block inputs and exclusive-ors them with the known Walsh code. The result will be a majority of either 1s or 0s, which is the original data bit. Even if errors occur along the RF path, typically either the number of 1s or 0s will be dominant over a 64-bit block. This

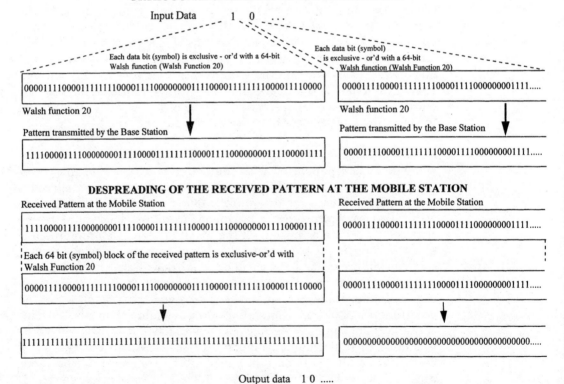

Figure 8.11    Orthogonal spreading/despreading.

**DESPREADING OF THE RECEIVED PATTERN AT THE MOBILE STATION WITH
INCORRECT WALSH FUNCTION**

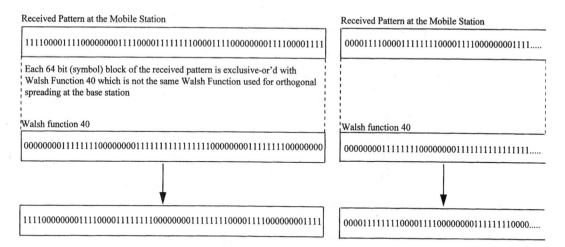

Inconclusive output - Equal number of 1s and 0s in the despread pattern.

**Figure 8.12**    Orthogonal despreading with incorrect Walsh code.

is considered as the input data bit by the mobile station. Figure 8.12 also shows the output when the despreading is done using an incorrect Walsh code. Due to the orthogonal nature of the Walsh code, exclusive-or-ing one Walsh code with a different one will result in an output consisting of an equal number of 1s and 0s. The receiver will not be able to make a determination of the input data bit/symbol.

**8.4.1.9  Quadrature spreading.**    Once Walsh spreading is done, the data is then quadrature spread with a base station specific PN sequence, also known as the short code. This gives it a base-station-specific identity and results in a quadrature phase shift keying (QPSK) output. In fact, all base stations use the same PN sequence, but each base station selects from among 512 possible offsets for its own identifying spreading code. This data is then transmitted to the mobile station.

The mobile station is able to differentiate between transmissions from different base stations due to the unique quadrature cover provided by each base station. Once the mobile station has locked onto its particular base station's transmissions, it is able to differentiate between the different logical channels transmitted by its base station based on the different Walsh codes used to provide a Walsh cover to logical channels. It is then able to extract information targeted toward it with the help of the mobile specific long code mask used by the base station.

### 8.4.2   Downlink broadcast channels

The downlink CDMA broadcast channel consists of the following code channels: the pilot channel, up to one sync channel, and up to seven paging channels. Each of these code channels is orthogonally spread by the appropriate Walsh function. Specifically, Walsh code W0 is used for the pilot channel, W32 for the sync channel, and W1–W7 for the paging channels. Walsh codes spread each of the channels orthogonally, enabling the mobile station to differentiate between the different channels.

**8.4.2.1   Pilot channel.**   The pilot channel is transmitted in every cell. This pilot channel is used by the mobile station to obtain initial system synchronization and to provide robust time, frequency, and phase tracking of the signals from the cell site. This signal is tracked continuously by each mobile station. The pilot channels are transmitted by each cell site using the same quadrature spreading code, but with different spread spectrum code phase offsets, allowing them to be distinguished. The fact that the pilots all use the same code allows the mobile station to find system timing synchronization by a single search through all code phases. This speeds up the cell acquisition process in the mobile station. The strongest signal found corresponds to the code phase of the best cell site. The number of possible codes which can be reused within a system is 512, however, care has to be taken so that no two neighboring cells use the same time offset since the PN codes are the same. The pilot channel data is a series of 0s and it is an unmodulated signal. Prior to being spread by the cell specific code the pilot channel is spread with a Walsh code of 0, to identify the channel.

**8.4.2.2   Sync channel.**   The sync channel uses the Walsh code W32. It uses the same PN sequence and phase offset as the pilot channel and can be demodulated whenever the pilot channel is being tracked. This sync channel carries cell site identification, pilot transmit power, and the cell site pilot PN carrier phase offset. With this information the mobile station is capable of establishing system time and knows the proper transmit power to initiate calls. The sync channel operates at a fixed rate of 1200 bps. The data is half-rate convolutional encoded with symbol repetition of 2 (i.e., the same encoded symbol occurs twice). This data is sent through a block interleaver and the output is spread with a Walsh code. It is then quadrature spread, which provides the channel with a cell-specific identity.

**Sync channel message structure.**   The sync channel is used to provide time and frame synchronization to the mobile station. Only one message, the sync channel message, is sent on the sync channel. It contains information about the base-station protocol revision level, the minimum protocol revision level supported by the base station (mobile stations that support revision numbers greater than or equal to this value can access the system), the system and network identification (SID, NID), the pilot PN sequence offset index, detailed timing information, paging channel data rate, and the CDMA channel number.

**8.4.2.3  Paging channels.**  These channels are optional and their number can range from zero to seven in a cell. They are covered with Walsh codes W1–W7. The paging channel can operate at a data rate of either 9600 or 4800 bps. The data is passed through a half-rate convolutional encoder and a symbol repeater. The block interleaver follows next. The output of the interleaver is a constant 19.2 kbps, which is modified with a long code. The long code is modified with a mask specific to paging channels. The mobile stations recognize the mask and the long code and are thus able to decode the information. This data is then given a Walsh cover followed by the cell-specific spreading common to all channels. Figure 8.13 shows the pilot, sync, and paging channel generation.

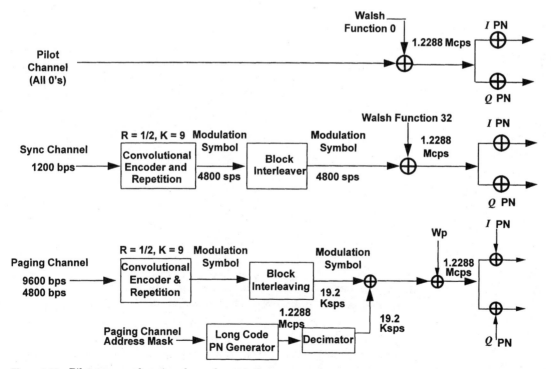

**Figure 8.13**  Pilot, sync, and paging channel generation.

Page messages contain pages to one or more mobile stations. Pages are usually sent when the base station receives a call for the mobile station and are typically sent by several different base stations. Orders are a broad class of messages used to control a particular mobile station. Orders can be used from acknowledgments to locking or preventing an errant mobile from transmitting. The paging channel has a special mode called a slotted mode. The slotted mode of paging channel operation is analogous to discontinuous reception (DRX) in GSM but with some differences. In this mode, messages for a particular mobile station are sent only in certain predefined slots which occur at certain predefined times. Through the access process the mobile station can specify which slots it will be monitoring for incoming page messages. These slots can occur from once every 2 seconds to once every 128 seconds. This capability allows a mobile station, which is operating in this slotted mode, to partially power down during slots other than its predefined slot. The order of page messages within a slot is also arranged in such a way that a mobile station has to listen to a portion of its slot, not the entire slot. This technique provides a very powerful method by which a battery-operated mobile station can conserve a considerable amount of battery energy when idle.

**Paging channel structure.** Once the mobile station has obtained information from the sync channel message, it adjusts its timing to correspond to normal system timing. The mobile station then determines and begins monitoring the paging channel. Typically, a single 9600-bps paging channel can support about 180 pages per second. Using all the seven paging channels on a single CDMA frequency assignment gives a capacity of 1260 pages per second. Paging channel messages convey information from the base station to the mobile station. Messages addressed to individual mobile stations can be addressed either by the ESN, IMSI, or TMSI. The paging channel supports the following messages:

- *System parameters message.* Information on the pilot PN sequence offset, system identification, and network identification is sent in the system parameters message. Registration information, such as zone identification along with information for the different types of registration, such as power-down, power-up, and parameter change is provided. Parameters, such as base-station class, number of paging channels, maximum slot cycle index, base-station latitude and longitude, and power control reporting thresholds are also present.

- *Access parameters message.* Information related to mobile station access is sent via this message. It contains information regarding the number of access channels, initial access power requirements, number of access attempts, maximum size of access message, values for different overload classes, access attempt backoff parameters, authentication mode, and global random challenge value. This information is used by the mobile station to modify its access procedure.

- *CDMA channel list message.* This message indicates the channel number of the CDMA channel.

- *Channel assignment message.* This message is sent to the mobile station to assign it a channel. The assigned channel can be a CDMA traffic channel, paging channel, or an analog voice channel. In CDMA, the base station assigns a paging channel to the mobile station to monitor in case there are multiple paging channels. The base station can use this mechanism to distribute the load across the paging channels. The base station can also direct the mobile station to acquire an analog system. If an analog voice channel is being allocated information about the system identification of the analog system, then the analog voice channel number, color codes, and so on, are provided. The mobile station uses this information to acquire the analog voice channel. If a CDMA traffic channel is being assigned information about the frequency, then frame-offset, encryption mode, and so forth, are provided. The base station can optionally request service-option negotiation or grant the service option requested by the mobile station.

There are several other types of messages dealing with authentication, obtaining information from the mobile station, general page messages, and orders.

## 8.5 CDMA Uplink Channels

The uplink CDMA channel is composed of access channels and uplink traffic channels. These channels share the same CDMA frequency assignment using direct sequence CDMA techniques.

Each traffic channel is identified by a distinct user-specific long code sequence and each access channel is identified by a distinct access channel long code sequence. The uplink traffic channel structure supports up to 62 different traffic channels and up to 32 different access channels. Each access channel is associated only with a single paging channel. These are the upper limits, and typically the base station supports fewer than these due limitations of interference and network resources. The long code sequence is $2^{42} - 1$ bits long.

The uplink CDMA channel has the overall structure shown in the Fig. 8.14. Data transmitted on the uplink CDMA channel is grouped into 20-ms frames. All data transmitted on the uplink CDMA channel is convolution encoded, block interleaved, modulated by the 64-ary orthogonal modulation, and direct-sequence spread prior to transmission.

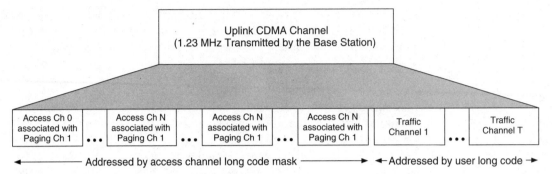

**Figure 8.14**   Structure of uplink CDMA channel.

### 8.5.1   Access channel

The access channel is an uplink-only, shared, point-to-point channel. It provides communications from the mobile station to the base station when the mobile station is not using a traffic channel. All access channel transmissions are at 4800 bps.

One (or more) access channel is paired with every paging channel. Each access channel is distinguished by a different long PN code. The base station responds to transmissions on a particular access channel by a message on an associated paging channel. Similarly, the mobile station responds to a paging channel message by transmitting on one of the associated access channels. The access channel is a random access CDMA channel. Multiple mobile stations associated with a particular paging channel may simultaneously try to use an access channel. A mobile station chooses an access channel by randomly choosing one from the set of access channels activated in that cell. It also chooses a PN time alignment from the set of available PN time alignments. Unless two or more mobile stations choose the same access channel and the same PN time alignment, the base station is able to receive their simultaneous transmissions. The base station controls the rate of access channel transmissions to prevent too may simultaneous transmissions by multiple mobile stations.

**8.5.1.1   Access channel message structure.**   An access channel transmission occurs in an access channel slot. The access channel slot length may differ from base station to base station. A mobile station, prior to initiating its access channel transmission, determines the relevant slot length. An access channel transmission consists of the access channel preamble and the access channel message capsule. The preamble consists of 96 zeros transmitted before the message capsule. Since the slot length can vary from base station to base station, the message capsule has a padding field which is used to pad the access channel message so that the entire transmission fills up the slot. The message body can be from 2–842 bits.

Access channel messages consist of registration, order, data burst, origination, page response, authentication challenge response, status response, and TMSI assignment completion messages. All access channel messages share some common parameters which can be classified as:

- *Acknowledgments and sequence number.* Parameters which acknowledge the most recently received paging channel message, message sequence number of the current message, indication of whether an acknowledgment is required for the current message, and more, are included here.

- *Mobile identification parameters.* A CDMA mobile station can identify itself either with an MIN, IMSI, ESN, MIN and ESN, or IMSI and ESN. Indication of type of address used by mobile station to identify itself, and the address itself are included here.

- *Authentication parameters.* If the base station has requested authentication parameters on access, the mobile station will include the authentication data, random challenge value, and the call-history parameter.

Apart from these common parameters, access channel messages also have message-specific fields. Registration messages have an indicator indicating the type of registration—power-down, power-up, distance-based, etc. If operating in slotted mode, the mobile station slot cycle index, then the protocol revision of the mobile station, station class mark, and an indicator of whether the mobile station can receive terminating calls can be used in case of roaming. Origination messages have most of the same fields as the registration message with additional fields, such as capability of mobile station (dual-mode analog and CDMA, CDMA-only, etc.), called party number, and service option fields.

**8.5.1.2  Access channel generation.**  The access channel supports fixed data rate operation at 4800 bps in 20-ms frames. Information rate of 4.8 kbps is input into a one-third-rate convolutional encoder which does channel encoding. The output from the convolutional encoder is input into a symbol repeater. A symbol repeater is used as a fast mechanism to input a constant bit rate into the block interleaver. Block interleavers operate efficiently when data is input at a constant rate. The block interleavers are optimized for a user information rate of 9600 bps. Inputting a rate of 9600 bps into the convolutional encoder results in a output of 28,800 bps of channel-encoded data which is then sent to the block interleaver. Since the output of the convolutional encoder for an access channel data rate of 4800 bps is 14,400 kbps, it is sent to a symbol repeater which repeats each symbol of encoded data once to give 28,800 bps of channel-encoded data. Each code-symbol output from the symbol repeater occurs two consecutive times. As discussed, a block interleaver is used to shuffle the data prior to transmission to overcome the effects of fast Rayleigh fading which can adversely affect long streams of data. Shuffling the data ensures that any information bits lost due to Rayleigh fading will not be sequential. CDMA uses a 32 rows × 18 columns (576 symbols) array for the block interleaver. Data is written into the interleaver by columns and output by rows. Since the block interleaver only shuffles the data and does not add to it, the output of the interleaver is the same as the input (i.e., 28.8 kbps of encoded data). A function called a 64-ary orthogonal modulator follows the interleaver. This function substitutes groups of 6 code bits for all 64 bits of one of the 64 available Walsh symbols. In this transmission all 64 Walsh codes are possible. The output from the orthogonal modulator is 307.2 kbps for encoded data. The orthogonal modulator's output is covered with a masked long code sequence which distinguishes any particular access channel from all the other channels the base station may receive. The long code is altered with a mask that includes such things as the access channel number (n), the corresponding paging channel in the downlink direction, the base station identification, and the current PN code offset. The last process is for each access channel to be quadrature spread with a cell-specific PN code, which assists the base station in differentiating transmissions from within its cell with transmissions from other cells/sectors. The entire user data of 4800 bps is transmitted in a 20-ms frame. Figure 8.15 describes the generation of an access channel.

### 8.5.2 Uplink traffic channels

Both the uplink and downlink traffic channels use a similar control structure consisting of 20-ms frames. These frames can be used for both traffic and signaling. CDMA supports two modes of transport of signaling information when a traffic channel is assigned. Both of these modes can be used in the uplink and downlink directions. *Blank-and-burst* signaling can be used to transmit signaling information. The operation of this mode is similar to that of AMPS. Whenever there is signaling information to be sent, one or more frames of primary traffic data, typically vocoded voice, is replaced with signaling data. The data rate is typically 9600 bps if rate set 1 is being used, or 14,400 bps, 7200 bps, 3600 bps if rate set 2 is being used. CDMA also supports another mode for sending signaling information known as *dim-and-burst*. This mode is possible because of the use of the variable-rate vocoder in CDMA. In this mode, the vocoder oper-

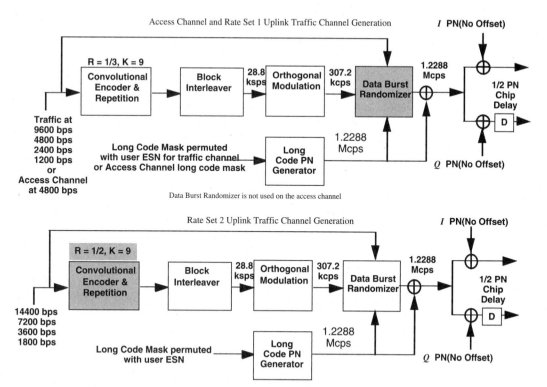

**Figure 8.15**  Access channel, rate sets 1 and 2 uplink traffic channel generation.

ates in one of its ½, ¼, or ⅛ modes, but the data rate is typically full-rate—9600 bps for rate set 1 and 14,400 bps for rate set 2. The bits saved due to not operating the vocoder at full rate are used for signaling. Typically, in this mode the vocoder's rate is limited only when it wants to transmit at full rate. In all other modes of operation, the vocoder rate is not limited since there will be bits remaining that can be used for signaling. Dim-and-burst signaling has an immense advantage over blank-and-burst signaling because the degradation in voice quality is essentially undetectable.

CDMA also provides for use of the dim-and-burst mode for sending primary and secondary traffic. For instance, primary data could be vocoded voice and secondary data could be a fax message. With the use of this mode, CDMA supports *simultaneous* voice and data via the same traffic channel.

**8.5.2.1  Traffic channel messages.**  There are essentially five types of control messages on the traffic channel: messages controlling the call itself, messages controlling handoff, messages controlling downlink power, messages for security and authentication, and messages eliciting or supplying special information from or to the mobile station.

**8.5.2.2 Uplink traffic channel structure.** Figure 8.15 describes the uplink traffic channel generation. The channel structure for an uplink traffic channel is similar to that of the access channel. The uplink traffic channel supports two rate sets, depending on the type of vocoder used. Rate set 1 contains four elements, specifically 9600, 4800, 2400, and 1200 bps. Rate set 2 contains four speeds, specifically 14,400, 7200, 3600, and 1800 bps. The mobile station has to support, mandatorily, rate set 1, while rate set 2 support is optional. The channel structure differs for rate set 1 and rate set 2.

Frame quality indicators are essentially CRC check applied to 14.4-, 9.6-, 7.2-, 4.8-, 3.6-, and 1.8-kbps traffic channel frames. This is then sent to a one-third-rate convolutional encoder in case of rate set 1 data, or to a half-rate convolutional encoder in case of rate set 2 data. Using a half-rate convolutional encoder for rate set 2 makes the protection for rate set 2 traffic frames weaker than that of rate set 1 which uses one-third-rate coding. However, lower protection is compensated because the rate set 2 vocoder is supposed to be of a better quality and have a more robust vocoding algorithm than the rate set 1 vocoder. The symbol repeater is next. It repeats the data as necessary, to result in an output of 28.8 kbps of encoded data. Orthogonal modulation occurs, which is similar to the access channel process. The orthogonal modulator's output is covered with a masked long code sequence, which distinguishes any particular mobile station's transmission from all the other mobile stations' transmissions the base station may receive. When transmitting on the uplink traffic channel, the mobile station uses one of two long code masks unique to that mobile station: a public long code mask derived from the mobile station's electronic serial number (ESN), or a private long code mask which is derived from the encryption and authentication process. Since each mobile station's ESN is unique, it follows that the public long code mask is unique for each mobile. It is called public because if the mobile's ESN is known then its long code mask can also be easily determined. The public long code mask is the unique code which differentiates mobiles in the network. It is the "C" in CDMA. A previous example discussed the analogy of a CDMA network as an office hall where users spoke different languages. The long code mask would be the language of the individual mobile stations. Finally, as is done for channels in the cell, each traffic channel is quadrature spread with a cell-specific PN code to enable the base station to determine transmissions from its cell. This cell-specific PN code is broadcast by the base station. Figure 8.16 describes the different processes involved in the channel generation scheme.

PCM Voice

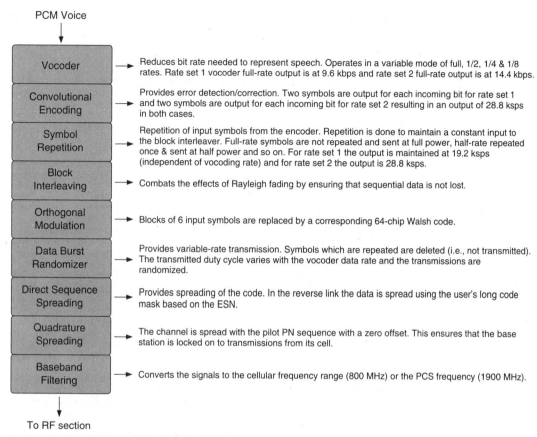

| | |
|---|---|
| Vocoder | Reduces bit rate needed to represent speech. Operates in a variable mode of full, 1/2, 1/4 & 1/8 rates. Rate set 1 vocoder full-rate output is at 9.6 kbps and rate set 2 full-rate output is at 14.4 kbps. |
| Convolutional Encoding | Provides error detection/correction. Two symbols are output for each incoming bit for rate set 1 and two symbols are output for each incoming bit for rate set 2 resulting in an output of 28.8 ksps in both cases. |
| Symbol Repetition | Repetition of input symbols from the encoder. Repetition is done to maintain a constant input to the block interleaver. Full-rate symbols are not repeated and sent at full power, half-rate repeated once & sent at half power and so on. For rate set 1 the output is maintained at 19.2 ksps (independent of vocoding rate) and for rate set 2 the output is 28.8 ksps. |
| Block Interleaving | Combats the effects of Rayleigh fading by ensuring that sequential data is not lost. |
| Orthogonal Modulation | Blocks of 6 input symbols are replaced by a corresponding 64-chip Walsh code. |
| Data Burst Randomizer | Provides variable-rate transmission. Symbols which are repeated are deleted (i.e., not transmitted). The transmitted duty cycle varies with the vocoder data rate and the transmissions are randomized. |
| Direct Sequence Spreading | Provides spreading of the code. In the reverse link the data is spread using the user's long code mask based on the ESN. |
| Quadrature Spreading | The channel is spread with the pilot PN sequence with a zero offset. This ensures that the base station is locked on to transmissions from its cell. |
| Baseband Filtering | Converts the signals to the cellular frequency range (800 MHz) or the PCS frequency (1900 MHz). |

To RF section

**Figure 8.16**  Generation of the uplink traffic channel.

### 8.5.3  Variable-rate data transmission

The uplink traffic channel structure has been discussed to this point. The way data is transmitted on this channel will now be examined. The data is transmitted on the channel via variable-rate transmission. The transmission rate is varied according to the input-information rate. When the transmit-data rate is 9600 or 14,400 bps, the transmission gate allows all encoded data to be transmitted. When the transmit-data rate is 4800 or 7200 bps, the transmission gate allows one-half of the output symbols to be transmitted, when the transmit-data rate is 2400 or 3600 bps the transmission gate allows one-fourth of the output symbols to be transmitted, and so on. The gating process operates by dividing the 20-ms frame into 16 equal length (i.e., 1.25 ms) periods called power control groups. Certain power control groups are gated-on (i.e., transmitted), while other groups are gated-off (i.e., not transmitted).

If the variable-rate data transmission scheme as described above is implemented, all mobile stations operating at the same data transmission rate will transmit data on the same power control group. At the base station, there will be certain power control groups having data, and other which will have no data. To evenly spread the data around within the entire 20-ms frame, a data-burst randomizing algorithm is used. This algorithm ensures that all gated-on power control groups are pseudorandomized in their positions within the frame and that every code symbol input to the symbol repeater is transmitted exactly once. Variable-rate data transmission reduces the average power in the uplink channel as it appears at the base station's receiver. This has the effect of reducing the interference levels throughout the CDMA system and improves system capacity and performance.

Transmission on the downlink is varied not by gating transmissions, but by reducing the power of the repeated symbols. Downlink transmissions use quadrature phase shift keying (QPSK) which requires a linear amplifier. Though not power efficient, it is not a problem in the base station. Uplink transmissions make use of QPSK (offset QPSK), using a power efficient, nonlinear, fully saturated C-class amplifier. This saves mobile phone battery life and provides long talk times.

## 8.6   Power Control

Power control is a valuable asset in any two-way communication system. It is particularly important in a multiple access terrestrial system where users' propagation loss can vary over many tens of decibels. Power control is of paramount importance for a CDMA system. The power at the cellular base station received from each user over the uplink must be made nearly equal to that of all others in order to maximize the total user capacity of the system. The system capacity is maximized if the transmitted power of each mobile station is controlled so that its signal arrives at the cell site with the minimum required signal-to-interference ratio. The bit error rate is too high to permit high-quality communications if a mobile station's signal arrives at the cell site with too low a received-power value. It the received power is too high, the performance of this mobile station is good; however, interference to all the other mobile station transmitters that are sharing the channel is increased and may result in unacceptable performance to other users unless their power is reduced. Very large disparities are caused mostly by widely differing distances from the base station, and to a lesser extent, by shadowing effects of buildings and other objects. Such disparities can be adjusted individually by each mobile subscriber unit simply by controlling the transmitted power according to the automatic gain control (AGC) measurement of the downlink power

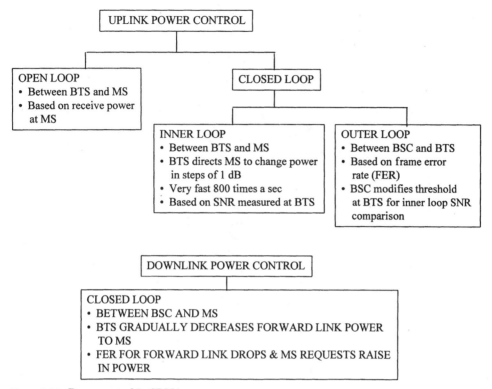

**Figure 8.17** Power control in CDMA.

received by the mobile receiver. Figure 8.17 lists the different types of power control used in CDMA.

### 8.6.1 Uplink open-loop power control

If the physical channel was completely symmetrical, measuring the power level of the signal received from the base station would determine the transmitter power level required for that user. This ideal situation is almost never the case since the downlink and uplink center frequencies are generally quite far apart. Measurement of total power received by a mobile is usually done by the automatic gain circuitry (MAGC) of a mobile receiver. Each mobile station attempts to estimate the path loss from the cell site to the mobile station. In CDMA, the pilot channel signals are used by all mobile stations for initial synchronization. The mobile station measures the power level of both the pilot signal from the cell site to which it is connected and the sum of all the cell site signals received at the mobile station. The downlink path signal strength of the mobile station is used by the mobile station to adjust its own transmitter power; the stronger the received signal, the lower the mobile station's transmitter power. This is called open-loop power control. It is open loop because there is no feedback information about power level change from the base station. Open-loop power control works independent of the base station. Reception of a strong signal from the cell site indicates that the mobile station is either close to the cell site, or has an unusually good path to the cell site. This indicates that relatively less mobile station transmitter power is required to produce nominally received power at the cell site from this mobile station transmission. In the case of a sudden improvement in the channel, the open-loop control mechanism provides a very rapid response in a period of tens of milliseconds. It adjusts the mobile station transmit level downward and prevents the mobile station transmitter power from being at too high a level. The sync channel from each cell contains data on the transmitted pilot channel power which the mobile uses to determine transmitted power. The general rule for adjusting the transmit power of the mobile station is:

$$Tx = -76 - Rx + Nom\_Pwr + Init\_Pwr + \text{Access probe correction}$$

Rx is the received power in dBm by the mobile station, the Nom_Pwr and Init_Pwr are parameters in the access parameter message, and the access probe correction is the first successful access attempt power correction.

### 8.6.2 Uplink inner closed-loop power control

Since propagation loss is not symmetric, primarily due to Rayleigh fading which depends strongly on carrier frequency and may differ considerably in the two directions, a closed-loop power control mechanism is used that varies the power transmitted by the mobile, based on the measurements made at the base station. Moreover, AGC by itself is not effective. Even after adjustment using open-loop power control based on AGC, the uplink transmitted power may differ by several decibels from one subscriber to the next. Closed-loop power control substantially mitigates the disparity between uplink and downlink power levels. It is typically implemented in the base station where the base station measures any user's received signal on the uplink and makes a determination if it's too high or too low by comparing it to a threshold. Then it sends a 1-bit command to the user over the downlink to command it to lower or raise its relative power by a fixed amount (1 dB). This type of power control is very fast and occurs 800 times a second. The dynamic range is ±24 dB of open-loop value and the total dynamic range (open and closed) is 80 dB.

Power control is more important on the uplink than on the downlink for two reasons. First, the power measurement is more complex, since for the downlink power is measured continuously on the unmodulated pilot signal. More important, power control is critical for achieving maximum uplink capacity and much less so for the downlink.

The propagation loss depends on the distance from the subscriber unit to the base station. However, it also depends on shadowing or blockage by buildings and other objects. As already noted, most of the long-term propagation loss due to distance is adjusted by an open-loop correction based on the AGC of the subscriber unit. Very rapid variations, shorter than about 1 millisecond, are mostly due to Rayleigh fading phenomenon that cannot reasonably be mitigated by power control. This leaves shadowing by objects and long-term fading that may be experienced by stationary subscribers. These are of long enough duration to be mitigated by the power control loop.

### 8.6.3 Uplink outer closed-loop power control

In the closed-loop power control process, the received signal as measured at the cell is compared to an adjustable threshold. The threshold determines the frame error rate (FER). Increasing the threshold reduces the frame error rate, thereby improving the quality of the speech. Reducing the threshold tends to increase the frame error rate. Typically, a system would attempt to maintain a frame error rate of 1 percent. Adjusting this threshold is referred to as "outer-loop power control." Unlike the power control scheme described in the previous section, this power control loop is controlled by the BSC/MSC/Vocoder network entities compared to the BTS. The standard does not specify outer-loop power control. A single threshold can be used for every unit in the cell or each subscriber unit can have its own threshold. Individual thresholds are not expected to vary over a range of more than a few dB. Individual thresholds will be beneficial since this will allow mobiles in extremely advantageous circumstances to have a lower threshold while providing a higher threshold to disadvantaged users. The use of individual thresholds significantly increases capacity.

### 8.6.4 Downlink traffic channel power control

The relative power used in each data signal transmitted by the cell site in response to control information is transmitted by each mobile station. In certain locations, the link from cell site to mobile station may be unusually disadvantaged. Unless the power being transmitted to this mobile station is increased, the quality may become unacceptable.

To accomplish the downlink power control, the cell periodically reduces the power transmitted to the subscriber. This process continues until the subscriber, sensing an increase in received frame error rate, requests additional power. The cell site receives the power adjustment requests from each mobile station and responds by adjusting the power allocated to the corresponding cell site transmitted signal by a predetermined amount. The adjustment is usually small. However, the down power control algorithm is not standardized. The standard specifies that the subscriber unit must monitor the quality of the down traffic channel and report this information back to the base station if told to do so. This is a closed-loop process similar to the uplink power control process. However, in the uplink direction, the closed-loop was based on maintaining the signal-to-noise metric at the proper level. The downlink power control process monitors frame error rate. As a result, the downlink power control process is substantially slower. When the 9600-bps transmission is used, the subscriber unit must inform the base station of the frame error count using a message defined in the standard (Power Measurement Report Message). The 14,400-bps transmission rate allows for a faster power control process. In this rate, a single bit has been set aside in every frame to be used as a *frame-erasure* bit. This bit is set by the mobile to indicate an erasure (an error) in the downlink traffic channel frame. The BSC considers the frame-erasure bit, and if the FER has crossed a predetermined threshold, it instructs the BTS to increase the forward-link power to the mobile station.

### 8.6.5    Impact of rogue mobiles

*Rogue mobiles* are mobiles that transmit at a power level much higher than ordered to. These mobiles that do not obey power control commands can degrade system performance. There are several methods proposed in the standard to counteract this behavior. A malfunction timer is implemented in the mobile station. This timer has a maximum length of 60 seconds. The timer should be reset periodically during the normal functioning of the unit. If the unit fails to function properly and does not execute instructions in the proper order, the malfunction timer resets will not be executed and the timer will run down as a result. When the timer runs down, the subscriber unit must disable its transmitter. The standards also define messages that can be used to order the mobile to disable its transmitter. These messages are called *lock orders*. Closed-loop power control, as discussed above, can be used to control the subscriber unit transmit power in the event that an amplifier malfunctions, but the phone still responds appropriately to power control commands.

## 8.7   CDMA Registration

In CDMA, registration is a process by which the mobile station notifies the base station of its location, status, identification, slot cycle, and other characteristics. The mobile station informs the base station of its location and status so that the base station can efficiently page the mobile station when establishing a mobile-terminated call. For operation on the slotted mode, which allows the mobile station to power down between its scheduled slots and conserve battery power, the mobile supplies the slot index to enable the base station to determine which slots the mobile station is monitoring. This scheme is also known as *mobile sleep mode* or *discontinuous reception* (DRX), similar to IS-54 and GSM. The mobile station also supplies the station class mark and protocol revision number so that the base station recognizes the capabilities of the mobile station.

The CDMA system supports nine different forms of registration:

- *Power-up registration.*   The mobile station registers when it powers on or switches from using the analog system.

- *Power-down registration.*   The mobile station registers when it powers off if previously registered in the current serving system.

- *Timer-based registration.*   The mobile station registers when a timer expires.

- *Distance-based registration.*   The mobile station registers when the distance between the current base station and the base station in which it last successfully registered exceeds a threshold.

- *Zone-based registration.*   The mobile station registers when it enters a new zone.

- *Parameter-change registration.*   The mobile station registers when certain of its stored parameters change or when it enters a new system.

- *Ordered registration.*   The mobile station registers when the base station requests it.

- *Implicit registration.*   When a mobile station successfully sends an origination message or page-response message, the base station can infer the mobile station's location. This is considered an implicit registration.

- *Traffic-channel registration.*   Whenever the base station has registration information for a mobile station that has been assigned to a traffic channel, the base station can notify the mobile station that it is registered.

As a group, the first five forms of registration are called autonomous registration since the mobile station initiates the registration in response to an event (i.e., without being explicitly directed to register by the base station). The base station can enable or disable any of the various forms of autonomous registration. It is expected that the operators will use the combinations of registration methods that are most effective for their networks. Some forms of autonomous registration have parameters that can be fine-tuned. The base station communicates the forms of registration that are active and the corresponding registration parameters via the system-parameters message. All autonomous registrations are enabled when the mobile station is roaming.

### 8.7.1  Roaming determination

For the purposes of roaming, there are system and network identification procedures defined in CDMA. A base station is a member of a cellular system and a network. A network is a subset of a system. Systems are labeled with an identification called the system identification or SID, and networks within a system are given a network identification or NID. A network is uniquely identified by the SID/NID pair. Figure 8.18 shows an example of a network with SIDs and NIDs.

The mobile station has a list of one or more home (nonroaming) SID/NID pairs. A mobile station is roaming if the stored pair does not match the one being broadcast by the base station. A mobile station is categorized as a foreign NID roamer if one of the stored SIDs matches with the network's SID. It is categorized as a foreign SID roamer if none of the stored SIDs match the network's SID. The mobile station, based on this information along with some parameters stored in its permanent memory, makes a determination about the type of service that will be available when it is roaming.

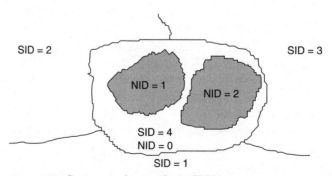

**Figure 8.18**   Systems and networks in CDMA.

### 8.7.2  Power-up registration

Power-up registration is performed when the mobile station is turned on, switched to an alternate serving system, or switched from using an analog system. To prevent multiple registrations when power is quickly turned on and off, the mobile station maintains a timer and only attempts power-up registration when the timer expires. This type of registration can be disabled via a systems-parameters message.

### 8.7.3  Power-down registration

Power-down registration is performed when the user directs the mobile station to power off. The mobile station does not perform power-down registration if it has not previously registered in the system. Power-down registration is not expected to be extremely reliable, as the mobile station may have driven beyond the range of the cellular system. It is expected to be even more unreliable for a portable, as a portable may be in a poor location, or it may have a poor orientation. Even though power-down orientation may be unreliable, a successful power-down registration allows an MSC to avoid paging the mobile station.

### 8.7.4  Timer-based registration

Timer-based registration causes the mobile station to register at regular intervals. Its use also allows the system to automatically deregister mobile stations that did not perform a successful power-down registration. The base station provides parameters to the mobile station that enable it to set the timer. The timer is also reset after each implicit or successful registration. To prevent all mobile stations from transmitting their timer-based registration messages at the same time and thus swamping the base station, CDMA provides a mechanism to stagger the arrivals of these registration messages. Mobile station timers can be a function of their paging channel slot index if they are operating in the slot mode, and if not, the mobile station chooses a pseudo-random value in between the parameters determined by the base station. In this way, different mobile stations will set their timers differently and register accordingly.

### 8.7.5  Distance-based registration

With distance-based registration, the base station sends its latitude, longitude, and a distance parameter in the systems-parameters message. When the mobile station starts receiving a new base station, the mobile station receives the new base station's latitude and longitude. The mobile station then computes a metric using these latitude and longitude values and those from the base station where the mobile sta-

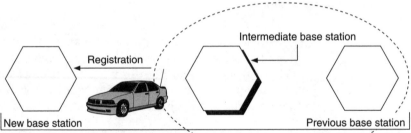

**Figure 8.19**   Distance-based registration.

tion last registered. If this metric exceeds the distance parameter from the base station where the mobile station last registered, the mobile will register. Upon registering, the base station becomes the center of a circle which is typically several cells in radius. The mobile station performs distance registration again only when it exits this circle. Figure 8.19 illustrates this concept.

### 8.7.6   Zone-based registration

With zone-based registration, a cellular system is divided into location areas or zones. The mobile station and the MSC both keep a list of the zones in which the mobile has recently registered. In CDMA, unlike GSM, a mobile station can be part of several location areas at the same time. When the mobile station enters a zone not on its list, it registers. Upon a successful registration, both the mobile station and the MSC add the new zone to their lists and set expiration timers on all other zones on their list. By keeping multiple zones on the list, the system can avoid multiple registrations along the border between zones. By setting timers on old zones, the MSC can avoid paging in the old zones after the timers expire. Zone-based registration is particularly powerful for defining boundaries between different sections of a cellular system or between systems.

### 8.7.7   Parameter-change registration

Parameter-change registration is performed whenever one of the mobile's internally stored parameters, such as station class mark or the paging slot index, changes.

## 8.8   Mobile Access

The entire process of sending one access message and receiving, or failing to receive, an acknowledgment for that message is called an *access attempt.* Each transmission in the access attempt is called an *access probe.* The mobile station transmits the same message in each access probe in an access attempt. For each downlink paging channel in the cell, there will be one or more uplink access channels from the mobiles in the cell. The access process may be initiated for an origination attempt, as a response to a page, or for a registration attempt. The access channel used for each access probe sequence is chosen pseudo-randomly from among all the access channels associated with the current paging channel. The first access probe of each access attempt is transmitted at a specified power level relative to the nominal open-loop power level. Each subsequent access probe is transmitted at a power level that is a specified amount higher than the previous access probe. The size of the power step increases and the time between the increases is controlled by a set of parameters the mobile recovers from the synchronization channel. The access probes do not continue forever, but are sent until a limit, determined by the system, is reached. This limit is announced to all the mobiles on the base station's synchronization channel. If the mobile does not receive an acknowledgment message, it reduces its RF power, waits a pseudorandom amount of time, and starts sending access probes again, just as before. The process continues until access is achieved, or until the maximum number of times the system will allow access attempts is reached. Once an acknowledgment is finally received on the paging channel, the base station assigns a traffic channel to the mobile station if needed.

## 8.8.1 Service configuration, negotiation, and multiplex options

During traffic channel operation, the mobile station and the base station communicate through the exchange of downlink and uplink traffic channel frames. The mobile station and the base station use a common set of conditions or characteristics for building and interpreting traffic channel frames. This set of characteristics, referred to as a service configuration, consists of multiplex options, transmission rates, and service option connections explained as follows:

- *Uplink and downlink multiplex options.* These control the way in which the information bits of the downlink and uplink traffic channel frames, respectively, are divided into various types of traffic, such as signaling traffic, primary traffic, and secondary traffic. Associated with each multiplex option is a rate set which specifies the frame structures and transmission rates supported by the multiplex option. The multiplex option used for the uplink channel can be the same as that used for the downlink channel, or it can be different.

- *Multiplex option 1.*  This applies to rate set 1. It provides for the transmission of primary traffic and signaling or secondary traffic. Signaling traffic may be transmitted via blank-and-burst with the signaling traffic using all of the frame or via dim-and-burst with the primary traffic and signaling traffic sharing the frame. Multiplex option 1 also supports the transmission of secondary traffic. When primary traffic is available, secondary traffic is transmitted via dim-and-burst with primary traffic and secondary traffic sharing the frame. When primary traffic is not available, secondary traffic is transmitted via blank-and-burst with the secondary traffic using all of the frame. Examples of primary and secondary traffic could be speech and fax. It is mandatory for the mobile station to support multiplex option 1. Figure 8.20 shows the different multiplex options for rate set 1.

- *Multiplex option 2.*  This applies to rate set 2. It provides for the transmission of primary traffic, secondary traffic, and signaling traffic. Signaling traffic may be transmitted via blank-and-burst with the signaling traffic using all of the frame, via dim-and-burst with the primary traffic and signaling traffic sharing the frame, or via dim-and-burst with the primary traffic, secondary traffic, and signaling traffic sharing the same frame. When primary traffic is available, secondary traffic is transmitted via dim-and-burst with the primary traffic, secondary traffic, and possibly signaling traffic sharing the frame. When primary traffic is not available, secondary traffic is transmitted via blank-and-burst with the secondary traffic using all of the frame. Support of multiplex option 2 is optional by the mobile station.

- *Uplink and downlink traffic channel transmission rates.*  These are the transmission rates actually used for the uplink and downlink traffic channels respectively. The transmission rates for the downlink and uplink traffic channels can include all rates supported by the rate set or a subset of the supported rates.

- *Service option connections.*  These are the services in use on the traffic channel. It is possible that there is no service option connection, in which case the mobile station and base station use the uplink and downlink traffic channels to send only signaling traffic and null traffic channel data; or there can be one or multiple service option connections. The services provided by the variable-rate vocoder and enhanced variable-rate vocoder are used for data and fax.

The mobile station can request a default service configuration associated with a service option at call origination, and can request new service configurations during traffic channel operation. If the mobile station requests a service configuration that is acceptable to the base

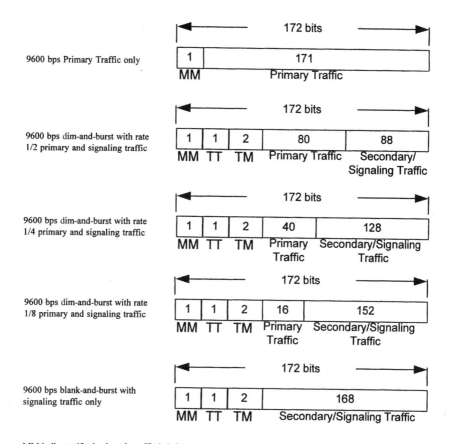

9600 bps Primary Traffic only

| 1 | 171 |
|---|---|
| MM | Primary Traffic |

9600 bps dim-and-burst with rate 1/2 primary and signaling traffic

| 1 | 1 | 2 | 80 | 88 |
|---|---|---|---|---|
| MM | TT | TM | Primary Traffic | Secondary/Signaling Traffic |

9600 bps dim-and-burst with rate 1/4 primary and signaling traffic

| 1 | 1 | 2 | 40 | 128 |
|---|---|---|---|---|
| MM | TT | TM | Primary Traffic | Secondary/Signaling Traffic |

9600 bps dim-and-burst with rate 1/8 primary and signaling traffic

| 1 | 1 | 2 | 16 | 152 |
|---|---|---|---|---|
| MM | TT | TM | Primary Traffic | Secondary/Signaling Traffic |

9600 bps blank-and-burst with signaling traffic only

| 1 | 1 | 2 | 168 |
|---|---|---|---|
| MM | TT | TM | Secondary/Signaling Traffic |

MM indicates if mixed mode traffic is being used, it is set to 0 if only primary traffic is being sent, 1 otherwise.
TT is a traffic type field which indicates if secondary or signaling traffic is being sent.
TM is traffic mode bit which indicates the mode of operation.

**Figure 8.20** Frame structure for downlink traffic channel for rate set 1.

station, they both begin using the new service configuration. If the mobile station requests a service configuration that is not acceptable to the base station, the base station can reject the requested service configuration or propose an alternative one. In turn, the mobile station can accept or reject this new service offering and propose its own. This process, called service negotiation, ends when the mobile station and base station find a mutually acceptable service configuration.

## 8.9 CDMA Air Interface Protocol Layering

Figure 8.21 shows a simplified, logical view of the CDMA protocol structure for the paging channel, access channel, downlink traffic channel, and uplink traffic channel.

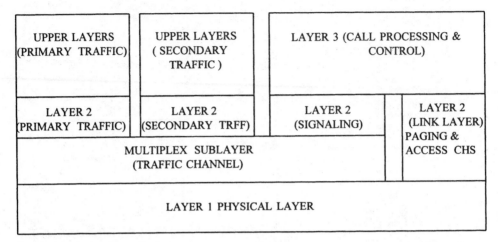

**Figure 8.21**   Protocol layering.

Signaling on all channels uses a synchronized bit-oriented protocol. The messages on all channels have a similarly layered format. The highest layer format is the message capsule that consists of a message and padding. Padding is used on some channels to make the message fit into a frame. An example of this is with blank-and-burst signaling on the traffic channel. If the message is less than a frame, the capsule is the entire frame; the padding bits extend from the end of the message to the end of the frame. The next layer format splits the message into a length field, the message body, and the CRC.

Layer 1 is the physical layer of the digital radio channel, including those functions associated with the transmission of bits, such as modulation, coding, framing, and channelization via radio waves. Between layer 1 and layer 2 is a multiplex sublayer containing the multiplexing functions that allow sharing of the digital radio channel for user data and signaling processes. For user data, protocol layering above the multiplex sublayer is service-option dependent. Typically, two higher layers are defined. Signaling protocol layer 2 is the protocol associated with the reliable delivery of signaling. Layer 3 messages between the base station and the mobile station, such as message retransmission and duplicate detection. Signaling layer 3 is the protocol associated with call processing, radio channel control, and mobile station control, including call setup, handoff, power control, and mobile station lockout.

In the mobile station, all of these layers are located in one physical piece of hardware. On the network side, these layers are distributed between several pieces of hardware which may or may not be located together. Acknowledgments are sent at the link layer. Responses are sent at the control process layer. To avoid excess signaling, link acknowledgments and control process signaling responses can be bundled into a single message. This is done by the mobile station, where the processing delay for the control process is very small. On the network side, the cell site can respond with a link layer acknowledgment, while the MSC can respond later with a control process response.

An end-user application, identified by a service option, can be viewed as plugging into sockets provided by the multiplex sublayer. A service option implies a particular layer 2 and all layers on top of layer 2. For example, service option 1, a rate set 1 vocoder, has a simple layer 2 and a vocoder as a single upper layer. Since there is no retransmission of voice frames, due to an error, the layer 2 for service option 1 does very little. A service option for a different user application, such as data, may have an entirely different set of upper layers. Some service options may have one or more layers providing identical services.

The multiplex sublayer and frame rate determination mechanisms are specified by the multiplex option. Multiplex option 1, the default multiplex option, allows both primary and secondary traffic. Thus, two service options can be defined that are optimized for different service options. A particular multiplex option may allow only a certain set of service options to be plugged into it.

## 8.9.1   Network signaling

Intersystem signaling to support automatic roaming, call delivery, and handoff is provided by IS-41 Rev. C. By using IS-41 Rev. C, different IS-95 systems can be networked together.

## 8.10   CDMA Handoff Process

Two types of handoff are supported in CDMA: soft handoff and hard handoff. *Hard handoff* is the traditional handoff mode, as discussed in the D-AMPS and the GSM chapters, where the mobile assists in the handoff process by taking measurements reports of neighboring channels and reporting them to the base station. In CDMA, hard handoff occurs between base stations having CDMA carriers with different frequency assignments. The hard handoff process for CDMA is similar to the handoff process of D-AMPS and GSM.

CDMA also supports another handoff process known as *soft handoff.* Soft handoff occurs between base stations having CDMA carriers with identical frequency assignments. Soft handoff allows both the original cell and a new cell to temporarily serve the call during the soft handoff transition. The soft handoff process transition takes place in three steps: (1) the mobile station is in communication with the original cell; (2) the mobile station is in communication with both the original cell and the new cell; and (3) the mobile station is in communication with only the new cell. This transition process reduces the probability of a dropped call, reduces handoff signaling by eliminating ping-pong effects which occur when the mobile alternatively hands off between the same cells (for instance, when traveling along the boundary of two cells), and eliminates the break in the handoff process. Hard handoffs are a break-before-make scheme, whereas soft handoff is a make-before-break scheme.

During call setup time, the mobile station is supplied a set of hand-off thresholds and a list of cells that are most likely to be candidates for handoff. While tracking the signal from the original cell, the subscriber searches all the pilot channels and maintains a list of those above the defined threshold. This list is transmitted to the MSC whenever it is requested, whenever the list changes by having a new pilot appear on the list, or whenever an existing pilot falls below a level that is useful to support the traffic. The list is transmitted in a Pilot Strength Measurement Message to the base station. The base station allocates a downlink traffic channel associated with that pilot and sends a hand-off direction message to the mobile station directing it to perform the handoff. For soft handoffs, the handoff-direction message lists multiple downlink traffic channels which have the same frequency as the current pilot. For hard handoffs, the frequencies will be different. The MSC begins the handoff process by assigning a CDMA channel at the

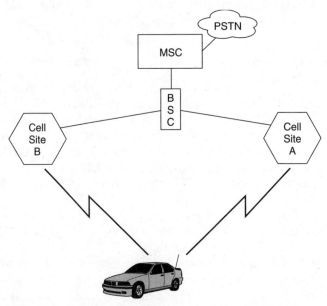

**Figure 8.22**   Soft handoff in CDMA.

new base station. The base station searches for and acquires the signal from the MS. At this point the base station commences transmission of the appropriate downlink traffic to the mobile station. All downlink traffic channels assigned to the mobile station will carry identical modulation symbols with the exception of the power control subchannel. Power control information is received from both cells. Both cells have to request a power increase for the subscriber to increase its power. Data from the mobile station is received by both cells and is forwarded to the MSC, where the best source is selected on a frame-by-frame (20 ms) basis, and is used to represent the data transmitted by the subscriber. Following the execution of the handoff-direction message, the mobile station sends a handoff-completion message on the new uplink traffic channel.

A similar process takes place when the mobile station moves from one sector to another within the same cell. In this process, called softer handoff, the mobile station functions in exactly the same manner as with soft handoff. The cell site, however, intercepts the handoff request, initiates transmissions in the new sector, and thus provides the parallel path as with soft handoff. The cell receiver actually combines signals from the antennae for both the sectors and demodulates based on the diversity combining of the signals. The MSC is notified of the activity, but it does not participate directly. No additional MSC/cell path is set up for softer handoff.

One of the critical components for soft handoff is the centralized vocoder/selector function. It needs to be centralized in order to be able to receive traffic from several base stations. It is conceivable that this function is resident at the base station, but such a configuration is not optimal since traffic would then have to be directed to the base station from the MSC. Some of the important functions of this vocoder/selector are listed as follows:

- It selects speech/data uplink from all cells involved in a soft handoff to pass along to the MSC. It has knowledge of the quality characteristics of the speech/data from all cells in order to select the best 20-ms frame on a frame by frame basis.

- It has a distribution function in the downlink direction. It distributes speech/data downlink to all cells involved in a soft handoff.

- It decodes uplink speech from $\frac{8}{13}$-kbps format to 64-kbps PCM and codes downlink speech from PCM to $\frac{8}{13}$-kbps format.

- It rate adapts and subrate multiplexes speech frames to fully utilize the transmission bandwidth of the wireline circuits.

- It provides a control function to handle all blank-and-burst and dim-and-burst messaging between the MSC and the mobile station.

## 8.11 Summary

CDMA-based cellular systems are a paradigm shift from the traditional approach to cellular systems. Larger bandwidth channels, frequency reuse of 1, use of multipath to improve voice communications, soft handoff, and so on, are some of the more radical concepts used in CDMA which run counter to traditional methods applied to cellular telephony. CDMA as a technology is very attractive to both operators and subscribers. For operators, it offers 10× capacity gain over AMPS and a larger coverage area which means that they need to deploy fewer cell sites to cover a given region. It also allows ease of deployment and subsequent frequency retuning as systems grow due to elimination of frequency planning. Internationally, in countries where a smaller chunk of RF spectrum has been allocated to cellular systems (in some countries as little as 5 MHz), CDMA is more attractive due to its high capacity. Essentially, a single 1.25-MHz channel can be reused in the entire system.

Subscribers have the benefit of low power transmissions, extended battery life with three hours of talk time and two-day standby, higher-quality voice transmissions (14.4/9.6 vocoder rates), reduction of interference and dropped calls, soft handoff (no handoff muting), and inherent privacy and security.

Currently there are commercial deployments of CDMA in the 800-MHz cellular band in Hong Kong, South Korea, the United States, and a few other countries. As more and more countries start deploying commercial CDMA systems around the world, CDMA will gain in popularity and acceptance.

# 9

# Personal Access Communication System (PACS)

## 9.1  Introduction

In some of the previous chapters we have discussed wireless technologies such as AMPS, D-AMPS, GSM/DCS 1800, and satellite services. In spite of several differences among these technologies, they are similar in one respect, they have been designed for ubiquitous nationwide outdoor vehicular mobile traffic. These technologies together can be grouped under what is called a high-tier PCS. There are several other wireless applications which we have not yet addressed, such as cordless telephones, wireless PBXs, and wireless pay phones. These applications are grouped under low-tier PCS. There are fundamental differences between the operating conditions of high-tier and low-tier PCSs that are explained in the following sections.

One way to determine if a high-tier or low-tier PCS technology should be used is by evaluating the region where the PCS service will be used by the subscriber. High-tier PCS can be classified as outdoor vehicular. These systems provide high-mobility, wide-area, two-way wireless voice communications. High mobility refers to vehicular speeds and wide area to nationwide and international coverage. Examples of digital cellular technologies which are used for high-tier PCS and which we have discussed are AMPS, D-AMPS, GSM, and CDMA.

## 9.2  Low-Tier PCS

Low-tier PCS encompasses wireless local loop, indoor residence (cordless telephones), indoor office (wireless PBXs), indoor commercial (for airports, malls, shopping centers, etc.), and outdoor pedestrian (for university campuses, shopping plazas, etc.) applications. Each of these applications has a different requirement and fulfills certain unique subscriber needs.

### 9.2.1   Wireless local loop (WLL)

In wireless local loop systems, the local loop is replaced by a wireless link. A wireless access fixed unit (WAFU) is usually located at a user's residence where it communicates to a PCS radio port via a wireless link. The wireless access fixed unit is connected to the telephone line in the residence which terminates at a typical telephone RJ-11 phone jack. As far as the end user is concerned, he or she uses a wire-line phone which behaves identically to that of a wire-line phone with a wired local loop. Thus, in this low-tier PCS application, there is absolutely no mobility at all. The wireless link is transparent to the user. Since the wireless local loop (WLL) is essentially a replacement for the wired local loop, it has to provide similar functionality to that of the wired loop. Thus, the voice quality has to be comparable to wire-line and the reliability has to be very good. The call-blocking probability should be minimal, and all the services that the user expects from the wired local loop should be supported on the wireless local loop. From a user perspective, the phone usage metrics would be closer to wire-line usage rates rather than cellular usage rates. The service also has to be cost competitive to wire-line local loop rates. Figure 9.1 shows a wireless local loop application. The wireless access fixed unit is permanently attached to the house. It communicates over the wireless link to the radio port and via an RJ-11 connection to the subscriber phone.

### 9.2.2   Indoor residence applications

One of the most common indoor telephony devices providing wireless access is the cordless phone. There are an estimated 60 million cordless telephones in the United States with roughly 15 million new units per year. One of the main reasons for the popularity of cordless telephones is the low cost of the subscriber equipment ($50–$150) and the mobility it provides. The wireless link in cordless phones is at the last stage of the telephone interconnection. The link between the telephone base set and the handset is wireless. Typically, this provides limited mobility within the residence and a few meters around the outside of the residence. The voice quality is not as good as wire-line telephones; eavesdropping is an issue, however, there are more expensive cordless phones which provide some sort of privacy; and battery life is of concern. Each cordless base station supports only one handset and the

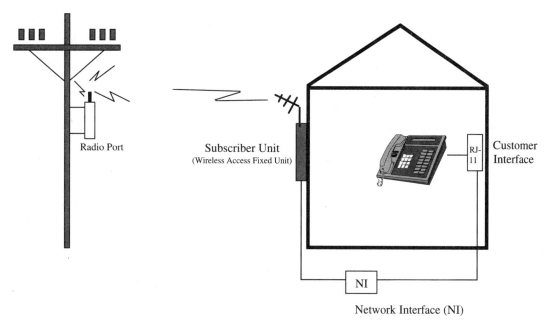

**Figure 9.1** PACS wireless local loop serving arrangement.

cordless base station connects to an RJ-11 wire-line telephone jack. Since cordless phones fall under the category of subscriber equipment, there is no difference from the network perspective if a particular RJ-11 jack is being used for a traditional wire-line phone or a cordless phone. The usage metrics for cordless phones are similar to those of wireline phones.

### 9.2.3   Indoor office applications

There is a need for wireless applications in office environments. Studies show that close to 40 percent of all calls are not answered in an office environment due to the called parties being away from their desks. It is obvious that there is a need for some sort of limited mobility in office environments wherein the user can be away from the desk but still be able to initiate and receive calls via a wireless handset. Wireless PBXs fill this need. Centrex services (PBX-like) provided by the local central office can also be used. Small wireless radio ports through which the handset communicates are installed in the office building along corridors, in conference halls, at meeting places, and so on. These radio ports are connected to a wireless PBX. This wireless PBX alerts both the user's desk phone and the portable handset in an attempt to complete an incoming call. The radio ports are required inside the building because of the difficult propagation environment of large buildings; multistory buildings with metal sheathing or thick, reinforced masonry are typical of this class and it is difficult to provide good RF coverage from radio ports on the outside. Characteristics of this type (indoor office) of application are:

- Two-way voice communications. Radio ports may restrict access to nonassociated users, if desired by the business owner. Voice quality should be comparable to wire-line.

- Most of the current office voice and low-speed data services and features should be supported.

- All business custom calling, messaging, and signaling services have user support.

- The ability to reach other users in the office via abbreviated dialing.

- Enhanced security for privacy, authentication, and prevention of unauthorized use.

- Call validation and billing verification.

- Handoff between radio ports as the user moves around the office building.

- Handoff between the wireless PBX and the macrocellular network as the user leaves the office while still involved in a call.

- Call blocking is acceptable since the user always has the option of using the desk phone.

- Concurrent ringing of the desk' phone and the wireless handset is typically required.

### 9.2.4  Indoor commercial applications

The applications which can be grouped under this category are wireless access in places such as airports, malls, and convention halls. The requirements are similar to those of indoor office with the exception that, since these are public places, the radio ports have to be connected to the public network. They also have to handle public users registering and deregistering from the system.

### 9.2.5  Outdoor pedestrian applications

Wireless service in small apartment complexes, residential neighborhoods, parks, and other recreational areas fall under this category. The radio ports would be mounted on outdoor electric poles, street light poles, and so on. They could, conceivably, also provide service within a residence. Handoff between radio ports is a must, along with good voice quality and low cost.

## 9.3    Characteristics of Low-Tier PCS

We have seen the different applications for low-tier PCS. All of these applications are characterized by the following attributes:

- *Low to medium mobility.*    Applications such as wireless local loop require no mobility at all, whereas indoor commercial, office, and outdoor residential applications require mobility, although not at speeds comparable to vehicular wireless.

- *Handoff requirements range from none to few per call.*    There is no requirement for handoff in the case of wireless local loop or indoor residential (cordless phones) applications, however, handoff is required for all the other low-tier PCS applications.

- *Voice quality comparable to wire-line.*    The primary competition to low-tier PCS is from wire-line telephony. To be acceptable as wire-line replacement to the user, the voice quality for low-tier PCS should be as good as wire-line telephony. This means that the vocoding rate has to be higher (to approximately 32 kbps) and the round-trip transmission delay has to be lower (to approximately <50 ms).

- *Lightweight, low-cost handsets.*    To be portable and to have a mass market the low-tier PCS handsets have to be lightweight and extremely low in cost. Being lightweight, the handset will remove some of the bulky components that are part of a high-tier PCS handset. In order to be low in cost, the RF interface has to be simpler so that the handset does not require expensive signal-processing parts.

- *Long talk time.*    This requirement, coupled with the lightweight requirement, necessitates low-power handsets. Since handsets cannot have bulky batteries, they have to have smaller, lighter batteries which have to last a longer time. This is only possible if the battery power consumption is much lower. This can be accomplished by having a very low average-transmit power (<30 milliwatts). Low transmit power equates to smaller cells.

- *Low-complexity radio signal processing.*    Since most of the low-tier applications have low to medium mobility requirements, there is no need for forward error correction and complex multipath mitigation. Rayleigh fading is frequent and space sensitive. Moving a few feet may improve the radio signal dramatically, when under a deep Rayleigh fade. Due to the fast movement of the vehicle in high-tier PCS, the user travels through several deep Rayleigh fades in a small amount of time. In this scenario, forward error correction and bit interleaving are very effective means in combating Rayleigh fading. But in a low-tier PCS environment the radio channel can be modeled almost as a quasi-static channel, due to slow movement of the user.

In such an environment, Rayleigh fading produces a multichannel that is either very good (where the FEC is not needed), or bad (where the FEC is needed). Because of the slow fading nature of the channel, bit interleaving to randomize error results in unacceptable delay. Error-detection schemes can be incorporated with fewer redundancy bits than error-correction schemes, thus requiring less bandwidth. Antenna diversity techniques can be used both at the radio port and at the handset to combat Rayleigh fading.

- *Adaptive equalization not required.*   Equalization is not necessary, since the cell size and delay spread are small.

- *Very high capacity.*   Due to the nature of some of the low-tier PCS applications which exist in high-user-density areas, such as offices, airports, malls, and apartment buildings, the low-tier PCS system has to support higher capacity. Also, the call-usage metrics, where talk time approaches that of wire-line phones for low-tier PCS, are different from those of high-tier PCS, which results in a need for higher capacity.

## 9.4   Characteristics of High-Tier PCS

Previous chapters discussed the technologies which are representative of high-tier PCS systems such as AMPS, D-AMPS, GSM, and CDMA. Although most of these technologies are different, they do share some common characteristics, as follows:

- *System capacity increases by increasing capacity of cell site.*   Due to the high cost of cell sites, the design considerations were to maximize users per MHz and to maximize the users per cell site. This expensive radio base station includes real estate, a very high steel tower with a strong support foundation, a temperature-controlled building, and backup batteries. Due to these high expenses associated with the cell site, high-tier PCS systems increase capacity initially by trying to increase the number of channels at each base station. These schemes are suitable for large cells but not for the smaller low-tier PCS cell sizes. Capacity of low-tier PCS is ideally increased by having a larger number of radio ports. This makes a case for low-complexity, low-cost radio ports.

- *High transmitter power consumption.*   Due to the need to cover highways running through low-population-density regions between cities, a high transmitter power requirement is necessary to provide maximum range from high antenna locations. Power of the system can be scaled down to accommodate low-tier PCS systems; however, the system was initially designed to operate at higher power and there is associated overhead, which adds cost and complexity to the radio ports and the subscriber units.

- *Low bit-rate speech coding.*  Typical coding rates vary from ≤14 kbps. Low bit-rate speech coding increases the numbers of users per MHz and per cell site. However, it also significantly reduces speech quality.

- *Inactive speech transmission.*  This increases the number of users per cell site. However, it reduces the speech quality because of the difficulty of detecting the onset of speech. The devices which take advantage of the voice activity factor to increase system capacity have a tendency to clip the beginning portion of a speech because of the nonzero delay needed for the device to detect the beginning of speech. Also, the low-tier PCS application users would be located in an airport terminal, a shopping mall, a train station, or in an office where there are plenty of background acoustic noises from public announcement systems, jet engines, passing trains, or other motorized vehicle sources. Since the speech detector cannot distinguish a user's voice from the background acoustic noises, these background noises will reduce the silent time periods of the communication channel.

- *High-complexity signal processing.*  Highly complex signal processing is required for speech encoding and for demodulation.

- *Adaptive equalization required.*  Due to the large cell sizes there will be larger delay spreads. Therefore, it is necessary to overcome the effects of delay spread such as intersymbol interference adaptive equalization.

- *FEC and bit interleaving to combat fading.*  High-tier PCS systems require these countermeasures against Rayleigh fading due to vehicle's high-moving speed. As already explained these countermeasures are not effective in a low-tier PCS RF environment.

- *Capacity is increased via sectorization.*  High-tier PCS systems typically increase capacity via sectorization. However, this scheme cannot be used for low-tier PCS because low height of radio ports and antennae surrounded by highly reflective urban environment will severely distort the signal and degrade the effectiveness of sectorization. Sectorization is possible for high-tier PCS because tall towers raise the antennae above the surrounding clutter, thus making sectorization effective.

- *High transmission delay.*  Typical round-trip transmission delays are about 200 ms. This delay in high-tier PCS technology results from both computation for speech bit-rate reduction and from complex signal processing, for example, bit interleaving, error-correction decoding and multipath mitigation associated with equalization or spread spectrum (CDMA). This high delay results in use of echo cancellors which further deteriorate speech quality.

The design objectives for high-tier PCS systems are evident and very different from those of low-tier PCS systems. Though high-tier PCS systems can be used for low-tier PCS applications, they will be more expensive, of poorer voice quality and more complex than solutions designed specifically for low-tier PCS applications. Figure 9.2 lists the main characteristics of the high- and low-tier PCS systems.

## 9.5   Cordless Telecommunications

Cordless phones provide limited mobility and are of low power. The market structure for cordless phones is very much different from that of high-tier PCS. Cordless telecommunication systems operate in an unregulated, open-market environment where system installation and

| | High-tier PCS (high mobility) | Low-tier PCS (low–moderate mobility) |
|---|---|---|
| Principal "designed for" application | Outdoor vehicular | Wireless local loop, indoor residential, office, commercial, and outdoor pedestrian |
| Base stations | Large and expensive | Small and inexpensive |
| Cell sizes | Up to several miles | <0.9 mile radius |
| Coverage | Nationwide | Zonal |
| Handset complexity | High | Low |
| Base-station complexity | High | Low |
| Antenna elevation | High (15 m or more) | Low (15 m or less) |
| Capacity | Low to medium | Very high |
| Principal means of increasing capacity | Sectorization, speech activity detection, power control | Install more base stations |
| Portable transmit power (avg.) | 125 mW | 25 mW |
| Talk time for portables | Short (~2 hours) | Long (>4 hours) |
| Vehicular service | >65 MPH | ~40 MPH |
| Voice quality | < wire-line | ~ wire-line |
| Expected usage | Low | High |
| Forward error correction and bit interleaving | Yes | No, uses CRC only |
| Adaptive equalization | Yes | No |
| Speech activity detection | Yes | No |
| Speech coding | <14 kbps | 32-kbps ADPCM |

**Figure 9.2**   Characteristics of high-tier and low-tier PCS.

frequency planning cannot be coordinated or planned and the driving design consideration is end-user cost and performance. Typically, these systems only specify the air interface and have to be connected to a local wire-line network or a private switching system such as PBX to be able to terminate calls.

### 9.5.1  Analog cordless systems

Analog cordless telephones are widely used in residential applications where the telephone cord is replaced by a wireless link to provide terminal mobility to the user. Analog cordless telephones in the United States have been in operation for quite some time. Initially, the frequency allocation for analog cordless was ten frequency pairs, five of them in the 49-MHz frequencies. They were paired with five frequencies near the 1.6 MHz in a frequency-division duplex (FDD) mode. This frequency scheme was not very effective due to wide separation between the uplink and downlink. This arrangement was changed in 1984 when the ten frequency pairs were reassigned to different frequency bands. Downlink transmission pairs were in the bands 46.6–47.0 MHz and uplink was in 49.6–50.0 MHz. The transmission bandwidth is 20 kHz and the effective radiated power (ERP) is 20 µW. Frequency modulation (FM) is used for the voice signal and the Federal Communications Commission (FCC) requires digital coding of the signaling for security. There are an estimated 60 million 46/49-MHz cordless telephones in use in the United States, and total sales are about 15 million units per year. There has been industry action to attempt to have more frequency pairs assigned due to congestion and deterioration of voice quality mostly due to the popularity of these phones. Large numbers of these phones are in operation in a limited space such as apartment complexes or high-rise buildings. This results in poor voice quality due to interference. In the United Kingdom, a standard similar to the one used in the United States was introduced. This standard allowed eight channel pairs near 1.7 MHz and 47.5 MHz and most cordless units were limited to accessing only one or two of these channels.

In Europe, an analog cordless standard known as CEPT/CT1 was introduced. This cordless standard provides for forty 25-kHz duplex channel pairs in the bands 914–915/959–960 MHz. The cordless unit is allowed to select any of the available 40 pairs for a particular call. The large number of channels results in better voice quality and low-blocking probability even in densely populated areas. Total sales of CT1 equipment, including its modified versions, are estimated at 2.2 million units and have increased to 2.7 million in 1996.

Japan has allocated 89 duplex channels near 254-MHz uplink and 380-MHz downlink for analog cordless telephony. The channel spacing is 12.5 kHz and the maximum transmit power is 10 mW. It differs from the CEPT/CT1 standard in that there are two dedicated control channels which are used to facilitate fast connections and conserve battery life. There is an installed base of about 20 million units with annual sales of 3 to 4 million units annually.

### 9.5.2   Digital cordless systems

While analog cordless systems' primary application was indoor residential, the intended applications for digital cordless systems are indoor residential, office, commercial, and outdoor pedestrian applications. Digital cordless systems will provide terminal mobility in these applications wherein users will be able to originate or terminate calls, while mobile in the coverage area, at pedestrian speeds. Some popular digital cordless telephone technologies are CT2, CAI, and DECT.

#### 9.5.2.1   CT2 common air interface (CT2 CAI).

CT2 is a second-generation cordless telecommunication system. The common air interface is an open air interface standard which provides interoperability between the base station and the handset. The CT2 standard was originated in the United Kingdom and was adopted by ETSI in 1992 as an interim European Standard for Cordless Telecommunications (I-ETS 300 131). The CT2 spectrum allocation consists of 40 FDMA channels with 100-kHz spacing in the range 864–868 MHz. The 100-kHz spacing reduces the bandwidth efficiency to half its value for CT1. The maximum transmit power is 10 mW with two-level power control which assists the base station in ensuring that the power levels of all handsets are within a specified range. This method contributes to frequency reuse. A call reestablishment procedure on another carrier after three seconds of handshake failure provides robustness against the degradation of the radio channels. The control bits are protected against errors. CT2 uses time-division duplexing (TDD) with a 32-kbps ADPCM vocoder. The speech and control data are modulated onto a carrier at a rate of 72 kbps and are transmitted in 2-ms frames. Each frame includes one base-to-handset and one handset-to-base burst. CT2 also supports the transmission of data up to 2.4 kbps through the speech codec and higher bit rates by accessing the 32-kbps channel directly.

The first commercial application of CT2 technology was the telepoint service introduced in United Kingdom in 1988. Telepoint service was a cordless pay phone kind of application which used only a few CT2 features. Base stations were located in places where people congregate, such as along city streets, shopping malls, train stations, and airports. Handsets registered with the telepoint provider could place calls only when within range of a telepoint radio port. Initially, only call originations were supported. Call terminations required manual registration and handoff was not supported. The user was limited to the currently used base-station range. CT2's success has been mixed. Telepoint was rolled out twice in the United Kingdom, but failed to attract enough subscribers to be commercially viable. However, it is very popular in Hong Kong and Singapore, with 200,000 subscribers in Hong Kong in 1995. CT2 systems are being deployed in other parts of Asia, such as Malaysia, Thailand, and China. There are about 1 million CT2 handsets in use in Europe. France Telecom has deployed CT2 widely and has about 100,000 subscribers.

A Canadian variant of CT2 called CT+ is being deployed in Canada. This variant provides some of the functionality which was missing in CT2, such as the mobility-management functions. Five of the 40 carriers are reserved for signaling. Each carrier provides 12 common signaling channels using TDMA. These channels support location registration, location updating, and paging, which enables the users of this service to receive a call. CT+ operates in the 944–948-MHz frequency band.

### 9.5.2.2 Digital European cordless telecommunications (DECT).

In 1992, the DECT, which is a digital cordless standard introduced by CEPT, was standardized. The primary applications for this technology were indoor office, commercial, outdoor pedestrian, wireless local loop, and wireless data (LAN) applications. It supports multiple bearer channels for speech and data transmission, handover, location registration, and paging. The DECT interface to the PSTN remains similar to that of a PBX of a wire-line phone. The frequency spectrum for DECT is the 1880–1900-MHz band. Ten carriers are allocated in this band. DECT uses TDMA and TDD with 12 slots per carrier in each direction. Refer to the next section for explanation on TDD.

A handset has access to all slots on all carriers. When a channel is needed, the least-interfered one is selected by the mobile terminal. Compared with CT2 phones, DECT doubles the transmission range and permits handoff between base stations. DECT base stations can allocate several time slots to a single user, if needed. This allows the subscribers to use higher bandwidth for certain applications such as data. The allocation of additional bandwidth to the user need not be symmetric. During retrieval of e-mail or database, most of the data is being sent only on the downlink direction with very little data being sent on the uplink. In such situations, it makes no sense to allocate equal bandwidth to both uplink and downlink channels. DECT allows unequal (asymmetric) bandwidth allocations to the uplink and downlink channels for such applications. The speech codec is a 32-kbps ADPCM vocoder with the transmission rate of 1152 kbps. Due to this high transmission rate adaptive equalization or antenna, diversity is required for certain applications. DECT is designed for low-cost, flexible operation in an uncoordinated environment. This means that the base stations need not be time synchronized. Synchronization is achieved by several schemes including additional fields for synchronization and parity. The base-station information such as base-station capabilities, paging, and system information is multiplexed onto the control channel of each active transmission.

Mobility and subscriber-location management in DECT networks are typically limited to the area serviced by the PBX. Within this coverage area, the system is able to keep track of the location of subscribers and page them for incoming calls. In a university campus type of environment, the coverage area could be the entire university that is served by a PBX. Within this area, handoff between base stations is possible. For interworking with the PSTN, DECT supports ISDN. ISDN terminations can occur at the DECT base station or in the handset. DECT systems are being shipped in Europe. They are used primarily for indoor applications. There were about 250,000 DECT sales in 1995.

## 9.6  Frequency-Division Duplex (FDD) versus Time-Division Duplex (TDD)

The wireless technologies discussed have been FDD (frequency-division duplex) systems. Essentially, an FDD system is one in which duplex (two-way) communication is achieved by using two simplex (one-way) communication channels. One channel is used for communication from the mobile station to the base station (uplink/reverse link) and another used for communication from the base station to the mobile station (downlink/forward link). These two channels are separated from each other in *frequency*. Most of the wide-area wireless technologies such as AMPS, D-AMPS, GSM, and CDMA are all frequency-division duplex systems. For instance, the frequency assignment for a U.S. cellular system is 824–849 MHz for the uplink simplex channel band and 869–894 MHz for the downlink simplex channels. The uplink and downlink channels are separated by 45 MHz. For each uplink channel, there is a corresponding downlink channel which is separated by 45 MHz from the uplink channel. These uplink and downlink channels are always allocated together for transmission of speech, and are used by the mobile and base station to achieve duplex communication.

Time-division duplex (TDD) systems, also known as ping-pong or time-compression multiplexing systems, use a different scheme to achieve duplex communications. Duplex communications is achieved by using a single transmission channel by alternating the direction of transmission. TDD systems are by definition burst systems and are not continuous transmission systems such as AMPS and CDMA. Communication between the base station and the mobile station occurs by information transmitted within time slots. Transmission on a physical channel is divided into individual time slots. Some of the time slots are used by the mobile station to communicate to the base station with the remainder used by the base station to communicate to the mobile station. The uplink and downlink channels are separated by *time,* not frequency as in FDD systems. To ensure that there is minimal delay, the transmission bit rate has to be at least 2 times faster than the coder rate. It is typically 2.25 to 2.5 times faster. More than twice the bit rate is necessary to compensate for the delay in switching from transmission mode to receiving mode and overhead associated with time slots. A TDD system can be described as a quasi-simultaneous, bidirectional flow, since one direction must be off while the other is using the frequency. A transmission rate in a digital speech system ensures that this off time is not noticeable during conversations. Figure 9.3 illustrates the difference in frequency use between FDD and TDD systems.

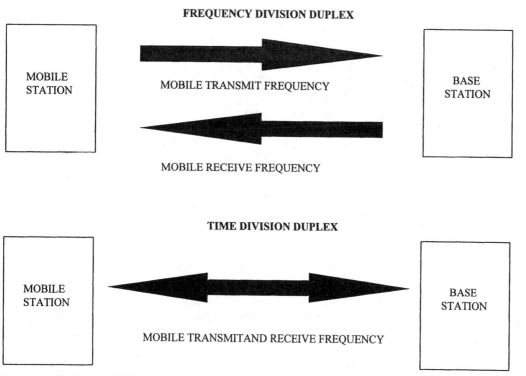

**Figure 9.3** FDD versus TDD.

### 9.6.1    Spectrum issues

Both FDD and TDD require the same amount of spectrum to serve a given area. In FDD, two bands of spectrum separated by a minimum guard band are required; whereas in a TDD system, only one band of frequencies is required. The advantage of TDD is that the allocation process for a wireless service is much simpler, since only a single band of frequencies is needed for a TDD system which is easier to find. Comparatively, two bands of frequencies are needed in a FDD system, separated by the required bandwidth, making it more difficult to find spectrum. Typically, the amount of unassigned spectrum in most countries is small and is generally not in paired groups. Thus, services requiring small amounts of spectrum, such as indoor commercial, are ideal for a TDD system. TDD systems can also be used in spectrum that is unlicensed (i.e., not regulated and does not have strict government controls on its use). Typically, frequency allocation for a wide-area service will come from an already existing service using FDD systems, thus mitigating the advantage of TDD systems for wide-area applications for high-tier wireless.

### 9.6.2    Radio design issues

The transmission rate on the radio channel plays an important role for a TDD system. Since a TDD system transmits at more than twice the rate of a comparable FDD system, it is much more constrained by the time delay spread. In a digital system, the time delay spread causes intersymbol interference. Higher transmission rates result in smaller bit and symbol duration making them more prone to intersymbol interference. If the design of a radio system calls for a radio channel symbol rate that is high enough to be limited by the time delay spread, then an FDD system can accommodate twice the number of users per transceiver than a TDD system. For instance, if the maximum possible transmission rate is 1 Mbps, and each user's information-generation rate is 100 kbps, then an FDD transceiver can support 10 users at 1 Mbps in the uplink and 1 Mbps in the downlink. A TDD transceiver can support only 5 users assuming negligible overhead, such as 1 Mbps transmission rate for uplink-only for ½ sec resulting in 500 kbps for uplink and 1 Mbps transmission rate for downlink-only for ½ sec resulting in 500 kbps for downlink. To support the same number of users as an FDD transceiver, two TDD transceivers will be required. Note that the amount of radio spectrum required for both systems is the same, only the number of transceivers required changes. If the radio system design is not limited by the time delay spread, then both FDD and TDD systems support the same number of users per transceiver. In the above example, the FDD transceiver would transmit and receive at 1 Mbps, supporting 10 users, and a TDD transceiver would transmit and receive at 2 Mbps, supporting 10 users.

A typical countermeasure against multipath is space diversity. This usually takes the form of two antennae at the base station and the mobile station in an FDD system. It is difficult for portables to incorporate two antennae in the handset and there are issues with customer acceptance. Since FDD systems use two different frequencies for transmit and receive which are not correlated, the fading characteristics will be different for uplink and downlink. The base station cannot accurately estimate the quality of the downlink based on the quality of the uplink. This makes diversity schemes more complex at the mobile station and the base station. Since the transmit and receive frequencies in a TDD system are the same, the base station can accurately estimate the quality of the downlink, based on uplink quality measurements, which makes it easier for the base station. Space diversity can be used only at the base station to improve the quality of the link without its need at the mobile station.

TDD systems are similar to FDD TDMA systems in the respect that they do not need a duplexer. Other FDD systems, such as AMPS and CDMA, will require a duplexer, since they are continuously receiving and transmitting.

TDD systems also require systemwide time synchronization of the base stations. The requirements for timing control are more stringent in TDD systems due to the use of the same link for transmit and receive and the effect of poor synchronization on overall system capacity.

## 9.7   Personal Access Communication System (PACS)

PACS is the United States standard for low-tier PCS. This standard is specified by the joint technical committee (JTC) of the United States T1 and TIA standards bodies. The PACS standard was formed by combining the Bellcore Wireless Access Communications Systems (WACS) and the Japanese Personal Handyphone System (PHS), which were both submitted to the committee as low-tier PCS systems. The PACS proposal was standardized in 1995. The key design requirements of the PACS standard were:

- Features comparable to wire-line access
- Quality, reliability, security, services and features, cost effectiveness, and the ability for rapid deployment
- Low-system deployment, subscriber-equipment costs
- Operating environment

- Small radio ports, low antennae, low power
- Moderate mobility support, and neighborhood roaming
- Wireless local loop support
- Evolution of services
- Wireless PBX and Centrex, expanded-data capabilities, and multi-media services

### 9.7.1   PACS system architecture

Figure 9.4 shows the different elements of the PACS system architecture. The A interface is between the different subscriber units supported by PACS and the RP (radio port). The subscriber units supported by PACS could be a portable handset for mobile voice applications, a data modem for data-only applications, or a wireless access fixed unit (WAFU) attached to a residence for wireless local loop applications. The radio port could be mounted outdoors on a utility pole for outdoor and WLL applications, or it could be wall mounted indoors for indoor office and commercial applications. The radio ports are connected by the P interface to the radio port control unit (RPCU). There are two kinds of RPCUs supported by PACS, depending on whether they incorporate the access-manager functions. The radio port control unit is responsible for controlling the radio ports, radio channel framing and formatting, RF spectrum management, multiplexing and demultiplexing, power control, and maintenance. The access manager is used for location management, paging, subscriber-profile management, and routing functions. The RPCU can be connected to a local central-office switch for public applications such as WLL, outdoor pedestrian, and Centrex applications.

PACS can interface to a PBX for private applications. The interface between the RPCU and the CO/PBX is the generic C interface if the RPCU does incorporate the AM functionality and is the switch C interface if the AM is physically a separate entity from the RPCU. The RPCU is, in that case, connected to the AM via a C interface specific to the AM. The AM is connected to a HLR/VLR via the SS7 network. The CO/PBX is the generic C interface.

### 9.7.2   PACS air interface

**9.7.2.1   Frequency assignment.**   The PACS air interface can operate in the entire PCS band (i.e., 1850–1910-MHz uplink and 1930–1990-MHz downlink). Channel bandwidth is 288 kHz with center frequencies spaced 300 kHz apart. The duplexing technique is frequency-division duplex (FDD) with 80 MHz separation between the uplink and downlink.

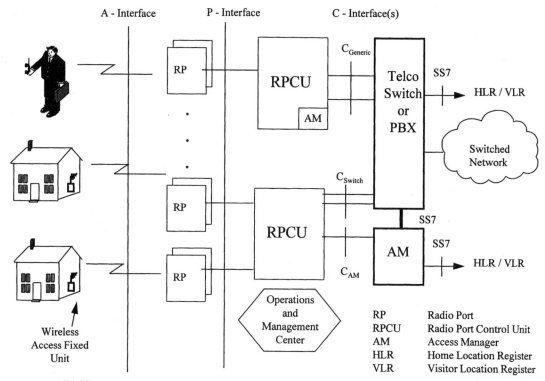

**Figure 9.4**  PACS system architecture.

**9.7.2.2  Access mechanism.**  The access mechanism is a unique time-division multiplex time-division multiple access (TDM) scheme. Time-division multiplex is a scheme where transmission occurs continuously, whereas transmission occurs only during active slots in TDMA. In TDM, transmission occurs continuously on all slots. If the time slot is idle, then the radio port inputs a unique filler in the time slot, which indicates that the slot is idle. In a TDMA scheme, the transmissions occur only when the unit has information to send. Nothing is transmitted on idle time slots. The TDM scheme is used for the downlink, and the TDMA for the uplink. Figure 9.5 shows the differences between TDM and TDMA downlink.

The PACS channel is divided into eight time slots of 120 bits each in a 2.5-ms frame. There is a time offset between the uplink and downlink bursts. The uplink burst is offset from the paired downlink burst by a one-burst interval plus 62.5 µsec. The bit rate over the air interface is 384 kbps at a symbol rate of 192 kbps. Thus, the modulation technique results in 2 bits per symbol. Both the radio port and the subscriber unit

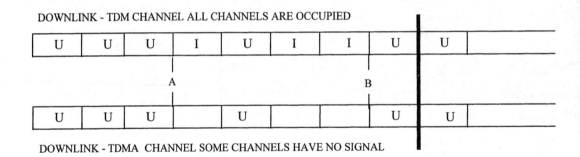

U = time slot in use
I = time slot is idle, idle pattern transmitted
Blank = no signal present

**Figure 9.5**    Slot occupancy example for an FDD channel in the TDM and TDMA downlink cases.

can have two antennae for receive diversity. Figure 9.6 shows the PACS frame structure for downlink and uplink channels.

**9.7.2.3  TDM versus TDMA.**  There are several advantages of using a TDM approach for the downlink in a wireless system, as is done in PACS. Some of the advantages of TDM, when it is used for the downlink as opposed to use of TDMA for downlink, are discussed in this section.

**Channel measurement process and timing.**  In order to determine the best frequency on which to communicate during call setup, and the best radio port for handoff—called automatic link transfer (ALT) in PACS, the subscriber unit must measure the frequency channels for signal strength and quality. The speed at which this can be done impacts the call-setup times, ongoing alternate channel quality measurements, and the ALT.

Figure 9.5 illustrates the difference in speed for TDM/TDMA channels, assuming that a subscriber unit begins to measure a given channel at time *A*, and that the channel-measurement process can be completed in one time slot. In the case of a TDM channel, the channel-measurement process takes just one time slot, since a signal is always present. The subscriber unit tunes to the downlink channel and measures the strength of the time slot. If receive diversity is used, then the subscriber unit measures one time slot using one antenna, and measures the next time slot using the other antenna. This process takes two time slots (assuming that the switching between antennae can take place instantaneously). If the channel is a TDMA channel, then

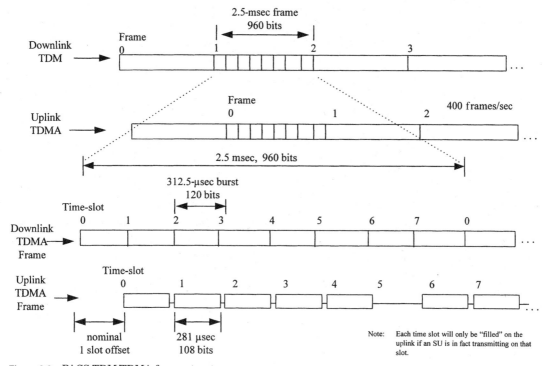

**Figure 9.6** PACS TDM/TDMA frame structure.

two time slots are required to take a single measurement, since there is no signal present at time *A*. If antenna diversity is needed then the subscriber unit has to wait until time *B* (i.e., three time slots later for a signal to appear). Thus, the total time required is six time slots compared to two time slots required for the TDM scheme. In the TDMA scheme, the time will vary according to the loading of the channel. If there are less channels in use, then it will take longer for the measurements to take place. If a lot of channels are in use, then the subscriber unit takes shorter time to make the measurements. It can be seen that a fully loaded TDMA channel is the same as a TDM channel. Thus, the best-case performance for a TDMA channel would be as good as a TDM channel resulting in the following:

- TDMA provides faster cell acquisition, call-setup and ALT process.

- In TDMA, it is quite possible that the subscriber unit may not be able to take measurements for a particular channel while involved in a call. This occurs when the subscriber unit has to receive during a particular slot and the only busy time slot on a different downlink TDMA channel happens to be at the same time.

- The TDMA approach is more complex. It requires systemwide port-to-port synchronization. Since all time slots in a TDM channel have active signals, synchronization is relatively easier.

- Cochannel interference is reduced in TDMA compared to TDM because of existence of time slots without any signal, although this reduction is only for as long as the system is lightly loaded. A fully loaded TDMA channel will be comparable to a TDM channel.

Overall, the TDM downlink channel has several advantages over the TDMA downlink channel. The advantages of a TDM/TDMA plan over FDMA are as follows:

- Channel combining at the radio port is simplified.

- Duplexer elimination at the subscriber unit. (A duplexer is not needed since the subscriber unit does not have to simultaneously receive and transmit due to time offset of receive and transmit time slots.)

- Relatively relaxed tuning requirements associated with TDM/TDMA.

- Fewer channels to synthesize and faster tuning with TDM/TDMA.

- Fewer transceivers to equip and maintain with TDM/TDMA.

- Increased versatility where the burst character of TDMA frees a subscriber unit to perform control functions, such as diversity control and quality assessment between communication bursts.

**9.7.2.4  Power requirements.**  In the downlink, the channel is a TDM channel with a maximum of 800-mW RF transmit power. In the uplink direction, the channel is a TDMA channel with a maximum of 200-mW RF transmit power per burst.

**9.7.2.5  Frame structure.**  Figure 9.7 shows the frame structure for the PACS air interface. The basic frame structure is of only 2.5 ms. It has the advantages of significantly reducing the transmission delay of speech when coded using low-delay vocoders such as a 32-kbps ADPCM vocoder. This allows low-delay speech vocoders in a 2-wire loop environment without requiring echo cancellers. The modulation scheme used in PACS is ¼ differentially encoded, quadrature phase shift keying (¼ DQPSK). This results in 2 bits per symbol.

Both the uplink and the downlink frames have eight time slots out of which seven can be independently assigned. One of the eight time slots is dedicated for alerting, broadcast information, and system information in the downlink direction. In the uplink, this time slot can be used for priority access, such as emergency 911 calls and call maintenance. The basic structure of an individual time slot is as follows:

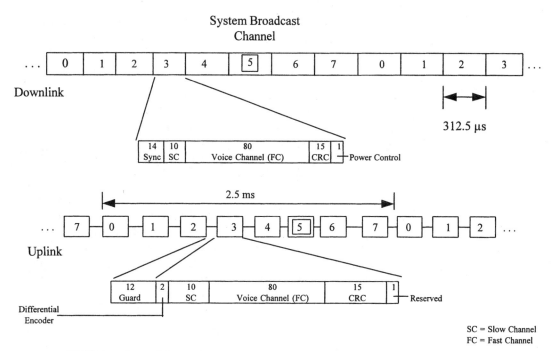

**Figure 9.7**   PACS time slot and frame structure.

- There is one fast channel (FC) in each time slot, which in both directions is used primarily for speech or data traffic. It consists of 80 bits. Since the frame rate is 400 frames/sec, this translates to 32 kbps. Typically, a 32-kbps ADPCM vocoder is used for speech encoding.

- A slow channel (SC) also exists in each uplink and downlink time slot. This channel can be used for signaling, maintenance, and measurement report. The bandwidth is 10 bits, which equates to 4000 bps.

- There are 15 bits in each slot which are error detecting coding applied to the fast- and slow-channel bits.

- In downlink-only there are 14 bits used for subscriber unit frame synchronization and 1 bit for power control which is not protected. This power control bit is used by the radio port to order the subscriber unit either to increase or decrease its transmit power.

- In uplink-only there are 12 bits for a guard time to allow the subscriber unit to turn on or off, 2 bits for differential encoding, and 1 bit which is reserved.

This totals to 120 bits. PACS also supports a superframe structure which can be used to support lower-bit-rate users. A 16-kbps user can be supported in PACS by having the subscriber unit transmit only during every alternate frame. Similarly, 8-kbps users and so on can be supported. PACS can also support more than 32 kbps per user by assigning multiple time slots in the same frame to the same user.

**9.7.2.6   Channel structure.**   PACS supports two types of channels: traffic channels and system broadcast channels.

**Traffic channels.**   There are two types of traffic channels supported in PACS, busy traffic channel and idle traffic channel. The busy traffic channel will carry user payload, either speech or data; the idle traffic channel carries a fixed signal and an idle channel indicator. The traffic channels can be assigned on a per-user basis whenever requested.

**System broadcast channel (SBC).**   The SBC is a dedicated channel occupying one time slot in the downlink and uplink directions. This time slot is permanently assigned for the use of the SBC. It is made up of three channels: system information channel (SIC), alerting channel (AC), and priority request channel (PRC). The system information channel is in the downlink direction and carries system-information messages used by the subscriber unit for cell selection and access. The alerting channel is also in the downlink direction and carries alert messages targeted toward the subscriber units. The priority request channel is in the uplink direction and is used for priority link-access request such as emergency 911 services in the uplink and responses in the downlink. The system broadcast channel is the same for all radio ports in a registration area. The SBC has capacity for paging/alerting support of 200,000 subscribers in a registration area with zero blocking probability. The system-information capacity at busy hour is 6.6 kbps.

PACS support discontinuous reception (DRX). This scheme allows a subscriber unit to listen for pages and alerts only during certain phases, enabling it to power down during the rest of the time, thus conserving battery power. When a subscriber unit registers with the network, the network assigns the subscriber unit a phase number and an alert value. The network and the subscriber unit remember these numbers as long as the subscriber unit is registered within the alerting area. Since the phase number and the alert value are negotiated in the encrypted mode during registration, the location identity of the user is kept private. The subscriber unit retains the phase number, even when the serving RP is changed, as long as the alerting/registration area remains the same. However, if the subscriber unit reregisters in the same alerting area, a different phase number and alert value may be assigned. The subscriber unit is assumed to have acquired sync with the serving RP. This means that the subscriber unit knows precisely when to "wake up" periodically in the alerting process. When the controller receives an incoming call for a user, it prompts all RPs in that alerting area to transmit the user's alert identification at its designated phase. During its allocated alerting phase, the subscriber unit wakes up and listens for any pages directed to it. It goes back to the sleeping mode if it determines that the alerting signal is not intended for it, or it takes appropriate action to receive the call if an alerting signal is intended for the subscriber unit.

**9.7.2.7 Security.** PACS supports both encryption and authentication as security procedures. It supports both the secret key methods discussed as part of IS-41 and public key schemes.

### 9.7.3 Port frequency assignment

RF frequencies can be assigned manually or through quasi-static autonomous frequency assignment (QSAFA). An autonomous frequency assignment technique is required for PACS due to large number of radio ports that will be needed to cover a given area. Moreover, the frequency assignments at the RPs may change as the environment changes, such as expansion of buildings and construction of new neighborhoods. Due to the high number of RPs and the need for frequent frequency changes at RPs, an autonomous frequency assignment procedure is useful.

**9.7.3.1 QSAFA method.** Channel planning is a concept conceived for an FDMA system. As the new generation of digital radios using TDMA schemes is introduced, channel planning implies the assignment of both frequency channels and time slots. In PACS, a service area is divided into many coverage cells with each being served by an RP. Service providers will decide the locations of RPs based on terrain profile, traffic, real estate, and so on. Every RP provides a fixed number of time slots in a given frequency channel for wireless connections.

Figure 9.8 shows the QSAFA algorithm. By using QSAFA, the system first runs an autonomous frequency assignment algorithm to assign the frequency channel(s) to each RP. Each RP receiver scans all the transmitter frequency channels and measures the received signal power for the frequency channel with the lowest-received power assigned for downlink transmission for that RP. In the case of multiple (N) frequency channels at a given RP, the N frequencies with the lowest-received power will be assigned to the transmitters. The measurement is performed by each RP independently and synchronously until the frequency reuse pattern remains the same as in the last iteration, or until a specified number of iterations is reached. Finally, the corresponding uplink frequency channel is assigned to that RP according to a duplex frequency channel pairing scheme.

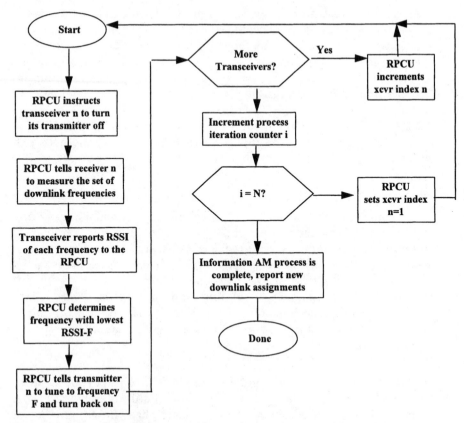

**Figure 9.8**    Automatic port frequency assignment.

When an RP is measuring the power, it turns its transmitter off. Hence, it is more convenient to do the interference measurement during the low-traffic hours to minimize service disruption. The same algorithm will be used to update channel assignments. This is necessary when, for example, a new RP is introduced into the system, a large structure is built, or the interior of a large building is changed substantially.

To initiate a phone call, a subscriber terminal first scans all the frequency channels to determine the optimal frequency channel and the corresponding RP. Then, the RP selects an idle time slot with the least interference to serve the requesting subscriber terminal. If all the time slots of the optimal RP are busy, another RP will be chosen.

### 9.7.4    Data transport capabilities

PACS supports four categories of data transport services. Individual message service (IMS), alternate speech and data, circuit-mode data service, and packet-mode data service. The capacity is greater than 28 kbps, and higher capacity is available by aggregating multiple 32-kbps channels. Data transport services have privacy and are transparent to the RPs. IMS provides private, reliable two-way messaging service and supports variable-length messages up to 16 Mbytes. Messages can be tagged as text, image, audio, video, or application. These messages can be sent as a call in the fast channel or during a voice call in the slow channel. Alternate speech and data allows a single call to have two modes. Signaling to the switch can be defined between the two modes. The circuit-mode data service provides a reliable real-time data transport service using link access procedures for radio (LAPR) protocol. It runs at full rate: ~29 kbps @ $10^{-6}$ error rate in a full-rate (32 kbps) channel. The LAPR protocol has automatic repeat request and is a modified V.42, optimized for the radio environment. The circuit-mode data service supports higher-layer circuit or packet protocols, such as data and fax modem, X.25, and TCP/IP. The packet-mode data service supports asynchronous shared access and provides variable-rate asymmetrical bandwidth which is useful for database retrievals, downloading files, and e-mail. It supports subscriber units using single slot (32 kbps) or multiple slots (256 kbps max).

### 9.7.5 Access procedure

The SU must attempt access only to a traffic channel that is explicitly marked available by virtue of the SYC and SC encoding. The RPCU ignores access attempts on all traffic channels except those explicitly marked available. Upon achieving synchronization with the marked-available traffic channel, the SU reads the contents of the FC in the next burst in the channel. After confirming that the contents of the SYC and SC indicates that this channel is available and after reading the complete port ID from the FC, the SU transmits an initial-access message to the RPCU. However, if the burst is received in error, the SU assumes that the characteristics of this traffic channel are poor. If the SYC and SC do not indicate this is the marked-available traffic channel, the SU assumes that the frame synchronization process has failed. In either case, the SU attempts frame synchronization again.

Upon receipt of an INITIAL_ACCESS request, the RPCU must either grant access with an ACCESS_CONFIRM message or deny access with an ACCESS_DENY message within 20 msec. If there is another unused time slot on the port, the RPCU should mark a new traffic channel as available within 20 msec of sending its ACCESS_CONFIRM message. Figure 9.9 describes the access process.

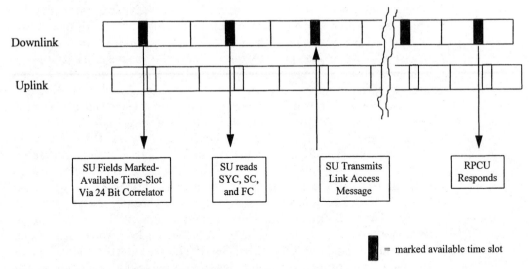

**Figure 9.9**   Access procedure for initial access and ALT requests.

### 9.7.6   Automatic link transfer (ALT) and Time slot transfer (TST)

ALT is analogous to handoff in high-tier PCS systems. One major difference is that in PACS the handoff is directed by the subscriber unit instead of the network. ALT and TST are two procedures used to maintain call quality as link quality changes due to changes in the interference environment. ALT allows a call in progress to be routed through different RPs. TST allows a call in progress to be transferred from one time slot to another on the same carrier frequency in order to mitigate changes in the uplink interference environment which is caused by arrival or departure of cochannel calls on the TDMA uplink.

In order to avoid the substantial network overhead which would be required for the radio access system to monitor the quality of every call in progress, the subscriber unit supports an ongoing measurement process and initiates ALT and TST wherever appropriate. Because of the use of TDMA for the uplink, the terminal will have time in which it is not actively transmitting or receiving information associated with the current call. During this period, the subscriber unit can make quality measurements of candidate radio channels. The subscriber unit compares quality information associated with the current channel with associate candidate channels in order to make decisions about ALT and TST. Decisions about TST are made based on quality information about the current uplink. Figure 9.10 describes subscriber-unit behavior. This information is not directly available to the subscriber unit. In order to allow for combined control of ALT and TST in the subscriber unit, PACS provides information about uplink quality to be given to subscriber unit. The subscriber unit makes use of uplink quality metrics fed back to it by the RP. The subscriber unit has an ongoing measurement process. When criteria for ALT or TST are met, the subscriber unit selects an appropriate channel/time slot and attempts an ALT/TST. Once the subscriber unit determines that an ALT/TST has to be done, it initiates the process by sending a message to the network using the newly selected channel. It is possible for the network to perform TSTs and some types of ALTs (e.g., ALTs between RPs connected to the same RPCU with only a brief switching delay). Other types of ALTs—ALTs between ports controlled by different RPCUs and ALTs between switches—will require a longer time to perform. For the former type, the network responds to the request by switching the channel, and informing the subscriber unit that the ALT is complete. For the latter case, the network informs the subscriber unit that the ALT is being processed. The terminal then returns to the original channel to continue the call while the network prepares the ALT. When the process is complete, the network informs the terminal, and the terminal switches to the new channel and informs the network.

A switch-based scenario for an inter–RPCU ALT is given in Fig. 9.11. This is an ALT between RPs connected to different RPCUs homing on the same switch and access manager. To initiate ALT, the SU temporarily suspends the voice conversation and requests an ALT from the network by signaling a marked available traffic channel on the new RF channel. It then returns to the old assigned channel and continues voice communication while the network prepares for the ALT. The

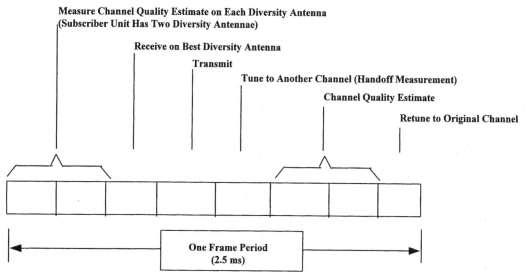

**Figure 9.10**   Time slot utilization in subscriber unit.

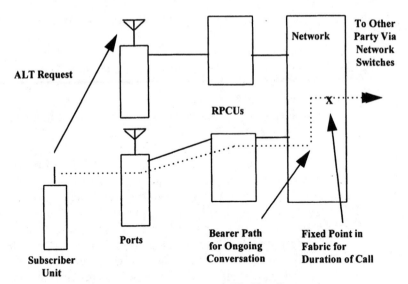

**a) SU Initiates Request To Network Via New Port**

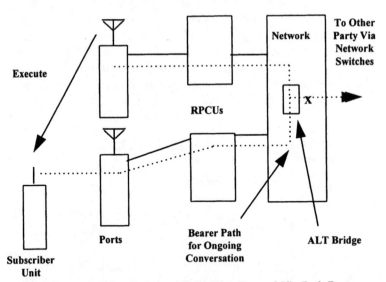

**b) Network Bridges Paths and Tells SU to Proceed Via Both Ports**

**Figure 9.11**    Inter-RPCU ALT.

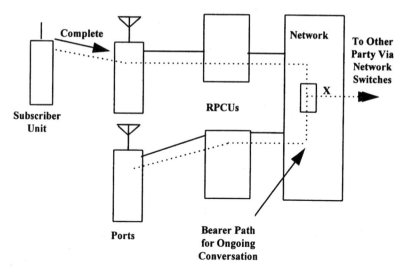

**c) SU Transfers to New Channel and Informs Network**

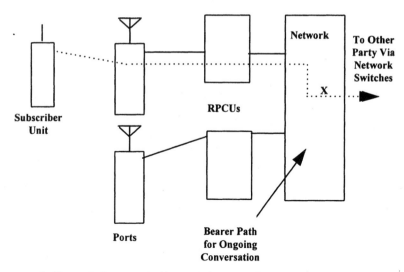

**d) Network Removes Bridge and Connects Conversation Directory**

Figure 9.11   (*Continued*)

access manager transfers the session privacy key to the privacy coder associated with new channel, and the switch inserts a bridge into the conversation path while bridging in the new port. Finally, the network informs the SU to execute the ALT via the old and new channels. Once the SU has made the transfer to the new port, it signals that is has completed the ALT. Then it resumes voice communication. The network can then remove the bridge from the path, and free up resources associated with the old channel at its leisure.

## 9.8   PACS Network Elements

PACS network elements are the subscriber unit, radio port, radio port control unit, and access manager described in the following sections.

### 9.8.1   Subscriber unit

Portable subscriber units are self-contained devices that perform all necessary functions to provide speech and/or data interfaces to the user. Fixed wireless access subscriber units are interface devices that serve to interconnect traditional wired customer premise equipment (CPE) to the air interface. The basic functions of the SU are as follows:

- *Transmission / reception of radio signals.*   The subscriber unit transmits and receives on the air interface. It should be capable of transmitting on either of its two diversity antennae as directed by the unit's controller.

- *Port selection.*   Upon power-up or movement from one coverage area to another, the subscriber unit selects the system broadcast channel associated with the best available receive RF channel. In making the decision, the subscriber unit verifies that it is allowed access, attempts to select its primary service provider and verifies that the RF channel is of sufficient quality to provide acceptable service.

- *Initial access.*   The subscriber unit seizes a marked available traffic channel in the TDM frame and transmits an initial access message to the RPCU in order to set up a single link duplex connection. Usually, the time slot on which the access request was made becomes the traffic channel for the call. However, a new time slot may be assigned at the time of the access request.

- *Priority access.*   Provision is made in the uplink of the SBC for users to notify the network of the need to make a priority call or emergency call (e.g., by dialing 911) when no traffic channels are marked available. Priority call access requests indicate the category of the priority access.

- *Quality-maintenance processing.*   The quality-maintenance processing consists of three components:

  1. Ongoing measurements and processing of measurement data allowing the subscriber unit to monitor quality.
  2. The trigger-decision mechanism, whereby the subscriber unit uses the processed measurement data to determine which action (ALT or TST) is required.

3. The choice of the new RF channel for ALT or the new time slot for TST is a process closely associated with the trigger decision.

In the quality-maintenance process, the phrase *error rate* is used by the subscriber unit to detect trouble situations. As part of the measurement process, the subscriber unit maintains a short list of channels ranked in order of mean received signal strength indication (RSSI).

- *Control of ALT and TST.*    Both automatic link transfer (ALT) and time slot transfer (TST) are normally controlled by the subscriber unit. The protocol, however, also allows an RPCU request to the subscriber unit to initiate an ALT.

- *Performing self-diagnostic and maintenance checks.*    The subscriber unit is equipped with self-diagnostic capabilities that are executed at power-up, such as power-on self test or upon receiving certain test commands.

### 9.8.2 Wireless access fixed unit

Figure 9.12 shows the architecture of a WAFU. It contains a subscriber-line interface, voice codec, clock source, digital modulator/demodulator, RF module, bus, and controller. The functions of each element in the figure are as follows:

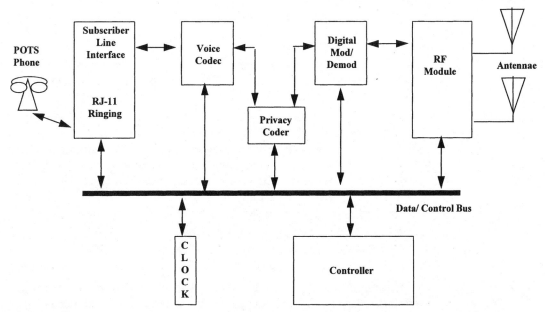

**Figure 9.12**    Wireless access fixed unit block diagram.

- *Subscriber-line interface.*   A direct interface to the analog telephone provides RJ-11 connectivity, ringing voltage, and analog to digital speech conversion.

- *Voice codec.*   Performs conversion between analog signal and digital signal at the appropriate bit rate.

- *Clock.*   Synchronizes the 32-kbps digital framing signal and other digital processes.

- *Digital modulator / demodulator.*   Transforms the data (bits) to and from a form suitable for transport over the air interface.

- *RF module.*   The front end of the digital radio component that includes transmit and receive radio circuits and that interfaces with the diversity-switched antennae.

- *Data / control bus.*   A common communications bus design that allows interconnectivity between components for maintenance, data control, and operational functionality.

- *Controller.*   Coordinates the operational functions of all other components.

If the subscriber unit is a portable SU then the subscriber-line interface function is replaced with a user interface function which has keypad, display, microphone, and speaker functions.

### 9.8.3   Radio port

The basic functions of the radio port are as follows:

- *Link between SU and RPCU.*   Whether in making a phone call or receiving an alerting message, the user's SU communicates with the wire-line network element (RPCU) through the RP that acts as an RF modem over the air interface.

- *Transmission / reception of radio signals.*   Each RP transmitter takes a framed and encoded signal from the modulator and converts it to the appropriate carrier frequency. The RP transmitter then transmits a modulated signal with a TDM RF channel structure to the SUs. The uplink TDMA bursts transmitted by the SUs arrive at the RP where the TDMA signal is synchronized, decoded, and converted to a line interface signal for transmission to the RPCU.

- *Channel coding / decoding for error detection and burst synchronization.*   The burst is synchronized between the RP and SU. Error detection is employed to allow mitigation of errors induced in transmission.

- *Synchronization.*   The RP extracts timing from the transmission facility (e.g., T1/E1) to synchronize itself to the network and may reference all transmit and receive frequencies to the same oscillator.

- *Termination of air interface.*   Message sets, signaling information, and operations data combine with the 384-kbps RF channel data to make up the transmission facility.

- *Self-diagnostic, remote test, and maintenance checks.*   The RP is equipped with self-diagnostic capabilities that are executed at power-up or upon receiving designated test commands. The RP collects quality indications, word-error indications, RSSI, and performance of RP circuitry information, and passes it to the RPCU.

- *Word-error indication (WEI) measurement.*   The RP performs WEI measurements and forwards detected-error indications to the RPCU for further processing.

- *RSSI measurement.*   The RP determines the strength of signals received in each time slot and forwards those measurements to the RPCU for further processing.

- *Quality-indication (QI) measurement.*   While in communication with the SU, the RP continuously monitors the receive signal quality of the radio link between itself by measuring the SU signal-to-impairment ratio. This information is used as the input to the uplink power control procedure and is transmitted back to the RPCU for processing.

As can be seen from Fig. 9.13 showing the functions of the RP, the RP performs mostly per-port functions, whereas a high-tier base station would typically do per-port and per-call functions. This results in increased complexity and cost of base stations, which is avoided in the PACS architecture. A single RP can support several channels.

### 9.8.4   P interface

The P interface connects the RP and the RPCU. The physical interface can be a T1 link where several RPs are connected in a drop-and-insert scheme. Alternatively, the interface could be a HDSL (high-bit-rate digital subscriber line) interface or a TV cable interface.

### 9.8.5   Radio port control unit (RPCU)

The RPCU provides management and control functions between the RP and the local exchange network. The functions are as follows:

- *Protocol conversion between the wire-line network and RPs*

   —*Multiplexing/demultiplexing.*   The RPCU provides multiplex and demultiplex functions between multiple RPs and the network.

   —*Transcoding between radio codec and wire-line codec.*   The RPCU performs transcoding as required by the C interface and transfers user speech and/or data through the external network.

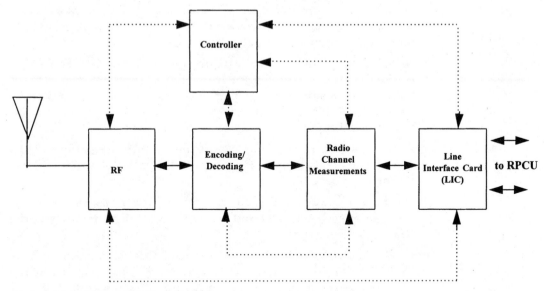

**Figure 9.13**   Radio port (RP) block diagram.

- *Registration and authentication.*  The RPCU communicates with various databases in order to determine the authenticity of an SU and to establish radio link security.

- *Encipherment/decipherment.*   The RPCU enciphers and deciphers the digital signals in each traffic channel as required for privacy. All user information and some signaling data are enciphered for each call using cipher keys. A unique cipher key is developed and agreed upon by the network and the SU during call setup. Encipherment and decipherment are performed by the SU and the RPCU for each call. Only the air interface is protected by this process.

- *TDM framing.*   The information to be sent to a given RP is formatted into an appropriate TDM stream by the RPCU. The RPCU multiplexes user information into the appropriate time slot. Symbols are always present, even in time slots which are not in use.

- *Broadcast channel formatting.*  The RPCU formats the system broadcast channel from its various components and multiplexes this formatted signal into a single time slot for RP transmission. The SBC provides operating information SUs within the coverage area.

- *Marking available traffic channel for normal calls.*  To enable access, the RPCU selects an unused time slot on that RP and marks it as available for the next normal call.

- *Marking available traffic channel for emergency calls.* If traffic channels are available, an emergency call will go through the normal process of call setup. However, if no traffic channel is available from all the nearby RPs for an emergency call, a process of marking available channels for emergency calls will also take place.

- *ALT and TST.* ALT and TST requests are generally SU-initiated processes. The RPCU functions, in conjunction with the network, effect the appropriate procedure (ALT/TST) upon request by an SU. Alternatively, the RPCU may request that an SU perform an ALT or TST under specific operational conditions (e.g., RP maintenance action may be necessary).

- *Managing RF spectrum allocation.* If the optional QSAFA frequency management scheme is implemented, the RPCU controls the overall QSAFA process operation for its served RPs in conjunction with other options of the network.

- *Performing word-error handling functions.* RP measurements of RSSI, QI, and WEI are delivered to the RPCU for each burst of every frame. The RPCU uses these indicators in determining the need for and to perform any mitigation of the effects of error segments.

- *Performance monitoring.* The RPCU determines and maintains system performance and call quality within its purview. The RPCU contains predetermined thresholds to monitor system performance/quality criteria. The RPCU tests this criteria during call setup.

- *Filtering, sorting, translating, and relaying signaling messages between the network, RPs, and SUs.* The RPCU acts upon signaling messages from the SU or the AM/switch by filtering, sorting, translating, or relaying these messages to their appropriate destination. They include call origination requests from the SU to the network, call delivery messages from the AM/switch to the SU, registration requests from the SU to the AM, alerting from the AM to the SU via the SBC, request for an inter-RPCU ALT, and priority access requests.

- *Exchanging signaling information (call setup, TST, and ALT) with the SUs.* The RPCU exchanges signaling information such as initial access messages to set up a call, messages to put the SU in the test mode, TST messages, and intra-RPCU ALT requests with the SU.

- *Power control.* In order to optimize the radio link performance, uplink power control is provided over the air interface. Downlink power control is not required. The RPCU provides frame-by-frame power control information to each served SU during each call.

- *Echo treatment.*    Echo treatment is not typically required by the air interface in conjunction with 32-kbps speech service. In the event that future, lower-bit-rate vocoders introduce the requirement, echo treatment should be implemented on the network side of the RPCU.

- *Maintenance and operations checks.*    The RPCU monitors radio link performance indications such as the RSSI, WEI, and QI, and initiates corrective action by use of the appropriate algorithm to optimize the air interface. These actions may include SU transmit power control, TST, and ALT. If an RP link transfer to an RP outside the control of the RPCU is required, an indication is sent to the AM to initiate transfer control to the appropriate RPCU and eventual RP. The RPCU also controls transmitter power on/off of the RP.

- *Interfacing with network operations systems.*    The RPCU provides the interfaces to such network systems as may be necessary to support PACS services. These interfaces are not part of this specification and will vary depending on the nature of the telephony network into which PACS interworks.

### 9.8.6    Access manager (AM)

The AM controls a telephony communications gateway between the PSTN and the PACS network. An AM controllable switch provides the voice path between the PACS network and the PSTN. This switch may or may not be dedicated to PACS; however, it must provide enough functionality to meet the needs of PACS under the control of the AM. A list of AM functions follows:

- Calls destined for PACS subscribers are delivered to the switch that serves the radio equipment in the area where the subscribers are active. It is the responsibility of the AM to direct the switch to deliver the call to the appropriate RP. The AM interacts with the database to determine the subscriber's radio location, status, alerting information, and terminating features. It then uses this information to control where and how to deliver the call to the subscriber.

- The AM controls the alerting process by first locating the SU and then by directing the switch to establish a voice connection to the appropriate RP, and alerts the subscriber. The first stage of alerting is a radio function that doesn't involve subscriber action. The second stage delivers the call with the complete incoming call parameters. The SU then provides the audible/visual indication to the subscriber.

- The AM works with the switch to provide originating service for wireless calls. The AM instructs the switch to associate the call origination with the subscriber. It queries the database for the subscriber's originating features and controls the switch to provide that set of features.

- The AM provides control over the actions required to maintain a seamless connection during ALT. This is a function that may be shared by multiple levels within the network. The RPCU is expected to provide this function for the RPs that it controls. The AM controls transfers between the RPCUs by controlling switch actions.

- The AM provides the storage, maintenance, access, and control of the data necessary to provide wireless service. The necessary databases include:

  —subscriber features (feature subscription)

  —dynamic subscriber data (associated with some features)

  —radio equipment configuration

  —alerting area mapping

  —terminal location

  —routing instructions

  —call-processing activity information

  —subscriber status

  —encipherment information

- The AM provides authentication and registration for the subscribers and the terminals.

- The AM provides operation functions and interfaces to operation systems for the proper management and administration of the radio network.

- The AM helps ensure the accurate billing of all calls originated from a PACS subscriber.

- The AM provides a level of control over the PACS switch to allow the offering of the various PACS services and features (i.e., ALT, along with the network-based features).

### 9.8.7   Network configuration

PACS architecture supports network elements access manager, ISDN/AIN switch, VLR, and HLR. There is considerable flexibility in the architecture as to the physical location of the access manager. The AM can be collocated with the RPCU. The advantage of this approach is that a single signaling transport mechanism can be used to communicate with the CPU, AM, and the switch—which leads to obvious savings in operating costs—although the disadvantage is that there has to be an AM along with every RPCU. The other alternative is for the AM to be stand-alone and connected to all the RPCUs within its area. The AM will also have a connection to the ISDN/AIN switch in order to control some of the functions of the switch. The advantage of this approach is that there is a one-to-many relationship between the AM and the RPCUs, leading to a need for a lesser number of AM. The final alternative is to collocate the AM with the VLR. In this way the RPCUs communicate with the AM via the ISDN/AIN switch. Operators may choose different architectures depending on their unique requirements. Although most of the signaling would be the same, the transport mechanism will change depending on the network architecture.

The RPCUs communicate with the AM for noncall-associated tasks such as registration, deregistration, enciphering, and ALTs. The RPCUs communicate with the ISDN/AIN switch for call control and ALT. The switch should be ISDN and advanced intelligent network (AIN 0.2) capable. AIN separates the call control from the service control and locates them in different network entities. The link between the RPCU and the switch would be ISDN BRI (basic rate interface). The switch does not have to be a mobile switch. In fact, there will be very minimal changes to a wire-line switch to support PACS. Most of the mobility aspects are transparent to the switch and are located in the RPCU and the AM. The call flows, assuming that the network architecture is the one where the AM is separately controlling several RPCUs, and has connections to the switch and the VLR.

## 9.9   PACS Numbering Overview

PACS supports the following two different numbering options:

- *Mobile identification number (MIN).*   Similar to today's cellular systems. MINs are programmed into the PCS handset. During registration, the handset makes the MIN known to the new system over the radio channel. The MIN also represents the number to be dialed.

- *Universal personal telecommunication number (UPT).*   Unlike MINs, UPT numbers, are assigned directly to people rather than to devices. They are device independent and allow users to be associated with one number rather than several, (e.g., home phone, office phone, and PCS handset). When a UPT number is dialed, the network first translates it into the address of the device serving the end user (e.g., MIN of the PCS handset) and then routes the call to that device.

For either type of number, the numbering formats are discussed in the next sections.

### 9.9.1  Geographic NPA-NXX-XXXX numbers

With geographic numbers, the NPA points to a geographic area and the NXX designates a specific switch. Calls to these numbers are always routed to the switch serving the NPA-NXX home switch for processing. This switch queries an AIN SCP for assistance.

### 9.9.2  Nongeographic NPA-NXX numbers

With nongeographic numbers, the NPA indicates the PCS. Within these NPAs, NXX codes are assigned to service providers. The service provider assigns XXXX values within these NPA-NXX codes to end users. Calls to these numbers are processed by an AIN switch and SCP, resulting in a query to the service provider's home database for routing instructions.

### 9.9.3  Nongeographic NPA database numbers

This scenario also uses nongeographic NPA code assignments for PCS; however in this case, the remaining 7 digits (NXX-XXXX) are assigned directly to subscribers. Routing a call to a nongeographic database number requires processing at an AIN switch and SCP to (1) determine which home database to query for routing instructions and subsequently (2) query this home database for these instructions.

## 9.10  Call Flows

### 9.10.1  Registration

A successful subscriber unit registration is shown in Fig. 9.14. Note that PACS-specific messaging is shown here, and the related IS-41 messaging between the AM, VLR, and HLR is assumed. For instance, during registration the AM will send an IS-41 registration notification message to the VLR which will forward it to the HLR. The HLR downloads the subscriber profile to the VLR and cancels the registration at the previous VLR. This messaging is identical to a typical IS-41 registration which is not shown in the call-flow design.

1. The SU receives the system information by reading the SBC. The SU reads the SBC upon powering up, when moving between RP coverage areas, and when instructed to do so by the network. If the SU has powered up, detected a change in the service provider ID, or received a reregistration instruction from the network, the SU begins the registration process. First, the SU selects an authentica-

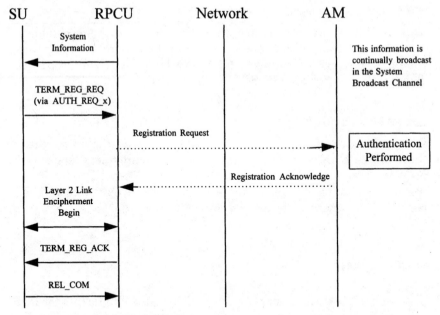

**Figure 9.14**   Successful SU registration.

tion method from the ones supported by the network. (This information is sent in the SBC.)

2. Upon acquiring a traffic channel, the SU sends an authentication request message to the selected RPCU. This message will have security, authentication, terminal ID or temporary TID, and other parameters.

3. The RPCU forwards the enciphered credentials and other information obtained from the TERM_REG_REQ to the AM.

4. The AM deciphers the credentials and performs the authentication. If the SU is authenticated, the SU registration procedure continues.

5. The AM assigns an alert identifier for the SU and sends it, with the required encipherment information, to the RPCU.

6. Layer 2 link encipherment begins between the SU and the RPCU.

7. The RPCU sends a TERM_REG_ACK with the alert identifier to the SU.

8. The SU responds to the receipt of the TERM_REG_ACK by sending a REL_COM message to the RPCU to release the associated radio resources.

### 9.10.2 Call origination

This call flow covers the most common scenario for call origination. The scenario covered is that of a registered SU making a nonemergency call. Other cases of call origination are essentially variations of this call flow and include emergency calls.

1. Upon acquiring a traffic channel, the SU sends an authentication request message to the selected RPCU. The authentication request message contains the CALL_REQ message.

2. The RPCU passes security credentials to the access manager.

3. The access manager deciphers the credentials, authenticates the SU, and determines that a call origination is requested. The AM assigns an RCID and also maintains a record of the RCID and encipherment information.

4. The AM sends the RCID, encipherment information, and decipherment call-processing information elements to the RPCU.

5. The RPCU receives the RCID, determines the associated ALT_DN for the call, and selects a network interface circuit for the call.

6. Layer 2 link encipherment begins between the SU and the RPCU.

7. The RPCU sends an RCID-ASSIGN message to the SU.

8. The RPCU sends a call-setup message to the network.

9. The network sends a message to the AM, indicating the called number.

10. The AM validates the request from the network and records the called number.

11. The AM signals the network to begin routing the call.

12. The network initiates the call setup.

13. The network may inform the RPCU that the call is proceeding.

14. If notified by the network, the RPCU sends a CALL_PROC message to the SU.

15. The network receives indication that the call has been routed, indicates to the RPCU that the route is complete and provides audible ring-back.

16. If notified by the network, the RPCU sends an ALERTING message to the SU.

17. The network receives an indication that the called party has answered, notifies the RPCU, and ceases sending ring-back.

18. The RPCU sends a CONNECT message to the SU.

19. The SU responds by sending a CONNECT_ACK message to the RPCU.

20. The RPCU acknowledges call connection to the network and notifies the AM of call connection.

### 9.10.3   Call termination

Before, the messaging between the AM, VLR, and HLR is assumed and not shown in the call flow since it is similar to an IS-41 call delivery. An incoming call that is delivered to an SU is described below:

1. The network attempts to terminate a call by notifying the AM of a call destined for an SU.

2. The AM initiates an SU alerting sequence by notifying all RPCUs within the registration area. This notification contains the alerting information that was associated with the SU at registration.

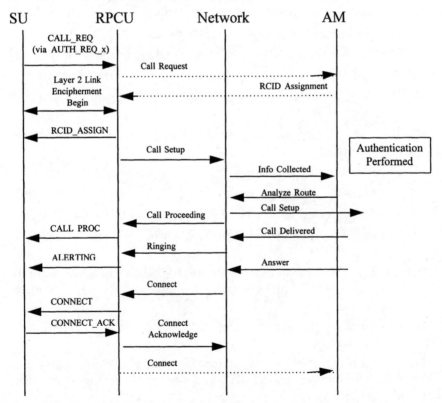

**Figure 9.15**   Call origination.

3. Each RPCU broadcasts alert messages for the given alert ID on its alerting channel.

4. When the addressed SU recognizes its broadcast alert, it acquires a traffic channel and sends an authentication request message to the RPCU. The authentication request message contains the ALERT_ACK message.

5. The RPCU selects a network interface on which it wishes to accept the call and sends the SU's security credentials to the AM with notification of the selected network interface for the call.

6. The AM deciphers the credentials, authenticates the SU, and determines that a call delivery is required. The AM assigns an RCID for the call and sends it to the RPCU while maintaining a record of the RCID and encipherment information. The AM notifies the network of the proper routing to complete the call. The AM passes on the network interface determined by the RPCU to the VLR. This number will be used by the PSTN to route the call to the proper RPCU.

7. Layer 2 link encipherment begins between the SU and the RPCU.

8. The RPCU sends the SU and RCID_ASSIGN message.

9. The network routes the call to the proper RPCU network interface and notifies the RPCU that routing is complete.

10. The RPCU offers the call to the SU by sending an INCOMING_CALL message to the SU.

11. The SU responds by sending an ALERTING message to the RPCU.

12. The RPCU notifies the network that the SU is ringing.

13. The network completes the circuit and provides audible ring-back to the calling party.

14. When the user answers, the SU sends a CONNECT message to the RPCU.

15. The RPCU notifies the network that the user has answered and that ring-back should be removed.

16. When the network acknowledges and the call is connected, the RPCU sends a CONNECT_ACK message to the SU and notifies the AM that the call has been connected.

## 9.11  Summary

The requirements for low-tier PCS are very much different from the requirements for high-tier PCS. These varying requirements play an important role in the design of the RF technology; and while high-tier

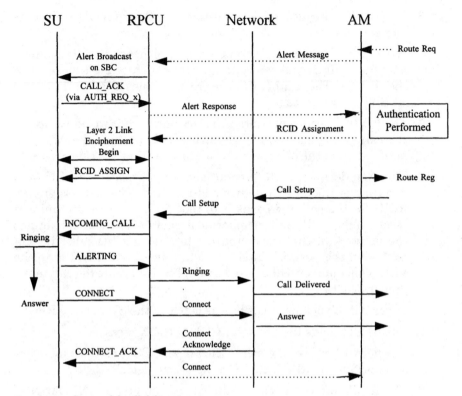

**Figure 9.16**  Successful call delivery.

PCS technologies can be used for low-tier PCS applications, they are clearly not optimized for such applications. There is a definite market niche for low-mobility applications and they will continue to coexist with high-mobility applications for well into the future. Currently, there is a great deal of activity in Europe regarding DECT and manufacturers are developing dual-mode DECT/GSM handsets. These handsets will allow subscribers to use the DECT network when in-building or in a campus environment, and in the cellular GSM network while traveling in a car. These kinds of applications are the wave of the future where a single dual-mode handset which can access both networks and provide dual-tier (low and high) PCS.

# PCS Business

# 10

# PCS Network Design

## 10.1  Introduction

This chapter provides a detailed overview of the considerations involved in the RF design of a wireless communications systems. This chapter will review the general principles which guide successful designs in today's high-capacity systems. The flow of general design and implementation activities for wireless systems today will be surveyed, recognizing basic phases in the process. A more in-depth technical basis for design will be discussed exploring the basic principles, tools, and practices. An example design is presented illustrating many considerations employed in designing a system to serve a large market.

## 10.2  The Wireless RF Design Philosophy

Wireless system design is much more than just radio transmission engineering. To be successful, a wireless system must adequately deliver the following:

- *RF coverage and penetration.*  Radio coverage must be provided over virtually all of the intended market area. RF signal levels must be sufficiently strong to penetrate buildings and vehicles to accommodate the usage styles of subscribers. The user should be able to communicate wherever reasonably desired, within the power output requirements specified by the FCC or other regulatory bodies.

- *Traffic capacity.*  There must be sufficient capacity to avoid significant blocking in any cell. The user should be able to communicate whenever desired, within acceptable call blocking parameters.

- *Mobility.* Call delivery and call continuity must be transparently maintained, regardless of the movement of the user within the system.

- *Quality RF environment.* Beyond coverage and penetration, the delivered signal must be free of interference to allow high-quality calls. This requires control of interference by frequency planning (in AMPS, TDMA, or GSM systems) or PN offset and cell-load planning (in CDMA systems).

The calling needs, such as messaging feature support, transparent integration with the PSTN, and both system- and subscriber-level economic concerns, are considered by the design engineers and are discussed in other chapters.

A successful PCS or cellular design must fully exploit the basic cellular concept of frequency reuse and strike a balance between coverage and traffic capacity. Interference must be anticipated and managed. Engineering design must be reviewed from a business perspective to ensure that the venture will be profitable within an acceptable period of time. The underlying perspective of RF design is easier to understand after a quick contemplation of the various types of systems which have been built during the recent era of mobile telephony.

Today, most of the first-generation cellular systems (AMPS, NMT, ETACS) are very successful, but all are experiencing severe congestion. Using the original analog technologies, the only way to increase capacity is to continue subdividing the market into smaller and smaller cells, and building additional base stations. However, now with increased public resistance on environmental and aesthetic grounds, it is becoming more and more difficult for the operators to successfully locate new base stations.

Congestion has driven the development of the second-generation digital technologies as discussed in earlier chapters. GSM and IS-54/IS-136 TDMA technologies offer greater capacity than the preceding analog systems through time-division multiplexing of multiple users on each radio carrier. Their digital formats allow additional calling features and improved security for both user and system operator. From an RF perspective, both U.S. TDMA and GSM systems are still frequency-channelized and achieve their capacity through simultaneous reuse of frequencies in multiple cells within the service area. The RF design of a TDMA or GSM system uses all the basic principles of an AMPS system with a few variations due to different levels of interference vulnerability, and in GSM the potential availability of frequency hopping to help reduce effects of interference and fading.

The spread-spectrum technologies, most notably CDMA, achieve substantial additional link gain due to their frequency spreading and trade off part of this windfall of gain to allow access by multiple users. The same wideband RF channel (or channels) are used in every cell, wherever needed throughout the system. All users share a common bandwidth. The user's data are distinguished and recovered by virtue of different orthogonal coding techniques and the use of timing offsets to distinguish them among signals of nearby cells, not by frequency separation. RF design of a CDMA system exploits the same propagation mechanisms as analog and TDMA/GSM systems, but code offset planning replaces the frequency planning required in AMPS/TDMA/GSM. However, power control is more critical than in AMPS/TDMA/GSM and requires special techniques.

The technologies mentioned above all were initially conceived for use in a macrocell environment, that is, where the cells are geographically large. Cell size ranges from a radius of many tens of miles in the rural areas to a mile or less in the large, metropolitan areas. However, some dense user environments generate so much traffic that a cell with coverage of even as much as one kilometer radius is hopelessly overloaded. This situation is easy to envision as wireless systems penetrate urban core areas and supplant the traditional use of wire-line telephones. Cell size must be reduced even further in very dense areas, with cells engineered to cover a few hundred or few dozen meters, and sometimes inside individual buildings. These tiny cells are termed *microcells*. It is difficult to engineer a system mixing both macrocellular and microcellular technology. The boundaries where macrocells and microcells meet, as well as the handoffs between the two environments, pose special problems. Special techniques must be used to achieve reasonable traffic sharing between the macro- and microcells in order to avoid mobility-restricting handoff problems, and to avoid and manage interference between the two worlds.

# 10.3  Modern System Design and Implementation: Phases

The design life cycle of a wireless system can be divided into several natural phases. Of course, the procedures of each operating company vary, as well as the approaches taken by different engineers, managers, and business developers. The circumstances of competition, site acquisition, business capital availability, hardware delivery schedules, and construction combine to impose many variations in the overall process. Nevertheless, there is a basic order to the development of a wireless system and each system goes through similar phases, although sometimes each phase has a "fuzzy" boundary. The RF system designer must see the big picture, working as part of a team to reach all objectives for the system. The design could be broadly divided into six phases. These phases are:

1. Preliminary design
2. Initial, detailed cell planning
3. Site acquisition
4. Construction
5. Initial system optimization
6. Commercial operation and growth management

Each of these six phases will be reviewed before discussing the more technical considerations.

### 10.3.1   Phase I: preliminary design

A workable preliminary design is developed for the system to allow business planners to conclude whether the system is economically viable during this phase. The preliminary design process may evaluate competing technologies (GSM, CDMA, etc.) and lead to a technology recommendation. The market is analyzed and divided into several zones for planning purposes, as shown in Fig. 10.1.

Each zone has its own distinct environment including terrain, vegetation and building types, traffic density, and other relevant factors. An orderly grid of cells is planned and interactively analyzed and modified until it appears that acceptable coverage will be obtained and no cell will intercept more traffic than it can carry within each zone. The basic configuration of the cells (omni or sector antennae, number of voice paths, etc.) is tentatively selected for each zone during this process. The initial grid layout is based on very high level assumptions about the

**Wireless Market**

Design Model Zone
- Dense urban
- Urban
- Rural
- Highway

**Figure 10.1**   Design zones.

coverage obtainable with one cell and the density of traffic in the zone(s). The focus is not on specific individual cells, but on overall system dimensioning.

Design could be based on coverage, traffic, or a composite as shown in Fig. 10.2. The anticipated traffic density is light and cells are constructed with the largest possible coverage footprints in some areas. This minimizes the number of cells required. The design in such areas is called *coverage limited,* and these cells are sometimes called *coverage cells.* In other areas, a large cell would intercept more traffic than it could possibly handle, so the cells must have smaller footprints to avoid unmanageable traffic loading. This requires a larger number of cells than what would otherwise be required. The design in these areas is called *traffic limited* or *capacity limited.* These cells are called *traffic* or *capacity cells.* Of course their footprints must mesh to produce a continuous fabric of coverage with no major gaps. There must be a reasonable degree of overlap coverage to allow handoff of calls from one cell to its surrounding neighbor cells.

Management needs to know how many cells will be required to serve a market long before detailed cell planning is possible. Engineers are troubled by having to make such guesses before there is any hope of working out a solid, detailed design. However, the need for such guesses is real. Capital must be committed, agreements signed with vendors to reserve manufacturing capacity, and an overall plan developed for system implementation.

Most initial guesses for planning purposes follow relatively simple assumptions, which should be clearly stated for comparison with reality as it later emerges. The number of cells in the system are bracketed by comparison with several limiting cases.

One notable case is the number of coverage cells required simply to cover the territory, ignoring traffic considerations. The total coverage

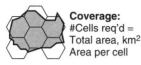

**Coverage:**
#Cells req'd =
Total area, km$^2$
Area per cell

**Traffic Capacity:**
#Cells req'd =
Total erlangs
Erlangs per cell

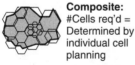

**Composite:**
#Cells req'd =
Determined by
individual cell
planning

**Figure 10.2** Coverage, traffic, and actual composite cells.

area of the market (in mi² or km²) is divided by the coverage size of the largest, normally configured cell. The result is the total number of cells required in the system.

Another case is the number of capacity cells required to handle all of the anticipated traffic with the simple assumption that traffic is where the cells are. An estimate of anticipated traffic provided or agreed to by the operator's marketing group is divided by the traffic capacity of the largest, normally configured cell. The result is the total number of cells required in the system.

The larger of the two cell numbers is then used as the number of cells required. However, real-life considerations suggest there is a better way. The coverage-only assumption is inadequate because traffic is often clumpy, and additional cells will be required to serve it. The traffic-only consideration also is often inadequate because traffic is so sparse in some areas that even large, coverage-limited cells don't intercept enough traffic. Consequently, the number of cells actually required in a system is usually larger than either the coverage cells or capacity cells.

The root-sum-square (RSS) rule shown in Fig. 10.3 was found to fit historical observations of the early development of a large sample of cellular systems. This method combines the results of the coverage and traffic assumptions, adding an extra margin to allow for the considerations noted above. For example, if a market needed 100 cells to achieve coverage and 100 cells were required to carry the anticipated traffic, the RSS rule would suggest planning for 141 cells. If 100 cells were needed for coverage and only 20 cells were needed for traffic, the RSS rule would suggest planning for just 102 cells.

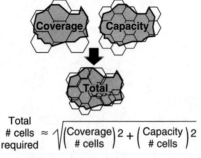

$$\begin{matrix}\text{Total} \\ \text{\# cells} \\ \text{required}\end{matrix} \approx \sqrt{\left(\frac{\text{Coverage}}{\text{\# cells}}\right)^2 + \left(\frac{\text{Capacity}}{\text{\# cells}}\right)^2}$$

Surveying existing cellular systems, the number of cells actually built seems to follow this formula

**Figure 10.3**  The root-sum-square rule.

Often, the preliminary design phase will include setting up several test transmitters and a driving test to collect signal strength data which characterizes the propagation behavior of the area and to fine-tune the propagation predictions upon which the grid development is based. In some cases, the preliminary design may even include a demonstration phase in which several actual model cells are installed and system performance is demonstrated prior to the commitment to build the overall system.

Although engineering information developed in the preliminary design will be useful in the later design phases, the main focus with the drive tests is to validate basic assumptions and to provide reasonable, high-level estimates of the system dimensions, such as the number of cells required, typical cell configurations, average traffic backhaul requirements, and switching resources required. These inputs feed the business analysis used by management and finance to determine the feasibility and develop workable strategic deployment plans for the system.

Figure 10.4 shows an example of a typical high-level RF network dimensioning plan. Note that the plan relates basic customer and traffic assumptions with the number of cells and radios required to provide the service. The plan forms the basis for initial expectations about capital, expenses, and profitability, although very tentative and preliminary. The whole process of business planning is discussed in detail in other chapters.

The preliminary design phase is finished when the management decision has been made to proceed with construction of the system, and the anticipated numbers of cells and other dimensioning details are accepted into the business plan.

| Year | 0/Start | 1 | 2 | 3 | 4 | 5 |
|---|---|---|---|---|---|---|
| Population | 1,000,000 | 1,000,000 | 1,000,000 | 1,000,000 | 1,000,000 | 1,000,000 |
| Penetration | 0.1% | 2.5% | 5.0% | 7.0% | 9.0% | 12% |
| # Subscribers | 1,000 | 25,000 | 50,000 | 70,000 | 90,000 | 120,000 |
| BH Erlangs/Sub | .100 | .05 | .04 | .03 | .028 | .025 |
| BH Total Erlangs | 100 | 1,250 | 2,000 | 2,100 | 2,520 | 3,000 |
| # of Cell Sites | 10 | 25 | 35 | 40 | 45 | 50 |
| Avg. Erl/Cell | 10 | 50 | 57 | 52.5 | 56 | 60 |
| Avg. Chan./Cell | 17 | 61 | 68 | 64 | 67 | 71 |
| Total Voice Chans. | 170 | 1,525 | 2,380 | 2,560 | 3,015 | 3,550 |

**Figure 10.4**   Example high-level network plan.

### 10.3.2 Phase II: initial detailed cell planning

This phase is sometimes indistinct from the final part of the preliminary design phase. Target locations are defined for each of the cell sites using the basic grids established in the preliminary design. Maps (search area maps) are created showing the desired ideal location for each cell along with an indication of how much leeway exists for relocation in case the ideal spot is not physically available. During this process, more detailed and technical preanalysis is usually completed for each cell to anticipate any troublesome aeronautical restrictions, nearby existing communications, or broadcast facilities which might interact with the cell, and to evaluate the immediate terrain and clutter features to help identify the ideal antenna height, cell configuration, and basic operating parameters.

The search-area maps and associated notes are delivered to the site acquisition team for their use in actual acquisition of the sites.

### 10.3.3 Phase III: site acquisition

During this phase, a group of site acquisition personnel search the market for usable sites that are as close as possible to the search-map targets. The acquisition of new sites is a continuing process which goes on at varying rates for the life of the wireless system. However, this specific phase is more narrowly conceived as the period of rapid acquisition of the initial round of sites—the sites necessary to begin commercial service, or to reach an arbitrarily defined level of system implementation.

Close interaction between site acquisition personnel and system design engineers is required during this phase. Each rooftop or other possible candidate location for a site must be reviewed carefully to identify any potential problems and to anticipate whether it is likely to deliver acceptable coverage and performance.

At each prospective site, acquisition personnel will be concerned with price and terms of lease or purchase, long-term availability, validity of title, special structural, foundation, or aesthetic considerations. Also considered are freedom from liability for safe disposal of hazardous waste which may have been previously deposited at the site, current zoning or land-use classification, and the likelihood of successfully obtaining any necessary regulatory approvals or waivers for use of the site. Other considerations are whether wireless operation is permitted for use at this site, whether it requires the issuance of a building permit, or whether it will be necessary to obtain a waiver or variance at public hearings. The attitude of the surrounding landowners and the neighborhood at large is critical, and ideal sites often are lost because of public opposition.

System designers will be concerned with RF integration at each prospective site—the impact of the coverage and interference produced by this site on the overall system, troublesome interference far beyond the boundaries of its intended coverage, and the sites impact on covering certain holes inside the intended coverage area. The site should be able to mount an antenna with a clear view of the area to cover and allow line-of-sight paths for microwave communication. Alternatively, access to leased T-1, DS-3, or fiber connections and AC power is desirable. Each prospective site must have 24-hour physical access possible for maintenance. Also taken into consideration will be other planned communications facilities (PCS, cellular, paging, land/mobile, radar, microwave, satellite, FM, or TV broadcast) on the rooftop or nearby site. The engineer will have to test the presence of troublesome EMI/RFI (electromagnetic interference/radio frequency interference) interactions. An additional check will be made to determine the presence of any AM broadcast stations as shown in Fig. 10.5 (either directional or nondirectional) close enough to require special AM measurements to demonstrate that the AM pattern has not been affected. If the AM broadcast pattern is affected, then installing a detuning skirt on the new antenna structure to reduce AM reradiation will be required.

It is desirable to set up a test transmitter and antenna in the precise location of the contemplated cell antennae and perform drive tests to evaluate the actual coverage if there are any lingering doubts about possible RF obstruction by nearby objects, or if the RF penetration characteristics of the surrounding area are questionable. Some wireless operators in dense metropolitan areas believe it will be necessary to test most of their sites in this way. This involves considerable effort,

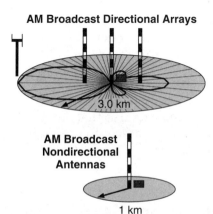

**AM Broadcast Directional Arrays**

3.0 km

**AM Broadcast Nondirectional Antennas**

1 km

**Figure 10.5** Proximity to AM broadcasters.

however it costs less than the discouraging realization that a site is inadequate immediately following the start of commercial service.

Both site acquisition personnel and system design personnel should be required to approve each site before its formal acquisition. A high level of cooperation and procedural integration is required to get things done. Both the acquisition group and the system design group must learn to think alike and absorb as much as possible of the other's concerns if the process is to meet its schedule. A detailed system and operating procedure is required to manage the collection of details on each prospective site. There must be an approval process so that everyone can quickly review task priorities on an ongoing basis.

Required filings and hearings should be scheduled if a site needs zoning changes. This requires legal counsel savvy in the local politics, as well as effective, articulate testimony with compelling exhibits to address any issues which arise in the process. Issues which may arise include the subject of human exposure to RF fields and possible biological effects; the general issue of aesthetics and clutter of the community by communications antennae; questions of risk to adjoining properties if severe weather or earthquake causes the antenna structure to collapse; the level of additional vehicular traffic in the area by system personnel while constructing and maintaining the site; whether the site will provide a possible hideaway for undesirables in the otherwise pristine neighborhood, and so on. Spokespeople knowledgeable about the system design and construction details will be needed to answer questions in the hearings.

System design personnel must complete the actual hardware specifications as the site details are finalized, with any last-minute changes recognized and included. This includes the final configuration of the site (such as omni or sector antennae, quantity of radios, antenna orientations, antenna height, and types of transmission line) as well as the operating parameters (such as RF power levels, base-station and switch data-fill values, default RF thresholds, and handoff parameters which will be needed by construction, installation, and optimization teams). The cell should be assigned its initial channels (or CDMA PN offsets) from the master frequency plan at this time. The transport facilities' planning and ordering will also have to be performed at this time. It is advisable to maintain all of this information in an orderly format—ideally in an on-line form so that any gaps or omissions can be readily spotted, and any deviations from typical defaults are found, justified, and corrected.

### 10.3.4   Phase IV: construction

The construction phase usually begins in a tentative way as soon as a decision has been made to pursue the site. Necessary preconstruction activities proceed on a contingent basis even if the zoning process will continue pending for several months or more. Detailed construction plans will be required as exhibits in the zoning process. There is considerable overlap between acquisition and construction phases.

If the antenna structure is a new tower or monopole, it will be necessary to do soil borings or investigations before the foundation can be designed. However, if the antennae are to be mounted on an existing building or tower, a structural analysis will be required. Structural analysis will also be required to determine that the host building can withstand the floor load imposed by the cell equipment.

A good contractor familiar with wireless site construction can generally complete all civil construction on a new stand-alone site on cleared ground within 30 days after permits are issued. This period can be reduced to less than a week for existing buildings. Tower erection and antenna installation are often done concurrently with civil construction.

When building new antenna structures as close as 3.0 kilometers to a directional AM broadcast antenna (i.e., one using multiple towers), it will be necessary to arrange for AM pattern measurements before and after the construction to conclusively demonstrate that the AM pattern hasn't been disturbed by the new wireless antenna structure. In cases where the wireless tower intercepts and reradiates enough energy to disturb the delicate AM pattern, it will be necessary to install a "skirt" or "sleeve" of wires on the tower and adjust it to cancel the reradiation as shown in Fig. 10.6, then repeat the after-construction measurements.

The wireless operator is financially responsible to the AM station if the AM proximity is overlooked and the wireless antenna structure

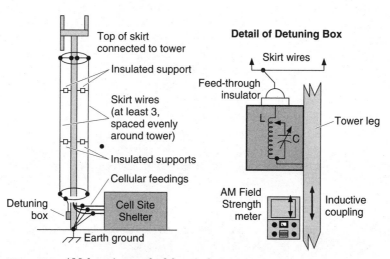

**Figure 10.6**   AM detuning method for wireless towers.

erected without taking the "before" measurements. The AM station can claim that the new wireless antenna is the source of all its real or perceived antenna system problems. Then, the difficult and expensive burden of mitigation or rebuttal falls on the wireless operator.

Double checks and reminders must be built into the process to ensure that any special zoning conditions and approvals are met and that any special promises or representations to the site owner or neighbors are fulfilled. The fewer unpleasant surprises during system optimization, the fewer the complications after commercial service.

The cell equipment is installed after physical construction of the site is complete. Leased-line or microwave facilities are activated and tested and overall commissioning and testing begins. This process should include bit-error-rate tests on the transport facilities; equalization and load testing of the battery plant, if used; internal diagnostic as well as external acceptance testing for the cell equipment; and sweep tests on the transmission lines and on the complete antenna system. Sites with towers should have tests to measure the effective ground resistance and a review conducted to ensure that proper grounding practices have been applied to protect all site equipment.

### 10.3.5  Phase V: initial system optimization

It might seem that the work is over with physical construction, installation, and acceptance testing completed. However, the most important step remains. Wireless sites are part of a complex, layered system, and the only way to ensure good performance is to painstakingly verify the final result. Independent checks of various hardware subsystems or of selected data-fill parameters cover only a tiny part of the overall operation of the system. Comprehensive optimization of each site is required, progressing from stand-alone testing to a final optimization of the site's relationships with all its neighboring sites.

The first step is to drive-test the new site for gross anomalies in coverage and to compare the actual with the expected coverage. If there are variances then the cause needs to be determined. The variances could be due to a defective antenna or transmission line, an unanticipated obstruction by nearby objects, or some simple misconnection within the site. The signal must be free of interference throughout the cell's intended service area. If interference is observed, the source must be determined. Solution for interference might require changes in the frequency plan (or CDMA offset plan). The signal levels on both uplink and downlink must be in balance.

The next step is to optimize coverage and handoff relationships with adjoining cells. This requires that the entire cluster of neighboring cells be operational. A drive route should be planned leading from each cell, or sector's coverage area, across the borders with all adjoining cells or sectors. Using a subscriber handset or specialized test equipment, a call should be set up and then carried back and forth to observe the handoffs. Drive tests help determine the handoff and hand-back boundaries. If these boundaries are not as expected then it could be due to data-fill on the original cell, data-fill on its neighbors, some propagation anomaly, or perhaps an unanticipated interference. Power control should be verified. Each of the wireless technologies (AMPS, TDMA, GSM, CDMA) has its own set of parameters or special considerations applicable to this testing, and each manufacturer has its own algorithms and features, yet the basic challenge is the same. It is more desirable to find a problem with an optimization team than to discover one later from hundreds of customer complaints or dropped-call logs. Intersystem handoffs require all of the above considerations with special attention and finesse.

The actual engineering time required to optimize a cell is estimated between 20 and 80 hours; 40 hours per cell has been used in projections for staffing in several large PCS systems. This has some major implications for operators, manufacturers, and consultants during the PCS build-outs of the coming few years.

There are a host of verifications to be completed in the higher-level switching layers of the system simultaneously with the RF optimization tests of cell clusters. In particular, call routing and translations must be carefully reviewed and verified.

### 10.3.6  Phase VI: commercial operation and growth management

Designing, building, and initially optimizing the system loom as major challenges for the new operator. This section focuses on some of the ongoing tasks and processes involved in managing the operation and growth of a system.

Before a system is placed in commercial service, traffic engineering is a very imaginative process. The actual number of customers is only a forecast; usage patterns and levels are unknown, and the uniform geographic distribution of the traffic is presumed. The exact shape of the traffic-interception zones of each cell is not yet determined, depending on final coverage and handoff parameters.

Traffic trends will emerge as commercial operations begin and the number of subscribers begins to climb from a few hundred to a few thousand. It is important to carefully monitor traffic statistics on each cell and sector and to anticipate overloads and blocking before they become recurring events. Additional channels must be added to cells in advance of this condition. Trending should be completed to anticipate when a fully equipped cell will begin to experience consistent overloading. The only way to create new capacity in a system is by cell splitting (i.e., shrink the coverage area of the overloaded cell and add additional cell(s) to subdivide its original coverage area). This involves site acquisition, construction, hardware installation, and optimization. There is a substantial delay involved, but advance warning is required to avoid a period of poor service waiting for the new cell to go into service.

Sometimes blocking can be anticipated purely from analysis of past trends in traffic on a cell. However, additional insight can be gained from the subscriber statistics at the sales and marketing departments (i.e., the new subscribers added monthly, the forecast for the next few months or years, and new groups of subscribers who are likely to have unique usage patterns).

In addition, pressures will come from customers, sales and marketing, and elsewhere to expand coverage into new areas. These expansions require resources, and as capacity increases in the system core, they compete for attention.

Building new cells requires capital resources and must be coupled with the operating company's budget process. It is likely that there will not be excess revenue to fuel rapid expansion of the system during the first few years of operation. Every improvement dollar must be justified and somehow found within the pool of available capital. Clearly, there is a role in the operating company for a network planner, a logistical coordinator to help interface between the traffic engineer, the finance department, and the sales and marketing groups to see that a growth plan is in place and the required capital is in the pipeline at the time it is needed. It is important to raise the issues so that all affected stakeholders are aware of the basis of the decision when resource contention exists. The decision of which cell to build is not a primarily engineering matter, although engineering can offer all stakeholders a valuable understanding of the performance consequences of the various alternatives. Engineering can also provide senior management the background and perspective it may need to build business justifications and effectively allocate additional capital for needed expansion.

## 10.4 RF Design Considerations

Having reviewed a nontechnical overview of the phases and events involved in system design and focusing on the events and some of the major concerns, the process with a focus on the technical factors and principles which shape the RF design will be discussed in this section.

The preliminary design is usually rushed by business pressures and occurs in what may seem to be a vacuum of solid technical detail. However, the inputs and assumptions of technical implementation must be developed with as much accuracy as possible.

### 10.4.1  Link budget

The link budget is the major foundation of the preliminary RF design. The link budget and the propagation analysis determine how much coverage can be achieved by an individual cell.

Link budgets always have been the dominant technique for evaluation of microwave path designs. However, their meaning and use may not be familiar to engineers from other disciplines. The choice of the word *budget* is no accident. A household or corporate budget traces revenue and expenditures showing where all the money goes. The income meets or exceeds the essential expenditures and everything is accounted for in a workable budget. An RF link budget traces RF power from the transmitter output connector through the entire path to the input connector of the receiver in much the same way. The link budget accumulates the various gains and losses of power along the path, and shows the level of RF power finally delivered to the receiver. The power received must exceed some minimum specified level if the signal is to be successfully demodulated and the embedded information extracted. Link budgets are always expressed in decibels for ease of computation. The transmitter output and receiver input are normally expressed in dBm (decibels above or below one milliwatt), and all gains and losses are dimensionless. If antenna gains are specified by comparison with an imaginary dipole antenna, the custom below 1000 MHz for cellular, then the radiated signal is expressed as effective radiated power (ERP), in dBm. If antenna gains are specified by comparison with an imaginary isotropic antenna, the custom above 1000 MHz for PCS and microwave systems, then the radiated signal is expressed as effective isotropic radiated power (EIRP), in dBm. Figure 10.7 shows the concept of a link budget and some of the commonly included losses and gains it accumulates.

There are actually two link budgets involved since wireless systems provide two-way communications—one from the base station to the subscriber's terminal, sometimes called the downlink or the forward link; the other from the subscriber's terminal to the base station termed the uplink or the reverse link. There is no advantage to be gained if one link has plenty of signal level while the other borders loss-of-signal. Satisfactory operation is required in both directions. Whichever direction has the more difficult-to-achieve link budget will become the weak link which dictates the overall performance.

As an example, Fig. 10.8 shows a typical link budget for a 1900-MHz GSM-based system. The parameters shown are not directly representative of any specific manufacturer's current equipment offerings, but are typical values for a conventional base station and a handheld subscriber terminal. Notice that the forward path (base station to subscriber, also called downlink) and reverse path (subscriber to base station, also called uplink) are individually analyzed.

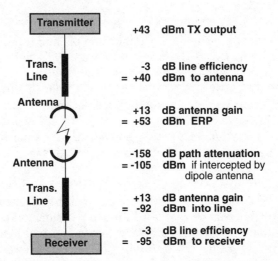

**Figure 10.7**    Link budget overview: tracing power flow.

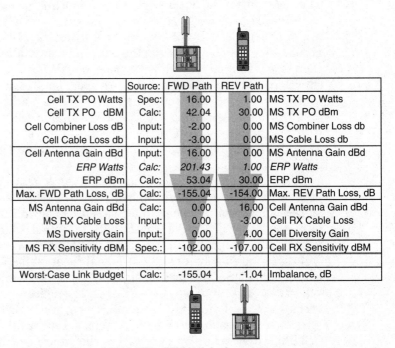

|  | Source: | FWD Path | REV Path |  |
|---|---|---|---|---|
| Cell TX PO Watts | Spec: | 16.00 | 1.00 | MS TX PO Watts |
| Cell TX PO   dBM | Calc: | 42.04 | 30.00 | MS TX PO dBm |
| Cell Combiner Loss dB | Input: | -2.00 | 0.00 | MS Combiner Loss db |
| Cell Cable Loss db | Input: | -3.00 | 0.00 | MS Cable Loss db |
| Cell Antenna Gain dBd | Input: | 16.00 | 0.00 | MS Antenna Gain dBd |
| *ERP Watts* | *Calc:* | *201.43* | *1.00* | *ERP Watts* |
| ERP dBm | Calc: | 53.04 | 30.00 | ERP dBm |
| Max. FWD Path Loss, dB | Calc: | -155.04 | -154.00 | Max. REV Path Loss, dB |
| MS Antenna Gain dBd | Calc: | 0.00 | 16.00 | Cell Antenna Gain dBd |
| MS RX Cable Loss | Input: | 0.00 | -3.00 | Cell RX Cable Loss |
| MS Diversity Gain | Input: | 0.00 | 4.00 | Cell Diversity Gain |
| MS RX Sensitivity dBM | Spec.: | -102.00 | -107.00 | Cell RX Sensitivity dBM |
|  |  |  |  |  |
| Worst-Case Link Budget | Calc: | -155.04 | -1.04 | Imbalance, dB |

**Figure 10.8**    Typical 1900-MHz GSM link budget.

**10.4.1.1    Key terms appearing in link budgets.**  In addition to the relatively straightforward factors such as antenna gains, transmission line (feeder) losses, and so on, link budgets can include less obvious factors related to specific features of the selected technology or the cell configuration. Often, system designers have particular insights concerning specific propagation modes or advantages/disadvantages of their specific technologies and system configurations which they express in customized terms. Several of the more commonly occurring terms are explained in the following sections.

**Maximum (allowable) path loss (sometimes propagation loss).**  Maximum path loss is the maximum amount of attenuation allowable for propagation between the transmitting and receiving antennae. The received signal level will be too low if more attenuation exists. If the actual attenuation is less, the received signal level will be more than sufficient and the extra signal level generally increases the reliability of the link, providing additional reserve against inevitable occasional fading. The amount of loss which can be tolerated is calculated by starting with the transmitter power output and adding and subtracting gains and losses of transmission lines, antennae, and any other hardware in the signal path. The required minimum signal level at the receiver is subtracted and the remainder is the maximum path loss. The path loss is numerically larger than any of the other terms in a link budget with a value usually between 100 and 200 dB.

The allowable maximum path loss for any given combination of transmitter, receiver, and other hardware may be simply determined as previously outlined. The maximum distance which the receiver can be from the transmitter before exceeding that amount of path loss needs to be determined. The laws of physics determine the result. The number of variables involved is so large that propagation prediction borders between art and science.

Later in this chapter the methods for propagation loss prediction will be surveyed, including deterministic and statistical models. Also examined are some commercial software tools for automating the prediction process, and basic field techniques for fine-tuning the statistical models. Errors of more than +/–10 dB in predicting propagation loss occur commonly in the real world. Statistically, if the standard deviation of the error is 8 dB, the prediction is better than average. The statistical techniques will be examined later in this chapter.

**Body loss.**  This is an allowance for the absorbing effects of the immediate surroundings of the subscriber handset. The most obvious component of this loss is absorption of RF energy in the tissue and clothing of the handset user. A typical value is roughly 3 dB.

**Building (or vehicle) penetration loss.** This is an allowance for the attenuating effects of the structure within which the subscriber unit is used, usually conceived as the difference in signal level between the environment at street level outside the building and the indoor locations where users may operate. It is common not only to characterize the average or median loss, but also to give a value for its standard deviation assuming a log-normal distribution.

The loss value depends entirely on the features and construction of the building or vehicle at a given frequency. Published penetration-loss estimates are interesting to read, but in the last analysis, measurements in Tokyo or Philadelphia are applicable only to Tokyo and Philadelphia. Differences in local building construction materials, architecture, and the general character of the neighborhoods have very pronounced effects. If at all possible, the figures used should be based on measurements made in the actual local market on a representative sample of typical buildings. The same is true of vehicles. Automobiles, pickups, and minivans can differ remarkably.

With the above serving as a disclaimer, some commonly reported values of penetration loss are from 12 to 20 dB for buildings with standard deviation of 4 to 7 dB respectively. Losses for vehicles are typically 6 to 10 dB with standard deviations of perhaps 4 to 6 dB, respectively. The statistical approach to predicting likelihood of penetration will be examined later in this chapter.

A blind, median value and standard deviation are not a very satisfying basis for predicting whether a metropolitan system is going to successfully penetrate specific buildings and centers. There is much interest in attempts to provide a prediction model which gives useful results based on the size and other characteristics of the building. Such models have been published, taking building size, the number of floors and on which floor the user is operating, and other factors into account. However, these models have not been widely used. Local applicability is still a question to be verified by measurements.

**Shadowing allowance.** Shadowing allowance is a special margin reserved to cover the possible additional losses a user may experience in obstructed locations. Conceptually, it is applied in much the same way as building-penetration or vehicle-penetration loss. However, it is usually specified as a median value without statistical considerations. Typical values can range from 6 to 15 dB.

**Diversity gain.** The link budget example includes diversity gain with none on the forward path and 4 dB on the reverse path. Diversity gain is an advantage obtained by receiving the subscriber's signal independently on two antennae at the base station. This type of diversity is called *space diversity* because the two receiving antennae are separated in space.

The purpose of space diversity is to reduce the prevalence of fading as shown in Fig. 10.9. The strength of the signal received at each end of the link rapidly varies up and down due to the random recombination of the direct signal and the signal reflected from a large number of objects and surfaces along the path as the subscriber moves about. The phase of a received signal is affected directly by the delay resulting from the length of the propagation path. In the real world, the signal propagates, perhaps directly, by reflection off of a host of objects in the immediate vicinity of the mobile and by diffraction over any obstructing objects.

As all these signal components arrive at the receiving end of the link, their phase and amplitude relationships are totally random. The signal heard by the receiver is the instantaneous vector sum of all the components. The signal level changes gradually due to obstruction and shadowing by terrain and nearby objects as a user moves about. Superimposed on this slower pattern of fading is a rapid flutter as the mobile passes through many locations where all the components cancel. The signal can range from 10 dB to 30 dB weaker than its average value during these fades. The fade locations are spaced at one-half wavelength apart, about 20 cm on cellular frequencies and about 7.5 cm on PCS frequencies. The fades are small from the user's perspective, affecting only a small percentage of the total area. They last only a small percentage of the time. These fades are called Rayleigh fades after Lord Rayleigh and his famous statistical distribution which fits this type of fading rather well. Rayleigh fading is shown in Fig. 10.10.

The signal recovered from each antenna will fade essentially independently of the other if the base station receives using two independent antennae separated by a short distance of 10–20 wavelengths.

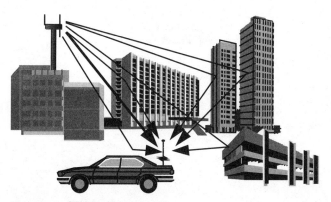

**Figure 10.9**  Multipath propagation.

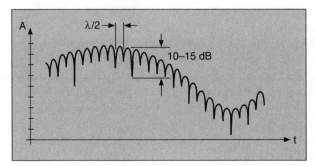

**Figure 10.10**   Rayleigh fading.

It is unlikely that the signal received on antenna A will be in a fade at the same instant the signal received on antenna B fades. The base-station receiver independently and instantaneously selects whichever antenna gives the best signal. Space diversity eliminates the effects of Rayleigh fades as shown in Fig. 10.11.

Space-diversity gain is not a gain in the sense of some fixed amount of increase in received signal level. However, it represents an additional immunity to fading which can only be characterized statistically through real-world tests or mathematical modeling. The effect is real, but not as easily measurable as an antenna gain or a transmitter power increase. Typical values range from perhaps 3 to 4 dB in typical cases, and as much as 6 dB using very sophisticated multibranch receiving correlators.

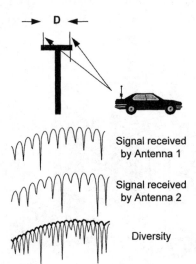

Signal received
by Antenna 1

Signal received
by Antenna 2

Diversity

**Figure 10.11**   Space diversity to combat Rayleigh fading.

Space diversity normally is applied only at the receiving end of the radio link. To transmit on two antennae using space diversity would not work against the fading mechanism since the two radiated signals and all their associated echoes would still be vector-summed at the single receiving antenna and Rayleigh fading would still result. Transmission using two antennae would introduce a new phase-cancellation effect producing cancellation of the transmitted signal in certain directions. This means that space diversity can be applied only at the base station because there is no room on a subscriber's handset for two separate receiving antennae several wavelengths apart (i.e., several feet apart).

At base stations where aesthetic or space considerations prevent using dual receiving antennae several wavelengths apart, it is still possible to obtain relatively effective receiving diversity by the use of polarization diversity in lieu of space diversity. This requires a special antenna including both the customary vertically polarized elements and a separate array of horizontally polarized elements mounted together in the same radome but with separate feed-line connectors. In an urban environment or in rugged terrain, the signal is scattered so much that there is substantial energy arriving in each linear polarization, both horizontal and vertical. At least one PCS manufacturer offers such an antenna commercially at 1900 MHz. There is no technological reason why this can't be done at 800 MHz; however, the radome would be wider than the traditional sector antennae.

**Frequency-hopping gain (GSM).** The GSM standard provides for frequency hopping even though all cell configurations will not allow it and not all manufacturers will support it. Frequency hopping is not employed in AMPS, IS-54 TDMA, or in CDMA. Frequency hopping can provide a beneficial antifading, anti-interference effect.

To understand the benefits of frequency hopping against Rayleigh fading, it must be realized that the fade results from a delicate combination of multiple signal components which happen to arrive in phases and amplitudes that add up to a vector sum which is very weak. The delicate random balance which produced the fade is destroyed and the signal returns to normal levels if the user physically moves. The same effect would occur if the user changed frequency, since all the involved signal paths are more than a thousand wavelengths long and they differ from one another by many wavelengths. For example, if the frequency changes by $\frac{1}{10}$ of 1 percent, all the path lengths expressed in wavelengths change by similar, but inverse amounts, and the new phase relationship is totally random. The fade does not exist on the new frequency.

The subscriber and the base station change frequency burst-by-burst, hopping through a predefined repetitive sequence of frequencies which can be fairly lengthy in GSM frequency hopping. If the subscriber is located where there is a Rayleigh fade on one frequency in the sequence, the fade will affect communications only during the time spent on that specific frequency. Communications will not fade during the remainder of the time. The powerful error-correction techniques (convolutional encoding, frame interleaving) compensate for the bad burst during each hopping sequence. This technique benefits both the uplink and the downlink better than space diversity.

The hopping technique works not only to combat fading but also reduces the effects of interference. Each related cell site is programmed with a different hop sequence so that cochannel or adjacent-channel collisions are not continuous but occur only during certain parts of the hopping sequence.

The net effect of frequency hopping is difficult to quantify except through actual field tests on loaded systems, or through detailed mathematical simulations. GSM manufacturers claim from 3 to 6 dB of overall link budget improvement from frequency hopping. Frequency hopping can make a big difference in performance and possibly even allow a more dense frequency plan to be applied and improves system capacity because GSM systems can give acceptable performance with carrier-to-interference (C/I) ratios with as little as 6 to 9 dB.

Frequency hopping is not feasible in all cell configurations. Mobiles are frequently measuring the signal levels they receive on the BCCH (broadcast combined control channel) frequencies of surrounding cells in a GSM system. These measurements are used in the mobile-assisted handoff process. Consequently, each cell must be configured to radiate a signal on its BCCH frequency at all times. This means no hopping can be applied in a cell with just one transceiver, since the transceiver has a full-time job on the BCCH frequency. However, in a cell with multiple transceivers, hopping is possible. One obvious method is to hold one transceiver always on the BCCH frequency, either with or without all of its time slots occupied by traffic. The remaining transceivers are configured to hop. This is called *transceiver hopping*. Another method is to keep all transceivers on steady frequencies without hopping but to switch the users' base-band inputs among the various transceivers. This is called *base-band hopping*. Transceiver hopping and base-band hopping are equivalent from the point of view of the RF signal and the user.

**Handoff gain (multiple-server advantage).** Some designers take note of the statistical nature of propagation and anticipate a coverage benefit in weak signal areas at the mutual edge of coverage of two or more cells.

In locations where the signal of one cell is weak, it is likely that one of the neighboring cells will be stronger and will help fill in the areas where the other cell fades. Composite coverage improvement values of approximately 3 dB are attributed to this factor. This rationale applies only where there are multiple, plausible signal sources and not at the edge of a system where the subscriber is leaving the fringe coverage of the last cell with no other adjacent cells to provide the benefit. Consequently, this factor is usually gracefully withdrawn from the link budget if challenged. In addition, CDMA proponents argue that this benefit can only be realized in the soft-handoff mode supported by CDMA systems, pointing out that the hard handoffs of all other technologies are not sufficiently fast acting to escape rapid fading, and thus CDMA-realize this benefit.

**Handoff margin.**  Handoff margin is the converse of handoff gain. Some designers recognize that the handoff process must include some hysteresis effects to prevent undesired ping-ponging. This requires a slight additional margin of signal strength, since handoff must be triggered while the signal is at least a few dB above the noise floor. Values for this factor are often shown as approximately 3 dB. If applied, this would essentially offset any handoff gain attributed to multiserver advantage above. Most designers simply ignore both handoff margin and handoff gain in their link budgets, but the topics are fervently debated in some quarters.

**Spreading gain.**  CDMA proponents point out the substantial gain they achieve by spreading the subscriber's low-bit-rate signal (8-kbps or 13-kbps vocoder) over a wide (1.23 MHz) bandwidth. This translates roughly into a 20-dB gain as compared to the channelized technologies such as AMPS or US IS-54 TDMA. There is a traffic-level-dependent offsetting loss due to the interfering energy from other users. The number of other users is deliberately restricted and user power levels are tightly controlled. Therefore, a very significant net gain is realized due to spreading.

## 10.4.2  Propagation physics

The physical principles of radio propagation are well understood. Close analytical calculations of wave propagation are possible for simplified, mathematically contrived cases involving free space, boundaries between different media, or for diffraction over concisely defined geometric shapes of homogeneous materials. The propagation from one real-world location to another involves so many variables and so many minor contributing signal paths that close numerical calculation of the final path loss and the received signal levels is not practically feasible. There are many unknown factors such as the dielectric constant of leaves, the precise geometry of the leaves and branches of the trees on the ridge between two sites, the precise effect of partial obstruction by a line of waiting cars and a freeway underpass. If indoor propagation is involved, the parameters involved in penetrating the building are even less well defined.

Despite these uncertainties, it is possible to obtain generally workable results by applying the usable parts of our physical knowledge using statistics to extrapolate measurements to help validate and fine-tune the physical and statistical models. Radio propagation prediction is about as reliable as modern weather forecasting.

**10.4.2.1  Modes of propagation.**  Certainly an accurate analytical solution based on physics is preferable to some loose estimate, assuming computation of the analytical solution is feasible. There are three ideal propagation situations which are easy to calculate with great accuracy. Unfortunately, these situations do not occur in the real world. There is almost always some contamination from other propagation mechanisms. However, these three situations are easy to recognize, easy to calculate, and can serve as tools for quickly estimating and determining the high and low limits for the anticipated propagation.

**Case I: free-space propagation.**  A free-space condition is when the transmitting and receiving antennae are surrounded by empty space with no other objects close enough to interact. In this mode, the only propagation effect is the natural spreading out or dilution of the transmitted signal at greater and greater distances from the transmitting antenna. A receiving antenna can be considered as having an equivalent capture area, in which it successfully intercepts the energy propagated toward it. The size of the equivalent capture area is a function of frequency and antenna type. In free space, the power intercepted by a receiving antenna is therefore inversely proportional to the square of the distance from the transmitting antenna as shown in Fig. 10.12.

The free-space path loss between two dipole antennae can be expressed in any of the following forms:

$$\text{Path loss, dB} = 32.26 + 20 * \log_{10} (\text{frequency, MHz})$$
$$+ 20 * \log_{10} (\text{distance, miles})$$

$$\text{Path loss, dB} = 28.13 + 20 * \log_{10} (\text{frequency, MHz})$$
$$+ 20 * \log_{10} (\text{distance, km})$$

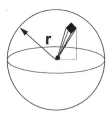

RF energy expands into
free space and the energy
intercepted by an antenna
decreases as $1/r^2$

**Figure 10.12**   Free-space spreading loss.

If both antennae are isotropic, the free-space path loss is:

Path loss, dB = 36.58 + 20 * $\log_{10}$ (frequency, MHz)

$$+ 20 * \log_{10} \text{(distance, miles)}$$

Path loss, dB = 32.45 + 20 * $\log_{10}$ (frequency, MHz)

$$+ 20 * \log_{10} \text{(distance, km)}$$

These equations are valid provided that (1) the antennae are at least
100 feet (30 meters) apart; (2) the first Fresnel zone (an ellipsoid-
shaped area between the antennae) is clear of any obstruction; and (3)
there are no reflections from any other objects nearby or along the
path. The first Fresnel zone is shown in Fig. 10.13.

The surface of the first Fresnel zone is actually defined as the collection of all points meeting the following path-length criteria. A ray from
one antenna, to the point in question, and another ray continuing on to
the antenna at the other end of the link gives a total length one-half
wavelength longer than the direct ray from one antenna to the other. At
its widest point midway between the two antennae, the radius of the
first Fresnel zone is:

$$d = \tfrac{1}{2}(l \times D)^{\wedge(1/2)}$$

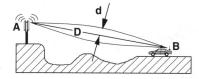

**Figure 10.13**   First Fresnel zone.

where    $d$ = the first Fresnel radius in meters
$D$ = half the distance between sites, in meters
$l$ = the wavelength at the signal frequency:
$l_{\text{meters}} = 3 \times 10^8/(\text{frequency, megahertz})$

Thus, the first Fresnel-zone radius on typical cellular or PCS paths is approximately 30 meters or 100 feet in its largest dimension at the midpoint of the path.

**Case II: partial cancellation by reflection.**   In most wireless mobile propagation, the subscriber is located in a vehicle or on foot with the receiving antenna a few feet above the ground. The base-station antenna is usually at least 20 times higher. Imagine that there are no major obstructions, although of course the first Fresnel zone may be in contact with the ground at some points. Under these conditions, there are two main signal components propagating to the receiving antenna on either end of the path. One is the direct ray between antennae; the other is a ray that reflects off the ground. The reflection point is closer to the subscriber than to the cell site in the proportion of their relative heights, typically about 20:1. That means the major reflection occurs only about 500 feet or less from the car on a 2-mile path over flat ground. The reflection with partial cancellation is shown in Fig. 10.14.

Reflection occurs at a very small angle, typically less than 1.0 degree. This is what is known as the *grazing incidence,* and this reflection is highly efficient. The reflected-signal phase flips almost exactly 180 degrees at the point of contact. This means that the receiving antenna intercepts a direct signal of amplitude 1.0 at some unknown phase angle, and a signal of amplitude 1.0 at a phase angle of almost exactly 180 degrees more. This would produce total cancellation except that the reflected signal has traveled just 1 or 2 inches further than the direct ray. This introduces enough phase difference to make this a partial, not a total, eclipse.

The apparent path loss between the two antennae is much larger than it would have been in free space because of this cancellation phenomenon.

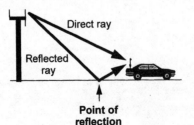

**Figure 10.14**   Reflection with partial cancellation.

Path loss, dB = $172 + 34 * \log_{10}(D_{\text{MILES}})$

$$- 20 * \log_{10} (\text{cell ht}_{\text{FEET}}) - 10 * \log_{10} (\text{mob. ht}_{\text{FEET}})$$

The received signal levels in free space and in the reflection mode is shown in Table 10.1. Table 10.1 is based on an ERP of 50 dBm transmitting at an 870-MHz frequency from a 200-foot base-station antenna to a mobile antenna of 5 feet. The comparison between received-signal levels in free space and in the reflection mode shows dramatic differences as the distance from the site is increased. In the reflection mode, the signal has faded into the noise at distances where a free-space path would still show a very high signal level. Notice the rate of signal decay with distance. In free space, the drop in signal level is 20 dB every time the distance is multiplied by 10, sometimes expressed "20 dB per decade." In the reflection mode, the signal drops roughly 34 dB per decade.

At first, this result seems to suggest that it is very difficult to cover a geographic area with a wireless system. In reality, the more rapid signal roll-off is beneficial, since it makes it possible to reuse frequencies relatively close together within the same market, thereby adding to system capacity. Performance actually suffers in sporadic cases where a subscriber happens to use a handset in a high location, for example, on a high-rise building, in an airplane, or on top of a major hill or mountain. Under these conditions, free-space conditions apply between that subscriber and his or her own serving cell as well as to several surrounding cells using the same channel. Meanwhile, in each of those cells, a subscriber traveling at ordinary elevations in the RF clutter is using a handset and experiencing roughly 34 to 40 dB per decade attenuation. The handset in the high location delivers interference to the calls in the surrounding cochannel cells.

**Case III: estimating obstruction losses.** The third situation of interest occurs when an otherwise clear path is blocked by a single large obstruction. An example is a mountain range between two sites at the level of the valley floor. This physical process is called *knife-edge diffraction.*

In this situation, the actual path loss between the two antennae is much greater than for free space, and may be significantly greater than for the reflection-cancellation mode. The path can be analyzed in seg-

**TABLE 10.1    Comparison of Free-space and Reflection-Propagation Modes**

| Distance, miles | 1 | 2 | 4 | 6 | 8 | 10 | 15 | 20 |
|---|---|---|---|---|---|---|---|---|
| Free-space | −45.3 | −51.4 | −57.4 | −60.9 | −63.4 | −65.4 | −68.9 | −71.4 |
| Cancellation | −69.0 | −79.2 | −89.5 | −95.4 | −99.7 | −103.0 | −109.0 | −113.2 |

ments to see if free space or reflection cancellation is appropriate, ignoring the obstruction's own additional attenuating effects.

The additional shadowing attenuation introduced by the obstruction can be estimated using formulas derived for the special case of knife-edge diffraction. Experience with microwave paths dating from the 1950s confirms that this fits the mountain obstruction situation fairly well, and it can even be applied to obstruction by urban buildings or objects relatively close to a cell site. Figure 10.15 shows the geometry for the knife-edge diffraction.

To estimate the additional shadowing loss due to the obstruction, parameter $n$ from the geometry of the path must be calculated first. Then the graph is referred to obtain the obstruction loss in dB. The sum of the shadowing loss and the otherwise-determined path loss is the total path loss. Other losses such as reflection cancellation still apply, but computed independently for the path sections before and after the obstruction. This method yields accurate results when diffraction is involved over sharply defined objects, and gives useful insight into the effects of antenna height changes and other options when general obstruction cases are encountered.

There are known analytical solutions to more than the three cases presented above. However, the cases are more and more complex and less likely to be encountered. The more esoteric models are appropriate to build into sophisticated algorithms in software tools, but are too cumbersome for the limited yields they produce in manual contemplation and calculation.

**Slow fading and fast fading.** The received signals on both uplink and downlink constantly fade and fluctuate if the subscriber is moving. The fades occur both slowly over a second or more and rapidly every few milliseconds as shown in Fig. 10.16.

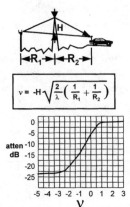

**Figure 10.15** Knife-edge diffraction.

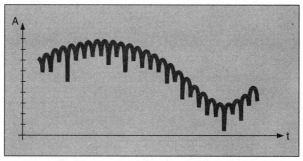

While multipath cancellation is responsible for the Rayleigh (fast) fading, the slower changes are due to obstructions and path geometry.

**Figure 10.16**   Slow fading.

The slow fades can be easily attributed to the effects of terrain considering the three propagation models previously presented. The fast fades occur because of multipath propagation which produces Rayleigh fading as previously described.

### 10.4.3   Propagation models

A simple, easy-to-use general model, which will predict useful cell coverage and how far the interference from that cell will travel into the layers of surrounding cells, is needed in the preliminary design phase for a new system. At this phase, the engineer is working with the average coverage of average cells and the most pressing question is how many cells to order from the manufacturer. It is irrelevant whether the model does not give reliable information for an individual cell, since specific coverage and performing-transmitter test measurements will be conducted before an operator leases space on a particular rooftop.

The class of models that has become popular in this situation is little more than mathematical curves fitted to observed-drive-test data. These curves can be sufficiently accurate for the initial planning purposes if drive-test data on several representative sites is available.

These tools are called *statistical models* or statistically derived models.

**10.4.3.1   Okumura-Hata, COST-HATA.**  During the 1970s, one of the first noteworthy systematic large-scale propagation statistical studies was published by Dr. Okumura and colleagues in Japan. The group collected and analyzed extensive volumes of data in dense urban and suburban environments surrounding Tokyo using a measuring lab in a motor home. Measurements were taken on virtually every signal source available (i.e., vertically polarized land mobile in various frequency bands, horizontally polarized television and FM broadcast stations, and other conceivable source). The resulting data were analyzed and curve fitted for trends with respect to every conceivable variable (i.e., base height, mobile height, frequency, type of terrain, and type of morphology, such as urban, suburban, and building type). The analysis reported in the paper was purely numerical and did not delve much into the underlying physics. However, the results confirm the free-space and reflection-in-clutter values of dB per decade.

More recently, the work of Okumura has been supplemented by further studies led by Dr. Hata and his colleagues. The European COST-231 committee has further refined the results for use with GSM systems at 1800/1900 MHz. Figure 10.17 shows an example of a measurement data at 800 MHz overlaid on the Okumura-Hata model's predictions. The curve appears to fit the data, although at any given point, the signal strength may be significantly above or below the curve. An observer with an understanding of physics will remember specific terrain on the drive-test path which is apparently responsible for some of the measured highs and lows. However, the curve is blind to such detail.

The actual formulae of the two models are as follows:

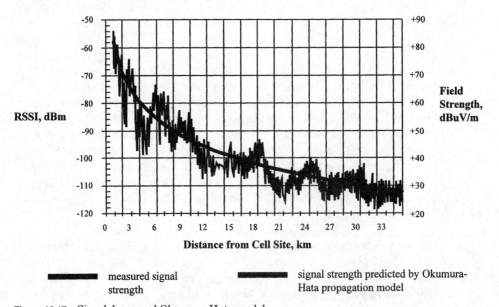

**Figure 10.17**   Signal decay and Okumura-Hata model.

**Okumura-Hata** (800-MHz band)

$$A \text{ (dB)} = 69.55 + 26.16 \log (F) - 13.82 \log (H)$$
$$+ [44.9 - 6.55 \log (H)] \times \log (D) + C$$

**COST-231-Hata** (1800/1900-MHz band)

$$A \text{ (dB)} = 46.3 + 33.9 \log (F) - 13.82 \log (H)$$
$$+ [44.9 - 6.55 \log (H)] \times \log (D) + C$$

$A$ = Path loss, dB
$F$ = Frequency, MHz
$D$ = Distance between base station and terminal, km
$H$ = Effective height of base-station antenna in meters
$C$ = Environmental-correction factor, dB

The environmental-correction factor for the two models with different RF environment is shown in Table 10.2.

There is a large environmental-corrections factor employed. It is advisable to compare the results with actual transmitter test measurements in the local market before selecting an environmental-correction factor. Figure 10.18 shows typical cell dimensions predicted for both cellular and PCS using these models.

**10.4.3.2 The Walfisch-Ikegami model.** The Walfisch-Ikegami model was developed for the dense urban environment where an almost unbroken fabric of buildings covers the area, divided by street canyons. The model uses a diffraction loss prediction technique treating successive blocks of buildings like a serrated diffracting screen. It draws statistical conclusions from the study of the range of possible diffractions. Its inputs are the dimensions and geometry of the building and street patterns. The model matches the measurement in the street canyon environment, but does not work well in a less rigidly structured suburban environment, since the pattern of regular diffraction does not exist. The application of the Walfisch-Ikegami model is shown in Figure 10.19.

**TABLE 10.2 Environmental-Correction Factor**

| Environment | Okumura-Hata | COST-231-Hata |
|---|---|---|
| Dense urban | −2 dB | 0 dB |
| Urban | −5 dB | −5 dB |
| Suburban | −10 dB | −10 dB |
| Rural | −26 dB | −17 dB |

| COST-231/Hata  f =1900 mHz. | Tower Height (meters) | EIRP (watts) | C, db | Range, km |
|---|---|---|---|---|
| Dense Urban | 30 | 200 | 0 | 2.52 |
| Urban | 30 | 200 | -5 | 3.50 |
| Suburban | 30 | 200 | -10 | 4.8 |
| Rural | 50 | 200 | -17 | 10.3 |

| Okumura/Hata  f = 870 mHz. | Tower Height (meters) | EIRP (watts) | C, db | Range, km |
|---|---|---|---|---|
| Dense Urban | 30 | 200 | -2 | 4.0 |
| Urban | 30 | 200 | -5 | 4.9 |
| Suburban | 30 | 200 | -10 | 6.7 |
| Rural | 50 | 200 | -26 | 26.8 |

**Figure 10.18** Okumura-Hata and COST-231-Hata typical results.

The model has been implemented in several of the popular commercial software tools in both two-dimensional and three-dimensional versions. It is useful for both macrocellular (cell antennae above buildings) and microcellular (cell antennae below building tops) environments.

**10.4.3.3 Statistical RF coverage and penetration considerations.** The Okumura-Hata and COST-231-Hata propagation models offer good predictions of systemwide factors, such as number of cells required. However, they are very poor predictors of the signal strength at a specific location. Slow fading and Rayleigh fading combine to dramatically influence signal strengths both above and below the best-fitting model curves. Figure 10.20 shows actual signal-strength variations along a specific path leading away from a site (jagged function), with the Okumura-Hata predicted values superimposed (smooth curve).

Half of the observed data will fall above predicted values and half below after selecting and optimizing a propagation model which fits the observed data well. One can arbitrarily add or subtract a fixed value in dB to the model and determine the relationship between the percentage of observations below the model and the value of the arbitrarily added level in dB. This manipulation is the same as adding or removing gain to the radio link.

This type of consideration implies that the statistics of RF signal variation can be used to adjust the transmitted power to achieve the desired level of service. This concept is valid even though the additional power and gain added to the link budget is limited. Figure 10.20 shows the probability of receiving a certain signal level at the cell edge or within a specific area.

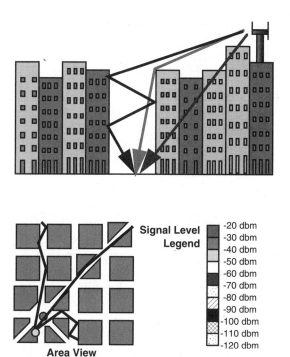

Signal Level
Legend

-20 dbm
-30 dbm
-40 dbm
-50 dbm
-60 dbm
-70 dbm
-80 dbm
-90 dbm
-100 dbm
-110 dbm
-120 dbm

**Area View**

**Macrocell**          **Microcell**

**Figure 10.19**  Walfisch-Ikegami model.

Uncertainty is derived from the attenuation caused by absorption and obstruction as the RF signal penetrates buildings and vehicles. The general solution is to (1) model the additional (excess) attenuation statistically, (2) determine the additional amount of signal level required to counteract the attenuation for an acceptable portion of the affected user population, and (3) design the system to achieve the required signal levels. Obtaining the increased signal level will require increasing cell antenna gain or height.

The first step in an analysis is to measure the statistical distribution of the excess attenuation in the type of building or vehicle you are trying to penetrate. The indoor-measured signal levels should be com-

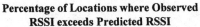

**Percentage of Locations where Observed
RSSI exceeds Predicted RSSI**

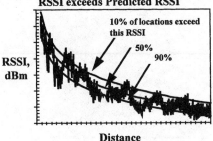

**Figure 10.20**  Coverage probability concept.

pared with the signal levels immediately outside the structure. The difference is the excess loss. The excess loss is usually found to follow a log-normal distribution about some average (mean) value expressed in dB for a large number of measurements. The scatter is characterized by a standard deviation expressed in dB.

If the outdoor RF level, mean excess loss, and excess loss standard deviation are known, it is possible to determine statistically the percentages of the indoor-user locations which will experience more than or less than an arbitrary loss in dB (e.g., the link budgets are completed for a new system and that the required receive-signal level on the downlink is –95 dBm). Measurements in the area of concern show a –95-dBm average, that is, 50 percent of locations in this area are above the required level and 50 percent are below it. The data scatters above and below this value. The standard deviation of the observations is 10 dB. The signal level to ensure that 90 percent of the observations receive at least –95 dBm needs to be determined. Figure 10.21 shows a statistical solution. Assuming the data follows the log-normal distribution, the cumulative log-normal curve is used to find an answer. The chart shows that 90 percent of the measurements will be included if the signal strength is increased by 1.29 standard deviations above its average value. The standard deviation is 10 dB, thus 1.29 standard deviations is 12.9 dB. The desired probability of service will be obtained if the signal level is increased by 12.9 dB. A more accurate reproduction of the cumulative normal distribution is presented for actual use in Fig. 10.22.

### 10.4.4  Propagation prediction tools

A whole family of radio propagation-prediction software tools has become available for use by wireless operators within the last few years. These tools use terrain databases to represent the earth's surface to any desired degree of accuracy. The programs construct a path profile and

## Cumulative Normal Distribution

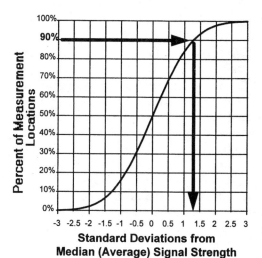

**Figure 10.21** Cumulative normal distribution example.

## Cumulative Normal Distribution

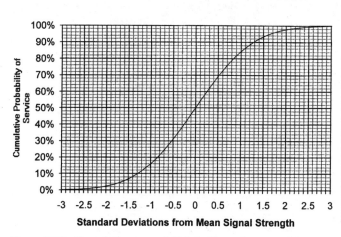

**Figure 10.22** Cumulative normal distribution reference.

| Standard Deviation | Cumulative Probability |
|---|---|
| -3.09 | 0.1% |
| -2.32 | 1% |
| -1.65 | 5% |
| -1.28 | 10% |
| -0.84 | 20% |
| -0.52 | 30% |
| 2.35 | 99% |
| 0 | 50% |
| 0.52 | 70% |
| 0.84 | 80% |
| 1.28 | 90% |
| 1.65 | 95% |
| 2.35 | 99% |
| 3.09 | 99.9% |

apply models of physical principles to predict the most likely path loss given the location and height of an antenna and the location and height of a receiving point. The program can calculate the signal strength at a series of points blanketing an entire area or market to produce maps readily showing subtle details of coverage or interference.

These programs manipulate data stored in layers as shown in Fig. 10.23. Input layers include terrain, land-use and land-clutter morphology, population or traffic data, and cell data, such as locations, heights, types and orientations of antennae, and masks which define what type of propagation model to use. Output layers include propagation results and a host of graphical displays of coverage, interference, and relational analysis for multiple cells.

**10.4.4.1  Terrain databases and resolution.**  The geographic resolution of such programs is determined primarily by the resolution of the terrain database. Digital terrain data is commercially available in a variety of resolutions for most of the industrialized world. Several techniques are available to create such data in countries where terrain data is not available in digital form. Manual digitization of paper topographic maps is the least expensive and least accurate of all. Satellite-altimetry data or aerial photography using stereoscopic techniques for altimetry can also be used and give better results at a higher cost. Sometimes the desired data exists for military purposes. Permission for its limited use may be obtained through political connections, or by presenting to the military the obvious benefits the new system will bring to the local economy.

Terrain databases are a series of elevation values corresponding to a grid of points on the ground. Figure 10.24 illustrates the three commonly encountered formats: geodetic, rectangular grid, and vector.

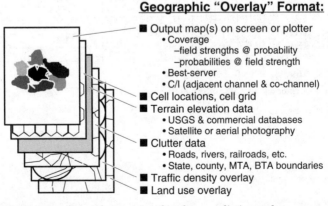

**Geographic "Overlay" Format:**

- Output map(s) on screen or plotter
  - Coverage
    - field strengths @ probability
    - probabilities @ field strength
  - Best-server
  - C/I (adjacent channel & co-channel)
- Cell locations, cell grid
- Terrain elevation data
  - USGS & commercial databases
  - Satellite or aerial photography
- Clutter data
  - Roads, rivers, railroads, etc.
  - State, county, MTA, BTA boundaries
- Traffic density overlay
- Land use overlay

**Figure 10.23**  General structure of modern prediction tools.

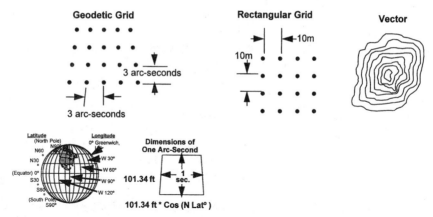

**Figure 10.24**   Common terrain database formats.

The geodetic format is where the elevation samples for every three arc-seconds of latitude and longitude are expressed in degrees, minutes, and seconds the same way time is divided into hours, minutes, and seconds. Adjacent points in a 3-arc-second geodetic terrain database are spaced 303 feet (93 meters) apart, north to south. The east to west spacing between points is the same as the north to south spacing for countries near the equator, but decreases to zero at the earth's poles varying in proportion to the cosine of the latitude. In North America, the east to west spacing between samples ranges from about 250 feet/84 meters in Miami, to 202 feet/67 meters at the western U.S.-Canadian border.

Terrain data in geodetic format is available in 30, 3, and 1-arc-second resolutions for most of North America and much of the world. The separation between points determines both the accuracy and the size of a digital terrain data file. A terrain of one degree square, the size of a typical city and surrounding area is roughly 3850 square miles. A 3-second terrain data file for this area will contain 1,440,000 discrete elevation samples. If each sample requires two bytes for storage, then each such area will require about three megabytes. The processing time to predict coverage of a single site with a 15-km radius using a typical program and workstation might average roughly five minutes at this density.

A 30-second terrain data file for the same area with its elevation samples ten times farther apart will contain 14,400 elevation samples and require less than 30-kbytes to store. The resolution of such a file will be so poor that major hills or ravines may be missed between successive elevation samples, and the propagation-prediction results will not be very useful. However, the processing time for a single site with a 15-km radius would be faster and probably less than 10 seconds.

A 1-second terrain data file for the same area will have elevation samples three times closer together than a 3-second file. A total of 12,960,000 elevation samples will exist requiring 26 megabytes for storage. The processing time for a single site with a 15-km radius will be increased to perhaps 15 minutes, but the resulting data will be highly detailed.

Terrain data is also available in an X-Y grid format with the distance between adjacent points set to a fixed value. Typical resolutions are from 10 to 100 meters. The available terrain data is in contour vector form in some other cases. Most of the better propagation prediction programs can use terrain data in any popular format. A conversion utilities to reformat the data into the desired format can be employed if your tool does not have this capability.

**10.4.4.2    Propagation prediction output formats and resolution.** The "carpet" of points where a program computes signal strength does not have to match the points in the terrain database. There is little to be gained by computing signal strength at points closer together than the terrain database, since interpolating the terrain database will not give information about any terrain features not contained in the terrain data. However, to reduce processing time or to avoid uselessly high detail in the output, it may make sense to compute signal strength at a density less than the density of the underlying-terrain data for large cells in rural areas.

Some prediction programs calculate signal strength for a grid of locations arranged geodetically, that is, spaced a specific number of arc-seconds apart. Other programs employ an X-Y grid with spacing in meters or feet. Both of these formats provide essentially the same resolution over the whole area. The calculated locations are evenly spaced everywhere.

Some programs compute signal strength along radials—straight lines leading away from a site. The angular spacing between radials might range from one-half a degree (720 radials) to three degrees (120 radials). A radial format produces a great density of data near the site. They are, however, far apart at the edges of the calculated areas. Thus, the resolution varies as a function of distance from the site. Some programs compute signal strength along radials and then interpolate the results to fill a geodetic or X-Y grid. The method employed and its implications must be understood when using a tool.

**10.4.4.3    Propagation models within prediction tools.** The propagation models used by the programs can be relatively simple using ray-tracing techniques to select modes of propagation. However, some programs use models which are quite sophisticated, including detailed radio-physics algorithms for reflection, diffraction, and scattering. Some programs allow the user to adjust the parameters of the model or even substitute their own models. Some include utilities to import drive-test measurement data, and perform statistical comparisons with the program's predicted signal levels at the same locations. This is useful for obtaining the general accuracy of the predictions in that area. It may also allow the user to recognize and even test desirable changes in the propagation model. Tweaking the model to fit drive-test data can cause problems in other environments if done indiscriminately.

Many of the software tools can also recognize databases of morphology—representing the type of land use or land clutter (LULC) at every point, and relating these categories to specific attenuations specified by the user. These attenuations are then automatically applied in all signal-strength or propagation-loss predictions generated by the program.

**10.4.4.4  Useful utilities and conveniences.** Most programs include routines for adding and manipulating user-site data. Using the graphical interface, the user can easily create a new site for computation, or edit the location, height, antenna type, or power of an existing site. Some programs offer nicely integrated cell-grid overlays which can be viewed on the screen and used by automatic routines for generating "straw man" sites at the exact target grid locations during early propagation studies while laying out a system.

Unfortunately, most of the programs maintain their databases of the user's sites in proprietary formats. However, some form of output utility is usually available to export as a flat file or some other common format and to import the same type of files. Newer programs are migrating their internal formats to industry-standard-database formats.

All programs include the capability of storing digitized antenna patterns so that the user may specify the type of antenna and its orientation. The program automatically retrieves and applies the correct antenna gain applicable to the path azimuth and elevation during propagation computation. Transmitted ERP or EIRP values are specified for each cell or sector so the program directly computes received-signal levels in dBm.

**10.4.4.5  Types of output plots and displays.** The tool programs can predict the coverage of an individual site or large groups of sites using the databases and propagation models described previously. A separate layer of data is generated to represent the coverage from each site. Multiple layers can be combined intelligently to display very useful analyses of the system.

The most common output is a composite coverage plot showing the extent of coverage of a wireless system without indicating which site (or sector) is responsible for the coverage at any given point. User-selected signal levels are usually represented by different colors. Such a plot makes it easy to spot holes in the coverage. Examples of this type of plot can be seen in the design example at the end of this chapter.

A multisite display mode is the serving-cell plot. It displays the coverage of each cell (or sector) in a unique color or pattern allowing quick, visual identification of the boundaries between coverages of adjoining cells and sectors. Such a plot is useful in anticipating traffic intercepted by each cell, planning and troubleshooting handoff parameters, and identifying unusual relationships such as small enclaves within one cell's coverage area where another external cell is stronger due to a quirk of terrain or morphology. Figure 10.25 shows an example of this type of plot generated with the MSI PlaNet tool.

Another useful display for a group of sites is a carrier-to-interference (C/I) plot. It shows the predicted C/I ratio at every point using different colors or symbols. The program knows to use the calculated best-serving cell at every point as the C value. The user must first provide the channel assignments for each cell so the program knows what cells are potential contributors of I at every location.

The C/I plot is a useful tool for anticipating interference problems in a system and evaluating various options as the frequency plan is manually shuffled in an attempt at improvement. It's important to realize that the C/I plot shows the best possible case. It assumes that the subscriber is always being served by the best (strongest) cell at any given point. A moving subscriber will drag before handing off if the system operator has not done a good job of optimizing handoffs. The user may continue to be served by the original serving cell long after passing the boundary where the handoff should have naturally occurred to its neighboring cell. This can take the user into some very hostile C/I environments and possibly cause interference or dropped calls.

Most programs provide utilities for viewing the actual-terrain database as a color-terrain elevation map or as a perspective-terrain view. Roads and other clutter detail may be superimposed. This is useful for visual orientation and familiarization with the area. The morphology database can be examined in the same way.

**10.4.4.6  More advanced features.** Some propagation tool packages include various traffic-engineering functions, and a few offer automatic frequency planning routines. Some users feel these features are the most useful and effective facets of the tools, while others rarely use them.

**Figure 10.25**  Example best-server plot.

**Traffic engineering features.** Propagation prediction tools have the unique capability to identify the strongest signal received at any location within the market, thereby attributing the coverage at a particular point to a specific serving cell or sector. This can be used by the tool to accumulate the traffic each cell is anticipated to carry if the geographic distribution of traffic is known or estimated. It is a relatively simple matter to provide algorithms within the tool to determine the number of trunks or voice paths required for each cell or sector. Figure 10.26 shows the concept of traffic accumulation by cell.

Several commercial tools provide this functionality. The most common approach is to allow the user to draw or define arbitrary polygons

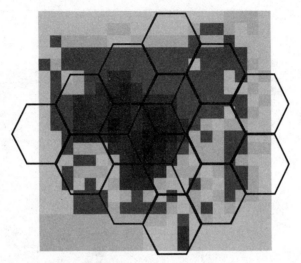

**Figure 10.26**   Traffic accumulation concept.

with a specific total number of busy-hour Erlangs spread uniformly through the polygon. Polygons can be overlapped or nested if desired to represent the specific traffic density expected in the market. The traffic density in the polygons can be subdivided by reference to a user-defined scheme of weighting factors indexed to the various land-use classes in the area (i.e., land fills generate a relative value of 1.0 while dense, urban centers generate a relative value of 10). An example plot, produced with the MSI PlaNet tool, showing such a traffic-density plot is shown in Figure 10.27.

It is important to realize that these automatic capabilities are no better than the traffic-distribution assumptions input into the model. However, there is no better or easier way to examine scenarios of system growth and cell splitting than with these techniques.

**Automatic frequency planning.**  System engineers have wished for frequency plans created without human suffering since the early days of AMPS cellular. Generating a workable frequency plan for a real system is difficult. Attempted adherence to an orderly grid, augmented with layer upon layer of ad hoc channel borrowing and exceptions, is the inescapable norm for large systems.

The major propagation tools offer routines for automatic frequency planning. These routines operate by (1) applying some sort of selection strategy for assigning channels to cells, then (2) testing the results to evaluate some measurement of overall C/I quality, and (3) changing the assigned channels where appropriate, and repeating the cycle. This pro-

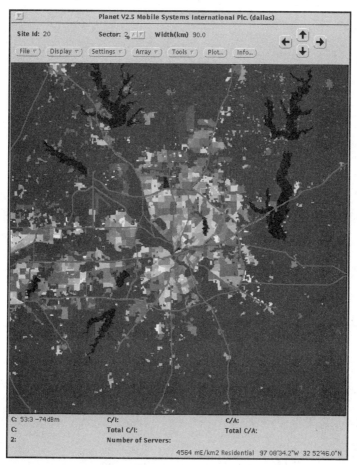

**Figure 10.27** Example traffic distribution.

cedure continues iteratively until the desired level of quality has been achieved or until it is obvious no further improvement is occurring.

There are many variables and considerations at each functional stage in any optimization scheme of this type. For example, the strategy of assignment of channels to cells may begin for a new system with a simple "stuffing" algorithm which implements some initial common distribution such as $N = 7$. Then, a "mask" may be defined to indicate which cells may have their assignments changed and which may not. The number of iterations or the target results may be specified. If the assignment routine is smart, it will also include the various taboos, such as minimum separation required between channels using tuned combiners.

Cyclic iterations to improve the frequency plan are possible only if there is some defined figure of merit or quality indicator to show whether one plan is better than another. Fortunately, the propagation tool can store C/I values for sample locations on a regular grid throughout the market and use some statistical measure of the overall C/I as the quality indicator. This can be made more appropriate by weighting the individual sample points or bins based on the traffic density in each. For example, it's more serious to have bad C/I at a spot in downtown than it is to have bad C/I in a cornfield.

Some users have complained that the automatic frequency planning routines available today are as cumbersome to use as the manual frequency planning process, so they continue to do things in the old manual way.

**10.4.4.7    Point-to-point and microwave path design.** Some tools provide path design features. A path-elevation profile is automatically generated and displayed with both endpoints defined and a frequency specified. In addition to showing the direct ray between antennae and the terrain profile along the path, the display includes the boundaries of the various Fresnel zones surrounding the central ray. Clearance or obstruction of the path can be quickly reviewed, and penetration of the Fresnel zone by obstructing objects may be identified. Arbitrary factors such as default tree height or known structures can be added.

Path-design tools are useful to the wireless operator since point-to-point microwave may be the most cost-effective method for interconnecting cells with the switch. It is useful to have a preliminary go/no-go conclusion on microwave at the earliest possible stage in site evaluation.

Automatically generated path profiles should never be used as the sole method of planning a path. It is highly advisable to have a field survey of the path completed by one of the competent consulting firms in the industry. Terrain databases and topographic maps have occasional errors. In addition, the latest maps typically do not show a man-made obstruction which was just completed weeks after publication. The cost of a path survey is far less than the cost of a surprise.

**10.4.4.8    PCS displacement of microwave incumbents.** One of the complications of PCS system design is the presence of existing microwave point-to-point systems operating within the frequencies now reallocated to PCS use. The FCC has provisions for microwave operators to relinquish their frequencies when requested by the PCS operator. The cost of replacing the facilities in some other frequency band is to be incurred by the PCS operator. This is a complex situation politically as well as technically.

The point-to-point microwave industry operates under a well-defined user coordination and interference resolution process with limited involvement by the FCC. The databases of existing paths and their technical details are maintained by private companies and independent consultants who perform the required interference calculations and public coordination notices whenever a new path is proposed. The best-known entity in the field is Comsearch, which maintains the most complete and most widely used database. Comsearch's database is available for use by customers and other consultants. Unless a wireless operator has experienced microwave-path coordinators within its in-house engineering staff, it will be more effective to subcontract this work to an entity already engaged in the complex interference analysis and case clearing.

**10.4.4.9  Survey of available propagation-prediction tools.**  Mobile Systems International's (MSI) PlaNet program is well known and is presently capturing a large part of the PCS planning market. PlaNet has all of the tools and capabilities described previously. It runs on a Unix workstation or server in an X Windows environment. There are many graphic enhancements and refinements for operator convenience. MSI is a U.K. company with offices in London, Dallas, and several other cities. MSI also provides consulting services.

Lunayach Communications Corporation (LCC) in Arlington, Virginia, produces a tool called CellCAD. CellCAD also features most of the functions described above. It runs on a Unix workstation or server. Prior to its development of CellCAD, LCC also developed a DOS-based tool called ANet. ANet required the user to manually manage site and propagation files, but substantially penetrated the U.S. cellular market in the early 1990s. CellCAD has eclipsed ANet for today's cellular and PCS systems. LCC also provides consulting services.

ComSearch in Reston, Virginia, produces a tool called MCAP with many of the features described above. While it has not gained as large a market share as PlaNet or CellCAD, it has followers. It, too, runs on a Unix workstation or server. ComSearch also provides tools and on-line access for its proprietary microwave-path database, as well as some PCS dimensioning tools. ComSearch also provides consulting services, especially in the area of microwave-incumbent issues.

CNet in Plano, Texas, offers a tool called WINGS with many of the features described above. It runs on a Unix workstation or server. In the late 1980s, it developed an earlier tool called SOLUTIONS which runs in a mainframe environment. It is still available for on-line use or licensed for in-house use, but does not include all the features and capabilities of WINGS.

TEC Cellular in West Melbourne, Florida, is one of the bright, new entrants into the tools arena with its WIZARD program. WIZARD runs on DOS/Windows machines and offers most of the functions described earlier, along with some innovative and highly efficient features. The use of PC platforms appeals to smaller operators and some consultants, not all of whom are Unix-experienced. TEC Cellular also offers system design consulting services.

AT&T offers a tool called Performance Analysis for Cellular Engineering (PACE). Originally focused primarily on collection and analysis of operational measurements and peg counts from its cellular systems, it soon gained a propagation-prediction module and has grown over the years into a well-integrated tool. It was recently transitioned from the DOS PC platform to the Windows platform. Although PACE's greatest appeal is to users of AT&T cellular network equipment, in the last few years it has gained a growing following within the general cellular community.

Communications Data Services (CDS) of Falls Church, Virginia, offers a tool called RFCAD aimed primarily at the public safety and broadcast industries. Although the propagation model is highly refined, it is not designed to work with the large numbers of sites in new PCS systems. However, CDS is a leading source of terrain data and topographic maps on CD-ROM, serving as an authorized distributor of terrain data for several U.S. agencies and several foreign governments. It also provides databases of towers, commercial sites, airports, and provides general database and GIS conversion services to wireless operators on a custom basis. Its sister company, Communications Engineering Services (CES) offers RF system-design engineering support, and also provides consulting services to resolve interactions between wireless sites and AM broadcast antennae.

Moffat, Larson & Johnson (MLJ) of Arlington, Virginia, provides a microwave-mitigation system called PathGuard, as well as various design and field-support tools. MLJ also provides general consulting services both in the RF system-design area and in more network-oriented aspects of wireless-system design.

Motorola provides a proprietary tool for internal use but does not sell or support its use by the market at large. The technical features are good but the user interface is more appropriate for knowledgeable internal users.

There are other companies and tools in the market and new tools are constantly being developed and introduced.

### 10.4.5  Propagation drive-test tools

The precise calculation or prediction of signal levels is virtually impossible except in very simple, pure cases. Measurement is much more reliable than prediction. However, measurement is physically more difficult than prediction. Useful measurements can be obtained from many different configurations of equipment, ranging from the simple to the very complex. A subscriber handset with a signal-strength indication calibrated in dBm gives useful information and first impressions to a knowledgeable user who might make observation notes using pen and paper. This is sometimes sufficient for simple conclusions regarding the need for a new cell in the area and so on. A larger volume of data is required to make accurate quantitative decisions required in designing wireless systems. Results are needed in graphical form for ease of analysis and understanding.

Several manufacturers market commercial tools for measurement-data collection to meet this need. The 800-MHz cellular industry is more mature and has a larger selection of such tools available for immediate purchase, but the 1900-MHz PCS industry is catching up fast.

The architecture of most of the tools is similar at the functional level. There are two key pieces of information for each measurement: signal level and location. Fig. 10.28 shows a conceptual diagram of a mobile measurement collector. Signal level is measured by a receiver and converted into a digital representation. Simultaneously, the test-vehicle location is determined usually by a navigation receiver. A magnetic compass and wheel-rotation sensors may be used to help bridge any gaps when the navigation receiver is unable to see satellites due to obstructions. The signal level and location are captured and logged

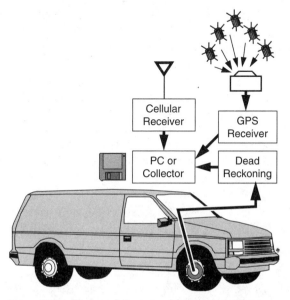

**Figure 10.28**   Basic architecture of drive-test systems.

onto storage media by a PC or PC-like processor for each measurement. The measurements are analyzed using software and appropriate plots are generated after collection. Some tools offer limited real-time analysis to confirm that the measurement data is being reliably collected during the drive tests.

**10.4.5.1  Purposes of field measurements.** Field measurements are taken for several reasons and each situation involves a slightly different focus. In the early phases of system design, tests are driven on sites which are believed typical but which may or may not themselves be candidates for possible lease by the wireless operator. Temporary portable test transmitters are installed for one or two days while the surrounding area is driven. The whole focus is to try to characterize the area and make any needed adjustments to the propagation model rather than to begin acquiring actual sites for use.

Later, individual sites are evaluated for suitability for actual lease or purchase by the wireless operator. These drive tests are to evaluate specific coverage of individual site(s) with an eye toward spotting any troublesome holes or shadows and identifying any potentially damaging interferences. This type of testing is highly desirable since effects of nearby obstructions can have a major effect on the propagation from a site. Such local obstructions are difficult or impossible to model using the standard propagation tools. Some wireless operators insist on drive testing all or a substantial majority of their new sites with the rationale that there is no other way to be sure of avoiding costly mistakes.

More drive tests are needed for network integration and optimization after the new cell is operable. The coverage of the new site is compared against the prediction and any deviations identified and studied for cause (e.g., bad antenna, local obstruction, and misprediction). The serving boundary of the cell is probed. Interference problems are identified if they exist. Handoff boundaries are identified and system parameters adjusted to obtain the best possible handoff performance. This type of drive testing includes both signal strength and tracking of call-processing events. The receiving apparatus is usually a subscriber unit with a data interface or a sophisticated tool using a subscriber unit as the functional core.

Measurements of all these types continue as needed after the system enters commercial service. Many operators conduct a general system field performance survey on a biannual or quarterly basis to identify seasonal factors and to resolve problems.

**10.4.5.2  Measurement considerations.** Measurements for different purposes may dictate the use of different hardware. The transmitter output power, antenna gains/ERP, receiver sensitivity, receiver bandwidth, and other features are all important. The sampling frequency of the measurements must be adequate to capture at least the slow-fading characteristics and to capture the Rayleigh characteristics if delay spread is a significant concern.

The area-characterization measurements early in the design process are mainly concerned with propagation loss so that signal strength is the primary variable. Continuous wave (CW), or "dead carrier" signal sources are the simplest to use. Although there may be interest in the degree of multipath propagation, there is no simple way to measure the delay spread. Although innovative analysis of the shape of the Rayleigh fading envelope using a CW receiver with adequately wide bandwidth may produce useful statistics. A special, modulated signal usually will be required and the receiver must have special demodulating and delay-tracking functionality if delay-spread information is desired.

In area characterization and site-evaluation measurements, it is frequently beneficial to have a receiver that can sample multiple frequencies simultaneously or fast enough to scan among several frequencies. This minimizes drive-testing time, assuming multiple test transmitters are available to equip several test locations simultaneously.

Measurements to observe actual system performance are concerned with signal strength and call-processing events and their associated messaging. This requires the use of a subscriber unit or some custom test set as the basic RF collection unit. Access, paging, and handoff messages can be captured and analyzed.

Both the measurement-collection platform and the postmeasurement-analysis platform must have sufficient speed and capacity to handle the volume of collected data.

### 10.4.5.3  Survey of available field-measurement tools

*1900-MHz PCS.*   The 1900-MHz PCS RF test-equipment market began in 1994. The market is dominated by simple CW test transmitters and receivers with a few exceptions coming largely from upbanding 900 MHz GSM capabilities. This will change rapidly as the PCS operators begin commercial service, and the existing drive-test products at 800 and 900 MHz will be upbanded and upgraded for PCS service. Some of the drive-test tools for the 1900-MHz PCS market are summarized next.

Rohde & Schwarz, and its U.S. affiliate, Tektronix of Beaverton, Oregon, offer a family of test transmitters and channel-delay-characterization equipment especially targeted toward the GSM PCS market. Although quite expensive and somewhat cumbersome physically, the system can derive both signal strength and delay-spread information which may be very useful in the early stages of system design and propagation model refinement. The test signal includes GMSK modulation and is ideally suited for GSM analysis.

MLJ of Arlington, Virginia, offers a CW test transmitter and receiver for 1900-MHz PCS use. The units are compact and relatively straightforward to set up and operate. MLJ also offers RF design consulting and network design consulting services for wireless operators.

Chase Electronics, a U.K. firm, offers portable test transmitters and receivers for CW tests at 800 MHz and up converters which allow operation at 1900 MHz for PCS. These units have been popular with some GSM operators.

Wireless Facilities, Inc. (WFI) with offices in New York and Chicago offers 1900-MHz CW test transmitters and receivers.

QualComm, the developer of the CDMA version adopted in the IS-95 standard, provides test transmitters and pilot beacons for characterization of areas and site evaluation for both 800-MHz and 1900-MHz systems.

*800/900-MHz CELLULAR.*   The 800-MHz RF test equipment marketplace is more than fourteen years old and many vendors on this list are in their third generation of measurement products. These vendors are working on products for 1900-MHz measurements as well. The functional equivalents of most of these products will be available at 1900 Mhz by the time PCS operators begin commercial service.

SAFCO Corporation in Chicago, Illinois, offers a two-tiered approach. For detailed system analysis, their OPAS collection tool and PROMAS office-analysis software can sample a large number of channels and characterize many sites simultaneously. The analysis capabilities are fairly extensive but the system is more expensive than lower-tier alternatives.

SAFCO's lower-tier tactical tool is the SmartSAM and SmartSAM Plus. These units use conventional subscriber mobiles as the RF front end. These tools are intended more for tactical use in shadowing call-processing activities on a system to collect information on dragged handoffs, interference, and other anomalies during actual test calls. The tools can be used for more general coverage analysis but do not scan as many channels as fast as the more expensive OPAS tool. SmartSAM Plus also offers US IS-54 TDMA analysis capability.

Lunayach Communications Corporation (LCC) in Arlington, Virginia, offers the real-time system analysis tool RSAT and RSAT 2000. These are tactical tools featuring call-shadowing capabilities directly competitive with the SAFCO products. RSAT 2000 offers US IS-54 TDMA capability and voice-synthesized reporting which is helpful in a one-person drive-test environment. LCC's earlier data-collection tool, CelluMate, provided similar capabilities but was limited to the AMPS environment. LCC also offers a portable AMPS receiver and data logger called the Walkabout, intended for data collection inside buildings.

COMARCO Wireless Technologies of Irvine, California, has developed advanced custom receiver and DSP design capabilities in military applications which it now brings to cellular. It has developed a series of sophisticated RF end-to-end testers (NAS-150, NAS-250, NAS-350) which measure not only the mobile RF environment, but also subscriber audio-loopback quality on mobile cellular links. The RF receivers have impressive channel-scanning rates and the collection software is well integrated.

GRAYSON Electronics of Blacksburg, Virginia, has developed CellScope, a PC-driven RF collection and analysis tool using a custom receiver. The user interface is easy to master and both the measurement and analysis capabilities are good.

Communications Data Services (CDS) of Falls Church, Virginia, provides MOBCAD, a PC software package capable of managing an instrumentation receiver (Anritsu or others) for data collection and simultaneously displaying the user's position on actual topographic maps while collection is in progress.

## 10.4.6   Site acquisition software tools

PCS markets will continue in a phase of almost explosive network buildup until year 2000. Some MTA networks will install more than a thousand sites. Merely identifying these locations will overrun the capabilities of the site acquisition practices used in the initial buildup of the AMPS cellular networks just a decade ago.

Planning and acquiring PCS sites could be substantially streamlined by better-integrated methods for identifying, qualifying, evaluating, tracking, and reporting sites in the field. During acquisition of a site, an operator must refer to many maps and databases to determine its desirability. The following list highlights the key information sources used in the process.

- Topographic maps to precisely identify site and determine coordinates, elevation

- Databases of "friendly" sites available for streamlined lease or acquisition

- Zoning maps, permitted-use maps, and so on, for use in intelligent selection of achievable sites

- FCC, FAA databases of existing towers and aeronautical obstructions

- FAA airport database

- FCC AM broadcast database (to determine whether AM measurements are required)
- Terrain data for use in propagation analysis
- Morphology and land-use databases for use in propagation analysis

In addition, various documents will be generated and must be managed in the acquisition process:

- Initial-lease agreement
- Site photographs for evaluation
- Tracking and approval forms for acquisition process

Mobile Systems International (MSI) of Dallas, Texas, offers a software tool called SATCAD which integrates all the databases, maps, and resources above onto a conventional laptop computer which can be carried in the field. The databases are displayed as layers on USGS topographic maps to allow immediate visual assessment and reaction to acquisition issues. The tool also captures digital photos of prospective sites and can import scanned or faxed documents to become part of the site record. An optional propagation-prediction capability allows rapid evaluation of sites and identification of coverage holes, areas of interference, and excessive overlaps, while in the field. The entire accumulated file on a site can be electronically transmitted back into the corporate database, eliminating paper transfer, delays for photo processing, and faxing or overnight shipment of paper maps. An industry standard open format is used for the database allowing easy interface with mainframe or workstation applications and powerful RF propagation tools such as PlaNet. Other software with the same core functionality undoubtedly will be developed by other vendors to help meet this widespread, burgeoning need.

### 10.4.7   Frequency planning

The degree of frequency reuse is generally determined by (1) the degree of inherent vulnerability to interference of the chosen technology, and (2) local propagation characteristics, which determine the degree of interference actually delivered. The required C/I ratio forces minimum cochannel separation and reuse factor.

In addition to the need to keep cochannel reuse separated by a minimum distance, there are two other types of complications in frequency planning. One restriction arises because of adjacent-channel interference, another because of restrictions on combining the outputs of multiple transmitters into a single antenna.

In all the channelized technologies (AMPS, NMT, ETACS), it is desirable to prevent a strong signal to occupy the channel immediately above or below the desired signal. If such an adjacent-channel signal ever grows significantly stronger than the desired signal, the analog technologies face the risk of dropping the call due to detection of the interferor's signaling, and the digital technologies are at the brink of increasing bit-error rate. For these reasons, it is not recommended to assign adjacent channels in the same cell or sector. When it is unavoidably necessary to assign adjacent channels in cells that are direct neighbors of one another, careful attention is required to be sure handoffs occur promptly near the desired boundary and do not drag. Dragging can expose the call to extremely poor adjacent-channel carrier-to-interference (C/I) ratios.

Combining multiple transmitters into one antenna is necessary at the cell sites in order to keep the number of antennae within aesthetic and economic limits. However, it is not feasible simply to connect all the desired transmitter outputs together in parallel. It is necessary to provide isolation, to prevent the output of one transmitter from passing back into another transmitter and causing intermodulation distortion. The required isolation can be achieved by any of three strategies, each of which has some implications for the frequency plan.

One method of isolation is to use tuned-cavity filters in the output path of each transmitter. The filters are highly frequency selective and severely attenuate any signals on frequencies other than the one to which they are tuned. Using this technique, it is possible to combine transmitters routinely 630 kHz apart (21 AMPs channels separation) and with careful attention survive 500 kHz separations (16 AMPs channels) under certain circumstances. However, this becomes the closest permissible separation between channels active on the same antenna at the cell.

Another method of isolation is to use hybrid combiners which get isolation by their unique circuit arrangement and not because of any frequency relationship. The drawback to this approach is that each hybrid combiner is a two-input/one-output device with an insertion loss of about 3.5 dB. Thus, it is not possible to "cascade" enough hybrid combiners to build up a workable number of AMPS channels without suffering tremendous attenuation. Sometimes a combination of tuned combiners and just one or two levels of hybrid combiners are used to achieve an unusually high number of channels at a single, busy AMPS cell site.

AMPS difficulties notwithstanding, GSM systems often use hybrid combiners since they have a relatively smaller number of transmitters per site (sometimes just two) and also because they cannot easily frequency hop if fixed tuned combiners are used.

A third method of isolation is to use multiple channel linear-power amplifiers. The signals of the various transmitters are combined at low power in a resistive network which does not suffer from intermodulation distortion, and then all signals are amplified together to operating power levels. The required amplifiers are expensive and critical to adjust, but this technique is being used successfully at 800 MHz in commercial network equipment.

The frequency plan finally adopted for a system must satisfy (1) cochannel C/I objectives at the cell boundaries, (2) freedom from adjacent-channel assignments in the same cell and careful deployment in adjoining cells, and (3) sufficient isolation between transmitters to avoid intermodulation interference.

There is no need for frequency planning with CDMA systems. They operate with a large number of users, typically 16 to 22 maximum, carried on a single RF channel which is reused in every sector and cell. Therefore, there is no frequency planning in the same sense as for the other technologies described above. The signals of individual cells and sectors are distinguishable because they are offset by differing numbers of bits of a pseudonoise sequence, the short PN code. Code planning serves the same role as frequency planning in the channelized technologies. There are a total of 512 discrete PN offsets available for assignment. The assignments are not totally unrestricted, since propagation delay between the cell and the mobile is significant in large cells and must be considered. This sets a lower limit for the number of offset steps required to ensure that no ambiguity can occur between any two given cells. However, the PN offset planning is considerably more relaxed than frequency planning. The PN offsets are a more plentiful resource than frequencies for the other technologies.

## 10.5 Example System Design

In this section, the principles described in the previous section are used to design a GSM network for the Dallas–Fort Worth BTA. This design was done using PlaNet tool from MSI. The modeling and assumptions that are input to the PlaNet tool are presented followed by the output in terms of the number of cells and transceivers needed for specific penetration and usage rates. The following sections show an example design of a 1900-MHz GSM system for the Dallas, Texas, BTA.

### 10.5.1 Modeling and design assumptions

A summary of the design assumptions is given in Table 10.3.

**10.5.1.1 Coverage zones.** The basic trading area (BTA) is divided into four coverage zones. These zones represent different RF and demo-

**TABLE 10.3   Modeling and Design Assumptions**

| | | | |
|---|---|---|---|
| Coverage criteria | Probability of coverage | Cell area | 95% |
| Margins | Fade margin | Outdoor | 8.6 dB |
| | | Shadow | 8 dB |
| | In-car margin | | 6 dB |
| RF signal level | Outdoor | All zones | −92 dBm |
| Thresholds | In-car | All zones | −86 dBm |
| Subscriber unit | Output power | Portable | 1 watt |
| Propagation modeling | Base models | COST-231 | |
| Traffic modeling | | | |
| | Penetration rate | Year 1 | 1% |
| | | Year 2 | 2% |
| | | Year 3 | 4% |
| | Subscriber usage | | 30 mE/Subscriber |
| | Grade of service | | 2% Erlang B |
| | Land-use weights (traffic spreading) | Commercial | 40% |
| | | Residential | 40% |
| | | Paved areas | 10% |
| | | Agricultural | 3% |
| | | Forest | 5% |
| | | Rangeland | 1% |
| | | Water | 0% |
| | | Open land | 1% |
| BTS assumptions | BTS EIRP | Suburban | 54 dBm |
| | | Rural | 60 dBm |
| | Antenna (type / gain) | | DB978H90 |
| | BTS antenna heights | Suburban | 30 meters |
| | | Rural | 90 meters |
| | Sectors / cell | All cells | 3 |

graphic environments. The zones are preliminary defined using population density derived from a zip-code-based population database. The zones are then fine-tuned using the USGS land-use database.

Dense urban zones consist mainly of center city, high-rise apartment complexes, large-scale, high-rise office complexes, and large-scale shopping complexes. This is reflected in the land-use database as commercial. Therefore, dense urban zones encompass city zip codes with a high concentration of commercial areas.

Light urban zones are comprised of small town centers, low-rise apartment complexes, low-rise office complexes, light industrial areas, or areas around strip shopping complexes. Light urban zones include densely populated zip codes with a population density of over 19,200 Pops/km$^2$ (7,500 Pops/mi$^2$).

Suburban zones are composed of single family homes and residential areas. Suburbs range in density from 2,500 Pops/km$^2$ (1,000 Pops/sq miles) to 19,200 Pops/km$^2$ (7,500 Pops/mi$^2$).

Rural zones consist of all remaining low-density areas.

Dallas has an unusually small downtown core which is largely unpopulated. For this reason, additional data would be required to identify the urban area. Due to its small size, the designers choose to ignore this area for their design and treat the area as part of the suburban area. The error in cell count due to this omission is expected to be within the margin of error of any design.

**10.5.1.2  Coverage strategy.** Three phases are considered for this design. Phase 1 represents the start-up network and consists of the Dallas and Fort Worth urbanized areas including major primary roads in the market center of the BTA. This footprint will allow for the user-friendly testing stage of deployment. In phase 2, most of the suburban areas will be covered in both markets. This will be sufficient for launch of commercial service. Total area and population coverage are achieved in phase 3.

**10.5.1.3  Coverage criterion.** For this design, 90 percent pops coverage was required throughout the entire BTA at 95 percent reliability over the cell area (85 percent at cell edge) for all areas.

**10.5.1.4  Margins.** The shadow-fading effect is a source of uncertainty when predicting the median signal level. For this design, the industry-accepted standard deviation of 8 dB was used.

In the absence of a fade margin the reliability at cell edge is 50 percent. The margin applied to increase the reliability to 95 percent over the cell area is 8.6 dB. Calculations are based on a single-server probability of coverage calculation. For the above calculations, the path loss-slope of 44.9 dB/decade and a standard deviation of 8 dB are used.

An additional 6 dB of loss is added to the fade margin to account for losses when the subscriber is located inside a vehicle. Although the standard deviation for this value is high, a mean value of 6 dB has become an unofficial industry-accepted standard for PCS 1900.

**10.5.1.5  Link budget.** The link budget for the design is given in Table 10.4.

The isotropic effective receive sensitivity level is the minimum signal level needed for 50 percent reliability (cell edge) of reception at the mobile station. Assuming a −102 dBm sensitivity and considering a body loss of 3 dB and an antenna gain of 2 dBi, the effective sensitivity level is: $-102 + 3 - 2 = -101$ dBm.

**10.5.1.6  RF signal-level thresholds.** The minimum signal power levels required to provide the desired coverage quality and reliability are given in Table 10.5.

**TABLE 10.4    Link Budget**

| | Suburban BTS—MS (downlink) | Suburban MS—BTS (uplink) | Rural BTS—MS (downlink) | Rural MS—BTS (uplink) |
|---|---|---|---|---|
| Transmit power | 39 dBm | 30 dBm | 41.5 dBm | 30 dBm |
| TX antenna gain | 18 dBi | 2 dBi | 18 dBi | 2 dBi |
| Body loss | 0 dB | 3 dB | 0 dB | 3 dB |
| TX feeder loss | 3 dB | 0 dB | 0.5 dB | 0 dB |
| TX combiner loss | 3 dB | 0 dB | 0 dB | 0 dB |
| Isotropic transmit EIRP | 54 dBm | 29 dBm | 60 dBm | 29 dBm |
| Isotropic transmit EIRP | 251 watts | 0.79 watts | 1000 watts | 0.79 watts |
| Receiver sensitivity level | −102 dBm | −104 dBm | −102 dBm | −107 dBm |
| Receive antenna gain | 2 dBi | 18 dBi | 2 dBi | 18 dBi |
| Diversity gain | 0 dB | 4 dB | 0 dB | 7 dB |
| Receive system loss | 0 dB | 0 dB | 0 dB | 0 dB |
| Body loss | 3 dB | 0 dB | 3 dB | 0 dB |
| Isotropic effective receive sensitivity level | −101 dBm | −126 dBm | −101 dBm | −132 dBm |
| Isotropic maximum free-space path loss | 155 dB | 155 dB | 161 dB | 161 dB |

**10.5.1.7  Propagation modeling.**  The base model used is the standard COST-231 which is supported worldwide and widely utilized for planning and implementing GSM-based Personal Communications Systems at 1800 MHz in Europe. COST-231 is a subgroup of the European Cooperation in the field of scientific and technical research whose charter is the development of propagation models for GSM and PCS systems. The COST-231 model is based on the measurement data developed by Okumura and considers the data above 1000 MHz unlike the Hata model. Due to the absence of measured data for unique clutter classifications, signal strength was predicted at mean levels. Signal variances based on specific clutter information were not evaluated in these propagation predictions.

**TABLE 10.5    RF Signal-Level Thresholds**

| | | Outdoor margin | Additional margin | Total margin | Isotropic effective sensitivity | RF signal-level thresholds |
|---|---|---|---|---|---|---|
| Outdoor | 95% probability | 8.6 dB | 0 dB | 8.6 dB | −101 dBm | −92 dBm |
| In-car | 95% probability | 8.6 dB | 6 dB | 14.6 dB | −101 dBm | −86 dBm |

**10.5.1.8  Traffic modeling.**  The goal of traffic modeling is to determine the number of cells and transceivers required to serve the traffic demand. The traffic-modeling methodology utilized is based on zip code population information. A penetration rate is then applied to the population to determine the number of subscribers. A usage figure is applied to the subscriber count to determine the total traffic demand per zip code. The total traffic per zip code is spread-based upon the land-use weights. The assumptions used for traffic modeling are as follows:

| | |
|---|---|
| Penetration rate | 1% in year 1 |
| | 2% in year 2 |
| | 4% in year 3 |
| Subscriber usage | 30 mErlang / subscriber in busy hour |
| Grade of service | 2% Erlang B |

### 10.5.2  Results summary

The output from the PlaNet tool based on the inputs from the previous section are summarized in the sections following.

**10.5.2.1  Area coverage.**  The composite coverage maps for year 1, year 2, and year 3 are shown in Figs. 10.29, 10.30, and 10.31, respectively.

In this design, 92 percent of the total pops of the BTA were covered to the required thresholds. Areas which did not meet the required threshold consisted mainly of rural areas where total coverage is unrealistic due to the need for excessive numbers of low-revenue sites. Although these areas are below threshold, there is still some degree of coverage in most of this area, however the reliability in these areas is lower than 95 percent.

In commercial, residential, and paved areas (roadways), 87 percent of the area received coverage in excess of the specified reliability. These three areas account for most of the total population.

In-car, portable coverage with 95 percent reliability was specified. However, the amount of outdoor, on-street coverage that results from the in-car design is approximately 68 percent of the total area. Commercial residential and paved areas (93 percent) were covered to outdoor levels. Typically, cellular systems were designed for outdoor coverage.

Mean signal strength in all areas was −77 dBm. The acceptable range for mean signal strength is approximately −85 dBm to −75 dBm based on this design's assumptions. Greater than this amount would indicate that the system had been overdesigned. Less than this amount would indicate that the system had too many coverage holes. The resulting values in this design are within this range and fall toward the conservative (overdesigned) end of the range.

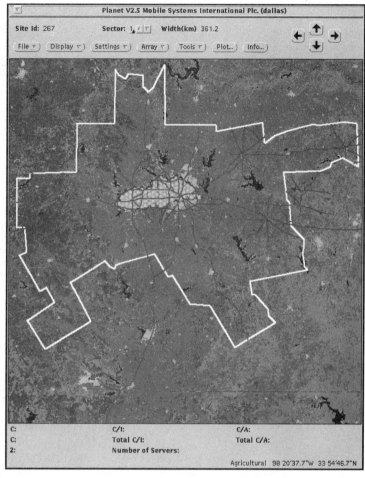

**Figure 10.29** Year 1 composite coverage.

**10.5.2.2 Cell count.** The final cell count for the Dallas–Fort Worth market was 249 BTSs with the yearly breakdown given in Table 10.6.

The bulk of the sites (86 percent) were required in the suburban areas; 14 percent of the sites were rural. At the time the design was completed, the urbanized areas of Dallas were considered too insignif-

**TABLE 10.6   Cell Count for the Dallas–Fort Worth BTA**

| Cell count | Suburban | Rural | Total |
|------------|----------|-------|-------|
| Year 1 | 92 | 0 | 92 |
| Year 2 | 102 | 5 | 107 |
| Year 3 | 21 | 29 | 50 |
| Total | 215 | 34 | 249 |

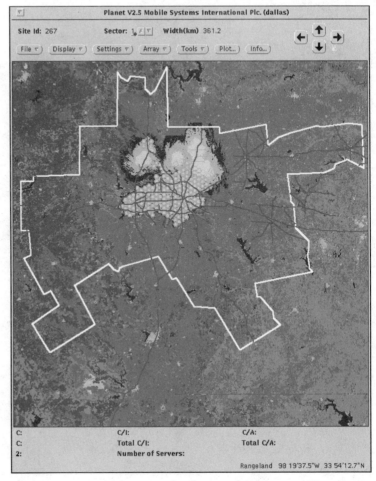

**Figure 10.30**   Year 2 composite coverage.

icant to warrant special attention. In a more thorough design, there may have been 5 to 10 urban cells in the downtown core. The cell radii used in this design by region are given in Table 10.7.

**10.5.2.3   Traffic analysis.**   In year 1, as few as 367 TRXs are required to provide capacity to meet the traffic demands of the market. This is as few as four radios per BTS. In order to ensure coverage in the case of a radio failure, it is common practice to specify a minimum of two radios per sector for all sectors. Using this requirement, 562 radios would be required in year 1. The worst-case sites require 7 TRX per BTS. In general, there were no·capacity problems which would require cell splitting to meet year 1 demand. However, the final design deployed should

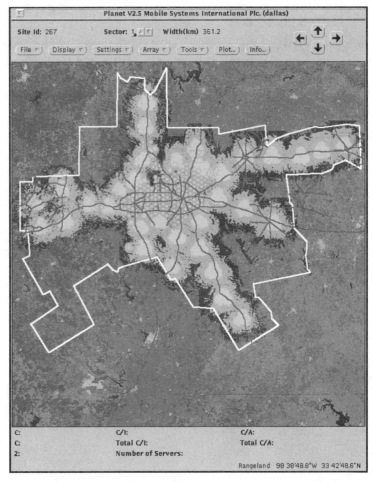

**Figure 10.31**  Year 3 composite coverage.

account for traffic at levels higher than year 1. On average, there were 253 subscribers per cell or 41 subscribers per TRX. As a reference, analog cellular is considered profitable at 20 subscribers per radio.

In year 2, 882 TRXs are required to provide capacity to meet the traffic demands of the market. This is as 4.5 radios per BTS. Using the two-

**TABLE 10.7  Cell Radii by Region**

| Region | Radius (km) |
|--------|-------------|
| Suburban | 2.6 |
| Rural | 15 |

radios-per-sector requirement, 1179 radios would be required in year 2. The worst-case sites require 11 TRX per BTS. The year 2 design required three cells (1.5 percent) to be added for capacity purposes. On average, there were 348 subscribers per cell or 59 subscribers per TRX. This is a substantial improvement in utilization.

In year 3, at least 1253 TRXs are required to provide capacity to meet the traffic demands of the market. This is an average of five radios per BTS, with a worst case of 14 radios after cell splitting. On average, there were 680 subscribers per cell or 114 subscribers per TRX. An additional 21 capacity cells (8 percent) were added at year 3, 4 percent penetration. This is consistent with previous capacity studies in GSM networks.

# 11

# PCS Economics

## 11.1  Introduction

The PCS industry structure was discussed in the previous chapters. Services that could substitute and/or complement PCS services were presented in Chaps. 2, 3, and 4. These competing personal communications services, depending on the extent of competition in the industry, could either expand and/or contract the market share from the PCS operators. The details of the different technologies, both RF and network access technologies that are available to operators, were also presented.

The industry structure impacts profitability since it defines the competitive forces to which it is subjected. It defines the pressures on the industry from its suppliers and buyers. These forces along with regulatory ones define the floor-rate profitability. The floor-rate profitability will drive the willingness of the nine potential operators to survive in the market.

In addition to these competitive forces, the PCS industry is also influenced by the dynamics of new technologies. Choice of technology could decide the availability of equipment for network operation, as well as capital and operating investment. Each of these technologies discussed in the previous chapters has its merits and demerits. No one technology has a cost structure that is better than the other in all market deployment scenarios. A case can be made for each technology to have a better-cost structure than all others for certain applications. A business case can be made to appear better than another, based on assumptions (either realistic or unrealistic) of network deployment parameters, such as network coverage, penetration rate, and equipment-sensitivity values.

From a business point of view, the operators will have to constantly weigh not only the capital investment and operating cost of different technologies, but also the availability of equipment. Availability of equipment will determine the cash flow and eventual profitability of each operator. Delay in deploying the network will make it harder to dislodge customers from their current operators, thereby increasing the operators' entry barriers. In addition, the operators will lose the increased operational and marketing experiences associated with the learning curve from operating a network.

As indicated previously, subscribers are not concerned with the aspects of technology and architecture. They care about cost, quality, and features provided by the services. There will not be a single victor among the different PCS technologies, but there will be several winners. Each of the different PCS technologies will survive and is better suited than others for certain applications and deployment scenarios. The classic dilemma for the new operators is shown in Fig. 11.1.

In order to avoid the religious warfare among the different technology players, this section will not concentrate on any one technology but will develop a typical PCS operator's business case. Most of the design and cost values used in this scenario might be GSM, not because of any inherent advantage, but more from the availability of information of about GSM equipment and its costs. The other tech-

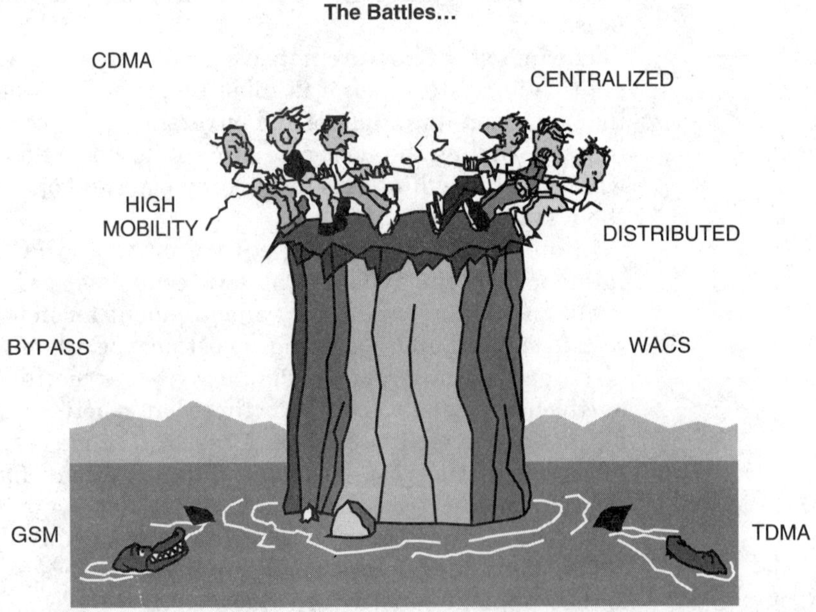

**Figure 11.1** Several winners but no single victor.

nologies being new, many of the technical and commercial parameters are "guest estimates" due to lack of sufficient operational experience. It appears that among the different technology, GSM is the only one with operational status. Developing a business case based on theoretical factors and estimates may not lead to proper analysis. An example is how Mercury Network in the United Kingdom grossly underestimated the number of cells required for initial basic-system coverage by a factor of 2. One of the factors that led to this gross underestimation is that GSM was relatively new and the technical parameters did not have the actual operational validity at the time its network was designed. The technical-design parameters had not been subjected to the rigors of a live environment. The accuracy of these nonoperational PCS technologies have the same accuracy as the GSM estimates of 1992. Whereas, CDMA has undergone several field tests, therefore CDMA's design parameters might have more validity than the GSM design parameters of 1992. However, CDMA still has not gone through the rigors of a high-capacity sensitive live environment.

The purpose of this section is not to compare the technologies, nor to sit in judgment on the merits of the different PCS technologies. This section is to provide PCS operators with the sources of revenue, costs, and show how the industry structure, with its current two cellular operators, impacts the business case for a new operator. The merits and demerits of the different technologies will be decided in the marketplace, and it is not the intent to fuel or confuse the ongoing battle among the different technologies. This section provides a business case based on one set of assumptions. It is not representative of the business case for any particular PCS operator. The data used to develop this business case is a typical composite of a middle-of-the-road outlook for cellular and PCS strategies for technology, pricing, interconnect, promotion, service offering, and other engineering and marketing data. The projections used are not forecasts for any PCS operators of what is about to happen, but one of the many possible outcomes of the inevitable disruptions caused by the emergence of PCS competition. Data is offered for the readers to understand the different elements that impact a PCS operator's business case and as a reasonable starting point for the user in an exploration of how the wireless market will evolve under a wide range of competitive, strategic thrusts and counterthrusts. This is done by analyzing a hypothetical, competitive environment encountered by a new BTA operator (BTAOp A) who has the license to operate the Dallas–Fort Worth BTA.

The business analysis in this chapter exclusively employed the Impulse Telecommunication Corporation's WIST business-analysis tool. WIST is a comprehensive economic-simulation tool that simulates a wide range of possible strategic scenarios for the wireless telecommunications market. The core of the economic-simulation model is found in the concepts and algorithms used to determine the rate at which consumers will buy and use the product or service, including the growth of the total market and the market share taken by each competitor. These models do not take revenue nor subscriber-penetration levels as input. Penetration levels are output derived by the underlying competitive and economic factors. WIST simulates the competitive market by using basic microeconomics principles of demand curves with the adoption rate determined by a diffusion rate time constant. This approach allows WIST the power to predict the market reaction to a wide range of business strategies adopted by the start-up PCS companies and by more established cellular companies.

The functional elements of an organizational structure required to build, maintain, and operate a successful wireless telecommunications network is presented in the next few sections. The new PCS operator will encounter a market environment widely different from the one faced by cellular operators in 1980s. The market area, the industry structure, as well as the key-decision variables such as price, packaging, distribution, and coverage are presented. This is followed by the technical and operational issues a new operator needs to plan and establish its business. This presentation is hypothetical and is not a representation of any particular operator. The purpose is to enlighten a new operator of the major elements that need to be focused on upon entering the business.

The marketing and technical inputs incorporated in these sections are inputs into the WIST simulation-modeling tool in order to obtain the financial pro forma. The business analysis using the WIST tool is based on a simplified model. The actual WIST tool has several enhanced features that provide a flexibility to model different scenarios and engage in "war game" and "what if" analysis. This tool is to wireless telecommunications players, operators, vendors, and consultants, as Quicken is for personal money management.

## 11.2   Organizational Structure

After raising of money for auction and the acquisition of a license, a BTA operator has completed the first and relatively easier part of the PCS operation. The next challenge is to build, maintain, and evolve a BTA network to compete with other entrenched players. The skill set of the auction team is raising money, analyzing, and anticipating (actually gambling or is it glorified game theory?) the moves of other auction participants. The auction team will need to be complemented by other skills required for building and operating a wireless telecommunications network.

Organizations can be structured by departmentalizing function, products, and matrix. Each has its merits and demerits, the success of each depends on the culture of the company, style of its management team, and its employees. A functional organization leads to increased efficiency and economic utilization of the employees, but it may also lead to situations of goal suboptimization as each department focuses on its own specialty or function at the expense of others. For example, the marketing functional group identifies certain area or coverage zones that are ideally suited to obtain new premium subscribers, while the site-acquisition functional group considers those same areas as troublesome because of the difficulty of acquiring sites there either due to zoning, expense, or aesthetic reasons. Usually, a team with representatives from the different functional groups will solve the problem of functional suboptimization and will improve communication between the different functional groups. Alternatively, a corporate planning group with the ability to draw on resources from the functional groups could also help improve communication between the different groups.

A generic organization structure for a BTA operator is shown in Fig. 11.2.

The number of employees in each group will depend on the tasks to be performed, its focus, bias, and the dominance of the individual leaders within the management team. At a microlevel, the number of employees in each department will be influenced by the number of subscribers, number of cell sites, and number of switches installed and planned for the network. The more the number of switches and cell sites, the greater the number of people in the operations department to build and maintain the network.

The organization structure of the preauction company is different from that of the postauction company. Preauction emphasized capital raising and gaming-theory skills while the postauction will emphasize

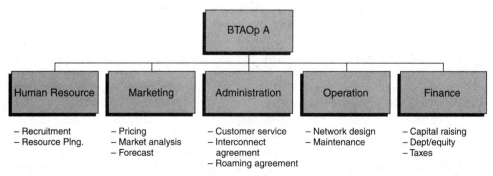

**Figure 11.2**  Organization structure of BTAOp A.

building the network and marketing the services. The organization will then evolve from its focus on building a network to operating a network. As with any organization, it will be impacted by both the external environment of changes in technology, political environment, and from changes in the internal environment from suppliers, customers, regulators, competitors, and ownership. These external and internal forces interact with each other to affect the organization as a whole. The adaptability and flexibility of the organization to future changes is essential for sustaining the organization. Such changes are much more rapid in wireless telecommunications than in industries such as steel or food.

## 11.3    Human Resource Department

In the next few years, six new wireless telecommunications networks will be built across the United States. The number of RF and other engineers needed to build a PCS networks will be enormous. Already, there is a shortage of engineers with RF and mobile telephony expertise. In addition to the RF engineers, there is a general shortage of people with knowledge of newer technology such as GSM and CDMA. Headhunters have been scavenging across the globe trying to find qualified personnel. They have not been very successful because most of the country has started building and/or expanding their cellular networks causing this shortage of qualified people. The universities, as usual, did not anticipate the demand for RF personnel. Therefore, they will not be able to supply enough engineers within the near future. It will be essential to retrain the existing work force in wireless telecommunications.

Human resource departments were once relegated to a second class status in many organizations. Previously, a line manager would have called the human resource department mainly for handling deviant employee matters. Today, the human resource department is invited and included in all major planning meetings. The shortage of people with wireless telecommunications experience along with increased legal complexities and regulatory requirements have enhanced the recognition that human resource personnel provide a valuable means for improving productivity. Increased awareness of the costs associated with poor human resource planning can lead to bad publicity and low morale, as well as unemployment compensation and training expenses. A weak human resource department can fail to keep up with compensation systems that attract and motivate employees. Also, a weak human resource department does not proactively prevent embarrassing discrimination lawsuits.

The human resources department forecasts future demand and internal supply for employees, based on input from other departments. Many vendors and operators are already heavily understaffed on the engineering front. Previously, an RF engineer worked in broadcasting for the same company for three to five years before moving to another company. With the advent of PCS, there is competition among vendors and operators for these engineers. The average length of employment of an RF engineer with a company has reduced significantly.

An operator without an adequate human resource department will most likely encounter potential delay in the network rollout. They will most likely have a less than optimal quality network. They will provide inadequate customer service and thereby impact revenue, survivability, and competitiveness of the organization.

## 11.4  Marketing Department

The marketing department markets the services to potential customers. To do this, it analyzes the market structure and determines the company strength and weakness vis-à-vis the competition. It determines the timing and availability of services and features that will be sold to the subscribers. Based on the demographic profile and competitive analysis, marketing will decide the best coverage zone for a network buildout. Marketing also develops and implements the distribution networks and pricing packages, and creates coherent advertising and promotion to maximize the revenue and profitability of the company. Based on its strength and anticipated competition, a marketing department will forecast the number of subscribers and revenue from which other departments determine their budget.

### 11.4.1  Market area

The Dallas–Fort Worth BTA is shown in Fig. 11.3.

There are 25 counties within this BTA. The different counties within this BTA are shown in Fig. 11.4.

The Dallas–Fort Worth BTA has 4.3 million people who account for approximately 45 percent of the Dallas–Fort Worth MTA population. This BTA covers 18,464 square miles of area, approximately less than 10 percent of the MTA's land area. It has a population density of 235 people per square mile. The demographic information for the different counties and the composite BTA and MTA is shown in Table 11.1.

The population distribution and other demographic information, such as per capita income, will determine the motivation of the MTA operators to aggressively compete in certain counties during the initial years. Approximately 45 percent of the Dallas–Fort Worth MTA's population resides in the Dallas–Fort Worth BTA, therefore, it is

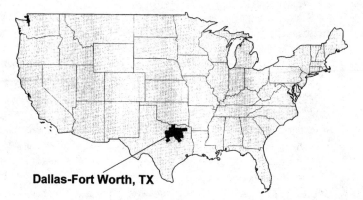

**Dallas-Fort Worth, TX**

**Figure 11.3**   Dallas–Fort Worth BTA.

likely the new MTA operators will focus first on providing coverage in this BTA before covering less dense BTAs within its MTA. Again, there are 25 counties within the Dallas–Fort Worth BTA. Some of these counties are more preferable for initial buildout than others. Collin, Dallas, Denton, and Tarrant counties probably will be included in the initial implementation because of their desirable demographic characteristics. These four counties together comprise approximately 80 percent of the population in less than 20 percent of the BTA's land mass. It has more than two-thirds of all the households in the entire BTA. The average house value in these counties is

**BTA  -  Dallas-Fort Worth, TX**

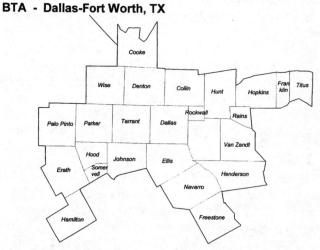

**Figure 11.4**   Counties of Dallas–Fort Worth BTA.

TABLE 11.1   Demographic Information for the Dallas–Fort Worth BTA

| | Population (in thousands) | % of MTA Pop | Area (sq miles) | Number of households (thousands) | Years of school | Median house value ($ thousands) | Per capita income ($ thousands) |
|---|---|---|---|---|---|---|---|
| MTA | 9,694 | 100 | 215,582 | 3,599 | 11.9 | 40.7 | 10.5 |
| BTA | 4,330 | 44.67 | 18,464 | 1,620 | 12.3 | 58.2 | 12.6 |
| Collin | 264 | 2.72 | 848 | 96 | 13.9 | 106.6 | 20.5 |
| Cooke | 31 | 0.32 | 874 | 12 | 12.2 | 47.7 | 11.6 |
| Dallas | 1853 | 19.11 | 880 | 703 | 13.0 | 79.2 | 16.2 |
| Denton | 274 | 2.82 | 888 | 102 | 13.6 | 89.1 | 16.1 |
| Ellis | 85 | 0.88 | 940 | 29 | 12.2 | 67.9 | 12.2 |
| Erath | 28 | 0.29 | 1,086 | 11 | 12.6 | 48.6 | 10.8 |
| Franklin | 8 | 0.31 | 286 | 3 | 12.0 | 46.8 | 12.4 |
| Freestone | 16 | 0.16 | 885 | 6 | 11.8 | 44.7 | 10.7 |
| Hamilton | 8 | 0.08 | 836 | 3 | 11.9 | 33.3 | 11.2 |
| Henderson | 59 | 0.60 | 874 | 23 | 11.9 | 53.6 | 10.7 |
| Hood | 29 | 0.30 | 422 | 11 | 12.5 | 73.8 | 15.0 |
| Hopkins | 29 | 0.30 | 785 | 11 | 11.9 | 43.6 | 11.0 |
| Hunt | 64 | 0.66 | 841 | 24 | 12.3 | 47.3 | 11.8 |
| Johnson | 97 | 1.00 | 729 | 33 | 12.2 | 61.1 | 12.1 |
| Kaufman | 52 | 0.54 | 786 | 18 | 12.0 | 56.6 | 11.6 |
| Navarro | 40 | 0.41 | 1,071 | 15 | 12.0 | 40.9 | 10.5 |
| Palo Pinto | 25 | 0.26 | 953 | 10 | 11.9 | 37.4 | 10.0 |
| Parker | 65 | 0.67 | 904 | 23 | 12.4 | 67.6 | 13.0 |
| Rains | 7 | 0.07 | 232 | 3 | 11.8 | 42.5 | 10.7 |
| Rockwall | 26 | 0.26 | 129 | 9 | 13.3 | 97.9 | 18.0 |
| Somervell | 5 | 0.06 | 187 | 2 | 12.0 | 55.3 | 11.9 |
| Tarrant | 1,170 | 12.07 | 864 | 439 | 13.0 | 72.9 | 15.2 |
| Titus | 24 | 0.25 | 411 | 8 | 12.0 | 44.4 | 11.2 |
| Van Zandt | 38 | 0.39 | 849 | 14 | 11.7 | 45.6 | 10.1 |
| Wise | 35 | 0.36 | 905 | 12 | 11.9 | 49.7 | 11.3 |

48 percent higher than the average house value of $58.2 thousand for the entire BTA. The average per capita income for these four counties is $17 thousand, about 35 percent higher than the average per capita income of the entire BTA. Thus, the focus of the initial system implementation for most new operators will be in Collin, Dallas, Denton, and Tarrant counties. Of course, within these four counties, further analysis based on zip codes and school districts can lead to additional microlevel focus.

## 11.4.2   Market structure

The PCS operators were licensed based on the MTA and BTA areas defined in Rand McNally's *1992 Commercial Atlas and Marketing Guide.* However, the cellular operators were licensed based on Metropolitan Statistical Area (MSA) and Rural Service Area (RSA). There could be several MSA and RSA areas within a MTA and BTA. The major MSA/RSA within the Dallas–Fort Worth BTA is shown in Fig. 11.5.

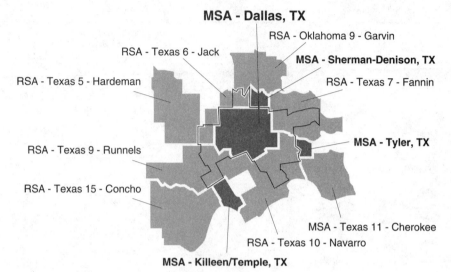

**Figure 11.5**    MSAs and RSAs of Dallas–Fort Worth BTA.

Counties are fully contained within the MTA/BTA boundaries. Each MTA/BTA has multiple counties within its area. The RSA and county boundaries, unlike the MTA/BTA and county boundaries, do not match. Potentially there could be multiple cellular operators in a county. Franklin county has seven cellular operators, though not more than two provide service in the same area. In addition, the same RSA could span multiple counties as in the case of RSA #658 (Texas 7—Fannin). Texas 7—Fannin RSA covers Franklin, Hopkins, Hunt, Rains, Titus, and Wood counties. Similarly RSA #660 (Texas 10—Runnels) covers some areas of Hamilton, Erath, and Somervell. In addition, there are counties where in some areas the service is provided by the MSA operator; while in other areas of the same county, service is provided by the RSA operators. Henderson county has both MSA and RSA areas. The service is provided in the MSA area by AT&T Wireless and Southwestern Bell, and in the RSA area service is provided by Sprint Cellular and GTE Personal Communications Services.

Table 11.2 shows the cellular providers in the different counties of the Dallas–Fort Worth BTA and their date of operation. The cellular operators in the different counties are the competitors for the new PCS operators in those counties. The date of operation indicates the start of these cellular operators and the associated head start over the new PCS operators. This information and the competitive strength of the other operators will help the new operators in planning and investing scarce marketing resources. This information and the demographic information will help a new PCS operator plan the network's deployment phases. All these types of information provide the competitive threat the new PCS operator is likely to encounter in each of the counties within the Dallas–Fort Worth BTA.

**TABLE 11.2  Existing Cellular Operator within the BTA**

| County | Cellular provider | Date of operation |
|---|---|---|
| Collin | AT&T Wireless | 3/86 |
| | Southwestern Bell Mobile Systems | 7/84 |
| Cooke | AT&T Wireless | 7/91 |
| | Southwestern Bell Mobile Systems | 7/90 |
| Dallas | AT&T Wireless | 3/86 |
| | Southwestern Bell Mobile Systems | 7/84 |
| Denton | AT&T Wireless | 3/86 |
| | Southwestern Bell Mobile Systems | 7/84 |
| Ellis | AT&T Wireless | 3/86 |
| | Southwestern Bell Mobile Systems | 7/84 |
| Erath | Lone Star Cellular, Inc. | 1/92 |
| | Southwestern Bell Mobile Systems | 1/91 |
| | Mid-Tex Cellular, L.P. | 8/91 |
| | Sprint Cellular Company | 5/92 |
| Franklin | K.O. Communications, Inc. | 4/92 |
| | Southwestern Bell Mobile Systems | 7/90 |
| | Sprint Cellular Company | 12/91 |
| | Peoples Cellular | 9/91 |
| | Etex Cellular Co., Inc. | 6/91 |
| | Lamar County Cellular, Inc. | 6/91 |
| | Century Cellunet | 8/91 |
| Freestone | Southwestern Bell Mobile Systems | 10/90 |
| | Sprint Cellular Company | 2/92 |
| | GTE Personal Communications Services | 4/91 |
| Hamilton | Lone Star Cellular, Inc. | 1/92 |
| | Southwestern Bell Mobile Systems | 1/91 |
| | Mid-Tex Cellular, L.P. | 8/91 |
| | Sprint Cellular Company | 5/92 |
| Henderson | AT&T Wireless Systems | 3/86 |
| | Southwestern Bell Mobile Systems | 10/90 |
| | Sprint Cellular Company | 2/92 |
| | GTE Personal Communications Services | 4/91 |
| Hood | AT&T Wireless | 3/86 |
| | Southwestern Bell Mobile Systems | 7/84 |
| Hopkins | K.O. Communications, Inc. | 4/92 |
| | Southwestern Bell Mobile Systems | 7/90 |
| | Sprint Cellular Company | 12/91 |
| | Peoples Cellular | 9/91 |
| | Etex Cellular Co., Inc. | 6/91 |
| | Lamar County Cellular, Inc. | 6/91 |
| | Century Cellunet | 8/91 |
| Hunt | K.O. Communications, Inc. | 4/92 |
| | Southwestern Bell Mobile Systems | 7/90 |

**TABLE 11.2    Existing Cellular Operator within the BTA (*Continued*)**

| County | Cellular provider | Date of operation |
| --- | --- | --- |
| | Sprint Cellular Company | 12/91 |
| | Peoples Cellular | 9/91 |
| | Etex Cellular Co., Inc. | 6/91 |
| | Lamar County Cellular, Inc. | 6/91 |
| | Century Cellunet | 8/91 |
| Johnson | AT&T Wireless | 3/86 |
| | Southwestern Bell Mobile Systems | 7/84 |
| Kaufman | AT&T Wireless | 3/86 |
| | Southwestern Bell Mobile Systems | 7/84 |
| Navarro | Southwestern Bell Mobile Systems | 10/90 |
| | Sprint Cellular Company | 2/92 |
| | GTE Personal Communications Services | 4/91 |
| Palo Pinto | AT&T Wireless | 7/91 |
| | Southwestern Bell Mobile Systems | 7/90 |
| Parker | AT&T Wireless | 3/86 |
| | Southwestern Bell Mobile Systems | 7/84 |
| Rains | K.O. Communications, Inc. | 4/92 |
| | Southwestern Bell Mobile Systems | 7/90 |
| | Sprint Cellular Company | 12/91 |
| | Peoples Cellular | 9/91 |
| | Etex Cellular Co., Inc. | 6/91 |
| | Lamar County Cellular, Inc. | 6/91 |
| | Century Cellunet | 8/91 |
| Rockwall | AT&T Wireless | 3/86 |
| | Southwestern Bell Mobile Systems | 7/84 |
| Somervell | Lone Star Cellular, Inc. | 1/92 |
| | Southwestern Bell Mobile Systems | 1/91 |
| | Mid-Tex Cellular, L.P. | 8/91 |
| | Sprint Cellular Company | 5/92 |
| Tarrant | AT&T Wireless | 3/86 |
| | Southwestern Bell Mobile Systems | 7/84 |
| Titus | K.O. Communications, Inc. | 4/92 |
| | Southwestern Bell Mobile Systems | 7/90 |
| | Sprint Cellular Company | 12/91 |
| | Peoples Cellular | 9/91 |
| | Etex Cellular Co., Inc. | 6/91 |
| | Lamar County Cellular, Inc. | 6/91 |
| | Century Cellunet | 8/91 |
| Van Zandt | Southwestern Bell Mobile Systems | 10/90 |
| | Sprint Cellular Company | 2/92 |
| | GTE Personal Communications Services | 4/91 |
| Wise | AT&T Wireless | 3/86 |
| | Southwestern Bell Mobile Systems | 7/84 |

It is clear that AT&T Wireless Systems and Southwestern Bell Mobile Systems cover most of the population in the Dallas–Fort Worth BTA. These two operators are clearly the major competitors. They have more than ten years of operational and marketing experience. Together they account for 430,000 subscribers in 1993 with AT&T Wireless having 40 percent and Southwestern Bell having 60 percent of the market share. The brand names for these two carriers are well established. AT&T Wireless has started bundling cellular services with their long-distance services. It is possible that Southwestern Bell will also bundle their cellular services with their local and other future services such as long distance with further deregulation. In the past few years, these two operators have upgraded their network to be digital ready (i.e., the network can be easily upgraded to digital without much transition pain). Each company has deep pockets with respect to advertising and promotion. In the last few years, each has shown creative-pricing packages to attract subscribers, and they have developed a distribution network which is better than those of many other operators in other MSAs. However, their service packaging and availability does not meet the needs of the consumer market.

In addition to the cellular operators in the Dallas–Fort Worth BTA, there will be two new PCS operators that are licensed to provide service for the entire Dallas–Fort Worth MTA. The Dallas–Fort Worth MTA includes the Dallas–Fort Worth BTA. These two operators are PCS PrimeCo (a partnership among Bell Atlantic, NYNEX, Airtouch, and US West), and Sprint Telecommunications Venture (a partnership among Sprint, TCI, Cox, and Comcast). Both of these PCS operators have nearly a nationwide footprint. They can virtually provide nationwide coverage. They, like the cellular operators, have deep pockets with respect to advertising and promotion. Both PCS PrimeCo and Sprint Telecommunications Venture (STV) can sustain long periods of adverse profitability in order to achieve market share.

The key competitors to a new Dallas–Fort Worth BTA operator will be the following:

1. AT&T Wireless Systems
2. PCS PrimeCo
3. Southwestern Bell Mobile Systems
4. Sprint Telecommunications Venture

There are several small operators who distinguish themselves by providing niche and localized services in addition to these four operators who have formidable resources in the technical, marketing, and financial areas.

This business analysis assumes there are five wireless telecommunications providers, though there could be many wireless telecommunications providers in the Dallas–Fort Worth BTA. Of the five, two are the existing cellular operators (CellOp A and CellOp B) as described in the previous section. The remaining three are the new PCS operators.

Two of these are the MTA operators (MTAOp A and MTAOp B) and the third is the BTA operator (BTAOp A).

### 11.4.3    Product and packaging

BTAOp A plans to provide a wide range of services to its customers. These include voice, supplementary, short message, class III fax, and packet-data services. These services will be of digital quality like those offered by the new MTA operators. The cellular and PCS operators will offer similar features, but PCS will be able to differentiate from cellular services by end-to-end digital quality, clarity of voice, and security of conversations. Although the cellular operators network is digital ready, it will be difficult to convert the network to digital due to the previously discussed legacy and historical weight of the existing customer base.

Price is a major factor that a consumer considers before subscribing to a service. There are several attributes of the product and services that the consumer prefers in addition to price. In many cases, consumers are willing to pay premium for certain desirable features. The major service attributes that a consumer prefers are digital, privacy of conversation, ability to send and receive data, ability to roam, and overall voice quality.

Consumer buying patterns in electronics goods have shown that consumers generally prefer digital products and services to analog products and services. There is an expectation that digital technology provides static-free voice quality and that phones will have a broad set of displays and push-button features.

Consumers are well aware of stories of cellular telephone conversation of the rich and famous being taped. These conversations have created embarrassing incidents that have caused many political figures to resign and have even embarrassed the U.K.'s royal couple. Digital technology allows embedded encryption techniques making it difficult to eavesdrop.

The proliferation of the Internet and other services has made the consumers acutely aware of the need for telecommunications networks to support data services. Many mobile professionals need to access e-mail and data that is resident on other networks. A product or service that can provide wireless data services will be desirable.

Consumers prefer to have a wireless telephone service when they are in other cities, and while they are in transit. Consumers also prefer that the network provides local and roaming mobility when they are outside of their home area.

Consumers are used to the voice quality of the landline. The landline voice is digitally coded with 64 kbps PCM. Most digital wireless telecommunications used compression to increase channel capacity.

Compression techniques impact the quality of the voice conversation. In addition to the inherent quality of voice, it is also related to background noise. In digital systems, background noise is minimized and the conversations are crisp and sharp.

The service attributes for the different competitors can be represented by an index value from 0.0 to 1.0. An index value of 1.0 to a service attribute represents that a typical prospective consumer considers a service attribute totally acceptable. An index value of 0.0 to a service attribute represents that a typical prospective consumer considers the service attribute totally unacceptable. The value of the service attributes described previously for each competitor is shown in Table 11.3.

Cellular operators have a clear advantage with respect to coverage due to their head start, while the new PCS operators have an advantage, due to their lack of a legacy, with respect to privacy and digital network. The new PCS operators can provide the latest technology to the subscribers without worrying about a transition plan for existing subscribers, due to the absence of historical baggage of older technology and an existing installed base.

**TABLE 11.3   Relative Product Strength Index of Different Operators**

| Operators | Factors | Consumer 1997 | 2010 | Business 1997 | 2010 | Emergency 1997 | 2010 |
|---|---|---|---|---|---|---|---|
| CellOp A | Digital | 0.00 | 1.00 | 0.00 | 1.00 | 0.00 | 1.00 |
| | Privacy | 0.00 | 1.00 | 0.00 | 1.00 | 0.00 | 1.00 |
| | Data | 0.00 | 1.00 | 0.00 | 1.00 | 0.00 | 1.00 |
| | Roaming coverage | 0.85 | 1.00 | 0.85 | 1.00 | 0.85 | 1.00 |
| | Voice quality | 0.85 | 1.00 | 0.85 | 1.00 | 0.85 | 1.00 |
| CellOp B | Digital | 0.00 | 1.00 | 0.00 | 1.00 | 0.00 | 1.00 |
| | Privacy | 0.00 | 1.00 | 0.00 | 1.00 | 0.00 | 1.00 |
| | Data | 0.00 | 1.00 | 0.00 | 1.00 | 0.00 | 1.00 |
| | Roaming coverage | 0.85 | 1.00 | 0.85 | 1.00 | 0.85 | 1.00 |
| | Voice quality | 0.85 | 1.00 | 0.85 | 1.00 | 0.85 | 1.00 |
| MTAOp A | Digital | 1.00 | 1.00 | 1.00 | 1.00 | 1.00 | 1.00 |
| | Privacy | 1.00 | 1.00 | 1.00 | 1.00 | 1.00 | 1.00 |
| | Data | 1.00 | 1.00 | 1.00 | 1.00 | 1.00 | 1.00 |
| | Roaming coverage | 0.00 | 0.75 | 0.00 | 0.75 | 0.00 | 0.75 |
| | Voice quality | 0.90 | 1.00 | 0.90 | 1.00 | 0.90 | 1.00 |
| MTAOp B | Digital | 1.00 | 1.00 | 1.00 | 1.00 | 1.00 | 1.00 |
| | Privacy | 1.00 | 1.00 | 1.00 | 1.00 | 1.00 | 1.00 |
| | Data | 1.00 | 1.00 | 1.00 | 1.00 | 1.00 | 1.00 |
| | Roaming coverage | 0.00 | 0.75 | 0.00 | 0.75 | 0.00 | 0.75 |
| | Voice quality | 0.90 | 1.00 | 0.90 | 1.00 | 0.90 | 1.00 |
| BTAOp A | Digital | 1.00 | 1.00 | 1.00 | 1.00 | 1.00 | 1.00 |
| | Privacy | 1.00 | 1.00 | 1.00 | 1.00 | 1.00 | 1.00 |
| | Data | 1.00 | 1.00 | 1.00 | 1.00 | 1.00 | 1.00 |
| | Roaming coverage | 0.00 | 0.75 | 0.00 | 0.75 | 0.00 | 0.75 |
| | Voice quality | 0.90 | 1.00 | 0.90 | 1.00 | 0.90 | 1.00 |

### 11.4.4    Market segment

The operators could focus on different market segments depending upon the strengths and weaknesses and market environment. The market segment is a distinct subset of the population, and is served by a distinct set of service offerings. Each market segment has distinct needs. The market segments differ in their consumption of services at different prices, therefore they are characterized by different demand curves.

It is assumed each of the five operators focus on the same three market segments in this business analysis. The three market segments are:

1. Business

2. Consumer

3. Emergency

The business segment is the total employed population of the market. This segment buys services because of business needs. They view wireless services as an efficiency tool to improve their productivity. The consumer segment buys services because it would be desirable, and not because the service is essential. This segment is very sensitive to price. The emergency segment is a subset of the consumer market characterized by infrequent use of cellular phone and low monthly price.

### 11.4.5    Coverage

The FCC requires that licenses must have a signal level sufficient enough to provide adequate services to at least one-third of the population in the licensed area within five years of being licensed. Also, two-thirds of the population in their license area must be covered within ten years of being licensed. Operators of a metropolitan BTA, such as Dallas–Fort Worth, will most likely cover the five- and ten-year FCC buildout requirement in a much shorter time frame due to competitive reasons as well as the densely localized population. BTAOp A must plan to build its network to provide competitive coverage comparable to that of the dominant cellular operators. There are critical relationships between penetration and market share with timing and extent of coverage. Penetration and market share increases with earlier market entry and wider coverage. Wider coverage allows subscribers to make and receive calls in a wider area.

The BTA network will be implemented in five stages as shown in Table 11.4. Initial implementation will cover the business areas and the major highways such as I-75 and I-635 that connect the business and residential areas. It will also cover residential areas that are demographically more likely to be subscribers to a new PCS service. This covers about 37 percent of the population and 3 percent of the land area.

TABLE 11.4   System Implementation and Coverage for All Operators

| Operators | Phases | Year | % of area | % of population |
|---|---|---|---|---|
| CellOp A | Initial coverage starts | 1985 | 3 | 37 |
| | Phase 1 starts | 1987 | 5 | 51 |
| | Phase 2 starts | 1989 | 10 | 76 |
| | Phase 3 starts | 1991 | 31 | 90 |
| | Final buildout starts | 1993 | 70 | 98 |
| CellOp B | Initial coverage starts | 1985 | 3 | 37 |
| | Phase 1 starts | 1987 | 5 | 51 |
| | Phase 2 starts | 1989 | 10 | 76 |
| | Phase 3 starts | 1991 | 31 | 90 |
| | Final buildout starts | 1993 | 70 | 98 |
| MTAOp A | Initial coverage starts | 1996 | 3 | 37 |
| | Phase 1 starts | 1997 | 5 | 51 |
| | Phase 2 starts | 1998 | 10 | 76 |
| | Phase 3 starts | 1999 | 31 | 90 |
| | Final buildout starts | 2009 | 70 | 98 |
| MTAOp B | Initial coverage starts | 1996 | 3 | 37 |
| | Phase 1 starts | 1998 | 5 | 51 |
| | Phase 2 starts | 1999 | 10 | 76 |
| | Phase 3 starts | 2000 | 31 | 90 |
| | Final buildout starts | 2009 | 70 | 98 |
| BTAOp A | Initial coverage starts | 1997 | 3 | 37 |
| | Phase 1 starts | 1998 | 5 | 51 |
| | Phase 2 starts | 1999 | 10 | 76 |
| | Phase 3 starts | 2000 | 31 | 90 |
| | Final buildout starts | 2010 | 70 | 98 |

The next stage of implementation will begin in 1998 and will cover 51 percent of the population and 5 percent of the land area by 1999. This stage will mainly target residential suburban areas. The subsequent implementation will cover 98 percent of the BTA population and 70 percent of the BTA land area. The system implementation and coverage for all five operators are shown in Table 11.4.

There are five wireless telecommunications providers competing in the three market segments in five coverage zones. This leads to seventy-five different elements as shown in Fig. 11.6.

Each shaded area consists of five competitors competing in the business-market segment as in coverage zone 1. PCS operators will compete with cellular operators only in coverage zone 1 (i.e., urban area in the early years). There is no competition to the cellular operators from the PCS operators in the rural areas. Effective competition for all segments throughout the BTA will occur in 1999 unless the PCS operators plans full coverage initially. Many large PCS operators are planning to provide almost full coverage in the initial few years due to their enormous financial strength and organization.

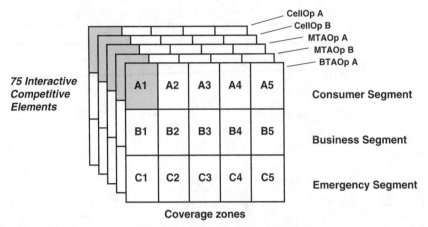

**Figure 11.6**   Competitors in different coverage zones and market segments.

### 11.4.6   Pricing

It is very likely that the new PCS operator with a new digital network will have a large idle capacity. Similar situations in other industries such as the long-distance market, had led to lower prices to recover fixed costs. This allows PCS operators to overcome the ten-year head start over the cellular operators. However, cellular operators are going to react, and probably will be proactive in lowering their prices. The advantage for cellular carriers is that their network is already amortized and they can afford to cut prices. Unfortunately, cellular networks for the most part are analog and do not have significant excess capacity. If the prices are lowered, it will increase the traffic load on the system and because the system is analog it is not designed to handle any significant increases. A sudden reduction in price by cellular operators with its accompanied increase in traffic load could lead to blocked calls causing a poor-service quality perception by subscribers. Furthermore, a cut in price will alter the revenue of the cellular operator. The total revenue could either increase or decrease depending on the elasticity of the demand curve.

Operators have different pricing plans for different subsegments within a market segment. The key elements in most pricing plans are the monthly fixed charge and the airtime charges. The high-usage business customer would be attracted by plans that have high, fixed, monthly charges and reduced price for airtime. The high, fixed, monthly charge includes some minutes of free airtime at peak and nonpeak hours. Different combinations of the fixed monthly charge and airtime charge are used to tailor the pricing plans for the different segments.

A way of comparing prices among the different operators is to develop a single effective price for each segment based on the different pricing plans. The effective price can be determined by plotting the various pricing plans and then determining the best-fit straight line through the lower envelope of all price plans as shown in Fig. 11.7.

The intercept of the line is the effective fixed monthly price and the slope is the effective airtime price. In addition, the prospective customer has several onetime charges, such as the price of the handset and activation charges. These onetime charges can be amortized over the life of the service assumed to be 48 months. These monthly charges of onetime fees are added to the fixed monthly charges. The prices used in the business analysis are offered by the different operators for the different segments as shown in Table 11.5.

BTAOp A has the most aggressive prices. This is partly to gain attention due to its late entry in the market. The head start of cellular operators is relevant in 1997, but may not be relevant in year 2010. It is expected that wireless telecommunications will be a commodity and as ubiquitous as the landline phones by the year 2010. Prices will be low and will be effectively the same for all wireless telecommunications operators. The monthly basic price falls to $15 and airtime charges fall to $0.05 per minute for all market segments for all operators. The effective prices might be the same for all operators but will be packaged differently depending on the segments focused by each operator.

## 11.4.7  Promotion

The current cellular carriers have a strong brand name and recognition. The BTAOp A should launch an aggressive advertising and pro-

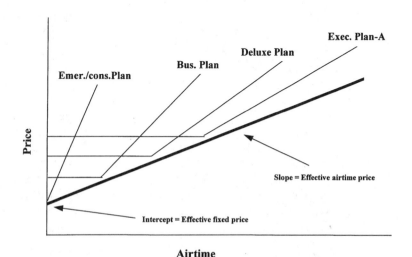

**Figure 11.7**  Computation of effective price.

**TABLE 11.5    Service Prices of All Operators**

| Operators | Prices | Consumer | | Business | | Emergency | |
|---|---|---|---|---|---|---|---|
| | | 1997 | 2010 | 1997 | 2010 | 1997 | 2010 |
| CellOp A | Monthly basic price | $30 | $15 | $40 | $15 | $25 | $15 |
| | Local airtime price | $0.30 | $0.05 | $0.30 | $0.05 | $0.35 | $0.05 |
| CellOp B | Monthly basic price | $30 | $15 | $40 | $15 | $25 | $15 |
| | Local airtime price | $0.30 | $0.05 | $0.30 | $0.05 | $0.35 | $0.05 |
| MTAOp A | Monthly basic price | $30 | $15 | $35 | $15 | $25 | $15 |
| | Local airtime price | $0.30 | $0.05 | $0.25 | $0.05 | $0.35 | $0.05 |
| MTAOp B | Monthly basic price | $30 | $15 | $35 | $15 | $25 | $15 |
| | Local airtime price | $0.30 | $0.05 | $0.25 | $0.05 | $0.35 | $0.05 |
| BTAOp A | Monthly basic price | $25 | $15 | $30 | $15 | $20 | $15 |
| | Local airtime price | $0.25 | $0.05 | $0.20 | $0.05 | $0.30 | $0.05 |

motion campaign to differentiate itself from the rest of the telecommunications providers. In addition to name awareness, it will need to conduct a campaign to increase awareness of PCS. Many of the people are not aware of PCS and might associate it with the same quality of service as cellular service. BTAOp A could benefit from the early entry of MTA operators who may have already helped increase PCS awareness.

In addition to the name and brand awareness campaign, there should be cooperative advertisements and channel-oriented promotions and a free or switch offering similar to but hopefully less expensive than the promotions of the U.K. PCS operator. A study of advertisements and promotions used by MCI and US Sprint in the first few years after deregulation of the long-distance market to attract those subscribers from AT&T might be beneficial to the BTA operator. Similarities exist between the MCI and US Sprint's environment and that of the BTA operator. Similar to MCI's low-cost offering, BTAOp A could go with a low-cost strategy, and/or similar to US Sprint's high-quality (fiber-optic network), the BTAOp A could emphasize a digital quality network. Unlike AT&T which was still shackled, and due to dominant-carrier status, could not quickly respond to MCI and US Sprint's offering in the early days of long-distance deregulation, the cellular operators will respond quickly both to low-cost strategy as well as the digital quality of the network. The cellular operators have already started lowering their prices and made their network digital ready.

The marketing expenses associated with advertising, promotion, sales commission, and so on can be grouped into fixed and variable expenses. The variable expenses can be expressed as a percentage of the revenue. The marketing expenses of the five wireless operators are shown in Table 11.6.

The cellular operators have a high-fixed overhead with a low-variable overhead, while the PCS operators have a low overhead with a high-variable overhead.

TABLE 11.6    **Marketing Expenses for All Operators**

| | CellOp A | CellOp B | MTAOp A | MTAOp B | BTAOp C |
|---|---|---|---|---|---|
| Marketing expenses | | | | | |
| Fixed overhead ($ thousands) | 50 | 50 | 2.5 | 2.5 | 2.0 |
| Variable overhead (% of revenue) | 10 | 10 | 15 | 15 | 15 |

### 11.4.8    Distribution

Initially, it will be difficult for the new BTAOp A to match the distribution network developed by the cellular operator in the last ten years. A strong distribution network and easier service-activation process are needed for the BTA operator to successfully compete with the entrenched cellular operators. The BTA operator's primary channel for service activation will be through selected mass merchandisers, such as Wal-Mart, so that mobile handset purchase and service activation will be convenient to customers. Service activation and mobile handsets will also be available via telephone and direct mail. BTAOp A will use business and residence service centers, and direct sales for corporate accounts. Direct sales and committed distributors will be required to help educate and differentiate the new operator's offerings from those of the cellular operators. An additional customer-care center will be available for direct contacts. This center will provide a single point of contact for service, equipment inquires, service and mobile handset activation.

Distribution in mass merchandise stores will be encouraged by over-the-air activation. The mass merchandisers require identical and prepackaged products. The products cannot have location-restriction information such as office codes and area codes. A large retailer such as Wal-Mart distributes its product to its stores from a central location. They will encounter an accounting and distribution nightmare if certain products are tagged to go to certain stores because they are pre-programmed with telephone numbers. Preprogramming telephone numbers and other location-specific information in the phones requires that these products can be sold only in certain store locations. This identification and tagging of products by the store's location creates central distribution problems for mass merchandisers. BTAOp A will program any location-specific information over the air, thereby allowing generic mobile handsets to be sold via mass-retail outlets. Alternatively, the customer, after purchase of the mobile handset, calls an 800-number service for activation.

The extent and quality of the distribution network will impact the market share for each operator. The distribution index for the BTAOp A compared with that of other competitors is shown in Table 11.7. Again, as with the product-attribute index, a distribution index value

**TABLE 11.7 Relative Distribution Strength Index for All Operators**

| Operators | Consumer | | Business | | Emergency | |
|---|---|---|---|---|---|---|
| | 1997 | 2010 | 1997 | 2010 | 1997 | 2010 |
| CellOp A | 0.90 | 0.99 | 0.90 | 0.99 | 0.15 | 0.99 |
| CellOp B | 0.90 | 0.99 | 0.90 | 0.99 | 0.15 | 0.99 |
| MTAOp A | 0.25 | 0.98 | 0.25 | 0.98 | 0.25 | 0.98 |
| MTAOp B | 0.25 | 0.98 | 0.25 | 0.98 | 0.25 | 0.98 |
| BTAOp A | 0.10 | 0.97 | 0.10 | 0.97 | 0.10 | 0.97 |

of 1.0 represents that the distribution network of an operator for that segment is such that it is optimal for the customer to buy the service from the operator. A distribution index value of 0.0 indicates that it is not conducive for a customer to buy the service from the operator.

The cellular operator has a better distribution network than the BTA operator because of cellular's entrenched status. Cellular operators have not focused on the emergency segment, therefore their distribution network for the emergency segment has low-index value. The BTA operator will initially be at a disadvantage with respect to distribution but the quality and extent of each operator's distribution network will be only marginally different by year 2010.

### 11.4.9 Market demand

The cellular market has outguessed every analyst's prediction. The growth rate in 1994 was 51 percent. The consensus among cellular-market analysts was a growth rate of 35 percent for 1994, a difference of 16 points or approximately 45 percent in error. Cellular forecasting of cellular subscribers has consistently been in error. The actual numbers have always provided a pleasant surprise, since the actual number of subscribers was much higher than that predicted by the analysts.

The number of cellular and PCS subscribers from 1996 onwards as forecast by Donaldson, Lufkin & Jenrette in their summer 1995 report is shown in Fig. 11.8.

The advent of new competition from the PCS operators will expand the overall wireless market thereby providing more subscribers to both the existing cellular and new PCS operators. The market penetration in 1994 was 9.29 percent. It is expected to be 40 percent by year 2004. The market will be large enough to sustain all the new players. However, some of the PCS customers will come at the expense of cellular. A large part of the expected growth will be due to digital and its associated features, such as quality, privacy, short message service. In addition, lighter, less expensive, and power-efficient mobile handsets along

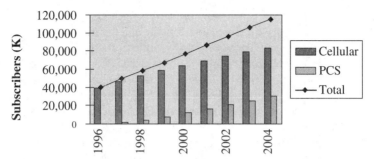

**Figure 11.8** Cellular and PCS subscriber forecast.

with packaging improvements will expand the wireless market from the early-adopter stage toward a mass market.

The price of services along with nonprice factors, such as marketing distribution channels, coverage, digital, and customer service, impacts subscriber demand. The penetration for each operator is determined as shown in Fig. 11.9.

The unmet demand for each operator is determined by subtracting the total number of subscribers from each company's mature-market penetration using that company's service-price offerings. The number of subscribers gained from unmet demand is determined by a market-segmentwide adoption rate characterized by a diffusion-time constant. Each operator approaches its unmet demand exponentially using the time constant, after some adjustments to reflect the nonprice, quality differentiation of the services offered by each operator.

Though price is a major factor that influences penetration, there are other factors that also influence market penetration. The nonprice factors could be divided into two parts. One is related to engineering and

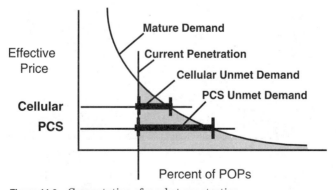

**Figure 11.9** Computation of market penetration.

the other to marketing. The engineering factors are grade of service, coverage, quality, and features. The marketing factors are the effectiveness of advertising, promotion, and distribution channels. The grade of service includes the traditional probability of blocking plus the probability of dropped calls. The value of this factor is determined by traffic engineering considerations. Coverage is another factor that a customer has preference. Customer's prefer broad-area coverage to narrow coverage. The quality of voice is another factor that determines the subscriber's choice of a network. Most digital networks compress the voice to increase capacity but in the process the quality suffers. In addition, the more valuable features, such as data and calling-line ID, would make a customer prefer one operator over the other. The relative-strength index for the nonprice factors were provided in previous sections.

The penetration level and subscriber-growth information for each operator is then determined by using the demand curve shown in Fig. 11.10, 11.11, and 11.12 for each market segment.

The analytical expression used for the demand curves is a modified rectangular hyperbolic function given below:

$$Q = K * \frac{1}{(P + P_0)^\alpha} - Q_0$$

Where: $Q$ is the percent penetration of served market
$P$ is the effective price in dollars per minute

The four parameters of this function (alpha, $K$, $P_0$, and $Q_0$) are constants and define the shape and elasticity of the demand curve. These parameters are given in Table 11.8.

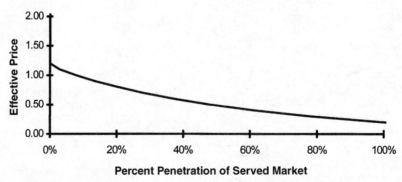

**Figure 11.10** Market penetration demand curve for consumer segment.

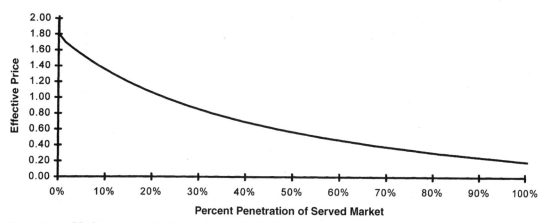

**Figure 11.11**   Market penetration demand curve for business segment.

## 11.5   Administration

The administration department of a wireless operation establishes interconnect agreements with LECs and IXCs and roaming agreements with other wireless telecommunications operators. It is also responsible for customer service and billing.

Wireless operating companies are assumed to provide interconnection for their subscribers to all other telephone networks. However, there are expenses and revenues attached to each inter-network call. Currently, cellular operators pay an access fee for all calls that terminate on their network, while LEC does not pay for calls from cellular operators that terminate on the PSTN network. CTIA has been lobbying the FCC to change this unidirectional payment of an access fee to bidirectional payment, whereby each operator pays the other for terminating calls on each other's network.

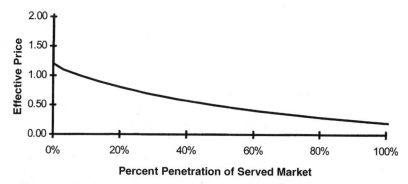

**Figure 11.12**   Market penetration demand curve for emergency segment.

TABLE 11.8   Parameters for the Penetration-Demand Curves

|          | Alpha | $K$  | $P_0$ | $Q_0$ |
|----------|-------|------|-------|-------|
| Consumer | 1.00  | 1.20 | 0.50  | 0.60  |
| Business | 1.00  | 1.00 | 0.50  | 0.44  |
| Emergency| 1.00  | 1.20 | 0.50  | 0.60  |

The business case assumes that an operating company that requests another company to complete a call will have to pay something for the service. These expenses or revenue is \$0.025 per minute and decreases every year by the inflation rate. Thus, BTAOp A pays LEC \$0.025 as an access fee for calls that terminate on its network while the LEC pays BTAOp A \$0.025 as an access fee for calls that terminate on the PSTN. These access fees are the same for all five operators.

Roaming revenues come from two sources, home subscribers when making calls while out of town, and out-of-town subscribers of other companies making calls while in town. Roaming clearly requires that the wireless telephone technology works in both companies. Roaming revenues from home subscribers employ a time-variant price per minute which is billed by BTAOp A and counted as revenue; however, some percent of it is paid by BTAOp A to the out-of-town company. This analysis assumes an \$0.80-per-minute charge of which 50 percent goes to the foreign operator. Similarly, roaming revenues from out-of-town subscribers use a time-variant price per minute. This amount is billed by the out-of-town company to its subscribers, and the amount passed back to BTAOp A is 50 percent. Thus, BTAOp A gets \$0.40 per minute for calls made by its subscribers outside its network. Roaming traffic is assumed symmetrical when the calls made by home subscribers exactly equal the calls made by visitors while out of town. This analysis assumes 3 percent of all home subscriber originated traffic is roaming traffic. The business analysis assumes that all wireless telecommunications operators in the Dallas–Fort Worth BTA have the same roaming charges and the same 3 percent of all home subscriber originated traffic is roaming traffic, as that of BTAOp A.

Customer service will coordinate repair and solve problems encountered by BTA Op A's subscribers. It will answer all billing-related questions of the customer. The customer-service center will provide a single point of contact for service and equipment inquires, service, mobile handset activation, service changes, mobile handset repair, reporting equipment thefts, and fraud.

The billing system used by BTA Op A provides customer usage in near real time. Instant authorization and credit-card verification is integrated. It has flexibility to provide instant sales promotion. It provides statistical information that can be used by the marketing department for fine-tuning pricing plans. The billing system will also provide account reconciliation with agents, dealers, and commissioned representatives, and for roaming charges on other companies networks.

The administrative expenses are included in the general and administrative (G&A) expenses. The G&A expenses include many others such as management overhead in addition to the administrative department's expenses. These G&A expenses for the five operators are given in Table 11.9.

## 11.6  Operations

The operation department builds and maintains the network. It provides RF and switch design, acquires the sites, relocates and clears the microwave paths, and builds the network. As indicated in the previous chapter, the design process involves a very iterative process that includes evaluation of competing technologies (GSM, CDMA); analyzing the terrain, vegetation, and building types; traffic density; and other technical and commercial factors. During the initial stages of the design the marketing and operation team work together because of the impact of design on both the marketing and operation of the network.

PCS licenses are required to coordinate their frequency usage with the cochannel or adjacent-channel incumbent fixed-microwave licenses in the PCS band. There are time limits for voluntary and involuntary negotiation periods. The negotiation for the relocation of microwave users can be a two-stage process. The first stage is the voluntary negotiation period. The voluntary negotiation period is for two years for all microwave users except public safety. The voluntary negotiation period for public safety is three years. During this period the incumbent microwave user is not required to relocate. This could force PCS operators to design networks around the microwave path which will limit the PCS operators flexibility, service, and coverage. The second stage is the mandatory negotiation period. This period is one year for all microwave users except public safety. The public safety users have been granted two years for mandatory negotiation. During the mandatory negotiation period, the incumbent microwave user must agree to relocate, provided the PCS operator meets FCC-mandated relocation

**TABLE 11.9  General and Administrative Expenses for All Operators**

|  | CellOp A | CellOp B | MTAOp A | MTAOp B | BTAOp C |
|---|---|---|---|---|---|
| G&A expenses |  |  |  |  |  |
| Start-up year costs | 2,400 | 2,400 | 3,000 | 3,000 | 2,400 |
| Fixed overhead ($ thousands) | 6,500 | 6,500 | 7,000 | 7,000 | 6,500 |
| Variable overhead ($/subscriber) | 50 | 50 | 60 | 60 | 50 |

obligations. The PCS operator must provide a comparable system to the incumbent microwave user. The definition of comparable system is not precise and can be interpreted differently by the PCS operators and microwave users. Many microwave users could use this opportunity to upgrade their current dated equipment to newer. The average cost for microwave relocation of a link is estimated at $250,000. This cost could increase to approximately $600,000 per link if a tower needs to be built. The number of microwave paths in the Dallas–Fort Worth area is difficult to determine. Industry estimates that there are more than 4,500 microwave paths that needs to be cleared in the United States. The business analysis presented in this chapter does not include the microwave relocation costs.

The operation department is traditionally the largest and is difficult to staff without proper planning because of the need for specialized RF skills. The number of people needed in this department is a function of the size of the network, (i.e., number of switches and cell sites). The traffic demand, grade of service, coverage plans, and choice of technology determine the number of switches and cell sites.

The size of the network depends on the coverage plan and traffic demand forecasted for the network. Traffic is determined by the airtime-demand curve. The airtime for each market segment arises due to call originations and terminations. The demand curves for each market segment for origination and termination are given in Fig. 11.13 to Fig. 11.18. These demand curves, similar to the penetration-demand curves are modified hyperbolic functions with the parameters shown in Table 11.10.

Equipment deployment is set initially by the need for areawide coverage. Later on, equipment deployment is driven by capacity requirement caused by traffic growth. Each network is assumed to serve all three market segments. Equipment is deployed in coverage zones

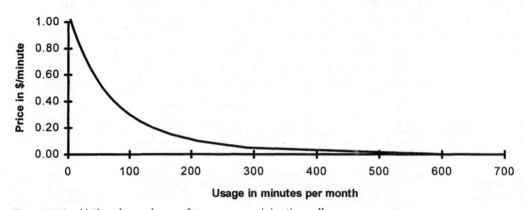

**Figure 11.13**   Airtime demand curve for consumer originating calls.

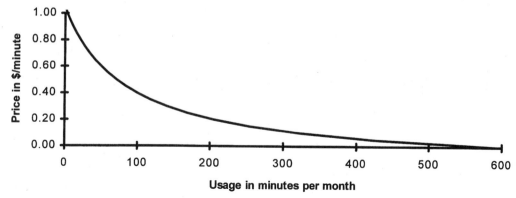

**Figure 11.14**   Airtime demand curve for business originating calls.

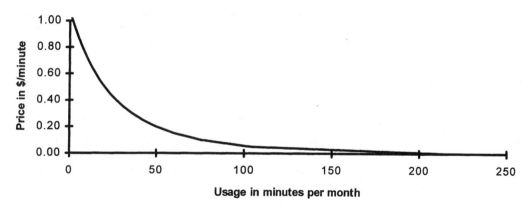

**Figure 11.15**   Airtime demand curve for emergency originating calls.

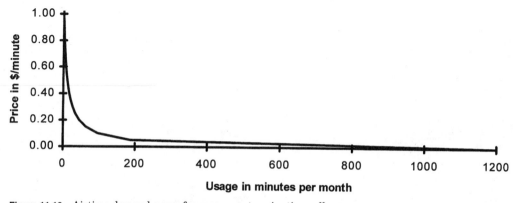

**Figure 11.16**   Airtime demand curve for consumer terminating calls.

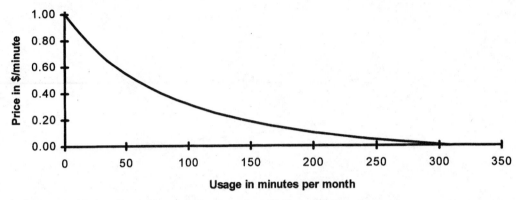

**Figure 11.17**    Airtime demand curve for business terminating calls.

defined in the previous section. Initially, the network is engineered for the defined-coverage zones. After the defined zones are covered, capacity growth is implemented first by adding more radios and T1 trunk cards to existing cell sites and switches, then more cell sites, and then switches.

The number of cell sites is heavily influenced by the choice of technology. However, as indicated previously, this study's intent is not to get into the middle of a religious warfare on technology. The purpose of this chapter is to enlighten the reader based on general network deployment rules. The size of the cells for the PCS operator and existing cellular operator are different due to the different technology and frequency. The cellular operators operate at 800 MHz frequency compared with PCS operators at 1900 MHz. The propagation characteristics at 1900 MHz are different from those at 800 MHz. The 800-MHz cell sites are larger. The cell-site size varies depending upon the terrain and radio environments. The urban areas, which are characterized by

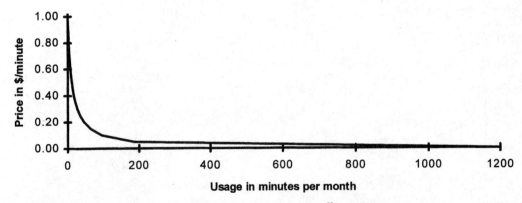

**Figure 11.18**    Airtime demand curve for emergency terminating calls.

TABLE 11.10    Parameters for the Airtime-Demand Curves

|  | Alpha | $K$ | $P_0$ | $Q_0$ |
|---|---|---|---|---|
| Airtime-demand curve |  |  |  |  |
| Consumer originating | 0.25 | 278 | 0.01 | 0.98 |
| Business originating | 1.00 | 102 | 0.15 | 0.84 |
| Emergency originating | 0.25 | 100 | 0.01 | 0.98 |
| Consumer terminating | 1.00 | 12 | 0.01 | 0.99 |
| Business terminating | 0.50 | 187 | 0.15 | 0.93 |
| Emergency terminating | 1.00 | 12 | 0.01 | 0.99 |

tall high-rise buildings and high-building density, typically have smaller cell sites. For more accuracy, the urban area is further divided into dense urban and urban. The suburban areas are the ones with one- or two-story residential buildings while the rural areas are the ones with open, agricultural land with sparse residential dwellings. Rural areas typically have large cell sites because of less obstruction to radio paths from tall buildings. The cell site sizes for 800 and 1900 MHz are given in Table 11.11.

The generic equipment infrastructure model is shown in Fig. 11.19.

The switch, cluster controller (BSC), and cell-deployment parameters are given in Table 11.12. The deployment parameters are used to expand the network as the traffic grows. As the traffic grows, more voice channels, cell sites, BSCs, and switches are added to the network. Growth initially takes place by adding switch ports and may later require additional switches if the maximum port capacity or maximum real time in terms of BHCA capacity is exceeded. The cell-deployment parameters are used to increase the system's capacity as demand grows by adding voice channels to existing cells, and then by adding cells. The BSC deployment parameters are used to deploy the BSCs. The maximum number of cells per BSC sets the systemwide average level at which new BSCs are added. In the case of cellular operators, all equipment vendors except Hughes Network Systems' cellular equipment does not have BSC in the network. The cell sites are directly connected to the switch. Therefore, this parameter is not applicable to the cellular network.

The cost of the switches, BSC, and cell sites used in the business analysis is shown in Table 11.13. These cost parameters, as with other

TABLE 11.11    Radius of Cells in Different Environments
at 800 MHz and 1900 MHz

|  | Urban (km) | Suburban (km) | Rural (km) |
|---|---|---|---|
| Cellular operators | 5.5 | 12.2 | 32.7 |
| PCS operators | 2.5 | 6.0 | 15.0 |

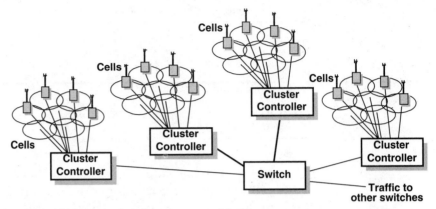

**Figure 11.19**  Equipment infrastructure model.

parameters, are estimates and are typical for the average of all vendor's equipment. The costs given in Table 11.13 include estimates for the fully engineered, furnished and installed (E, F, and I) package of goods and services, including spares, land, buildings, and power arrangements. The per-channel cost can be estimated as the linear "best fit" through the actual cost curve (a plot of total cost versus number of channels) for any vendor's specific configuration. The actual physical-equipment packaging for cell sites may often provide channels only in blocks of 8 or 16 or some other number at any single cell site. The linear per-channel cost model represents approximate real-equipment costs using a best-fit model.

The amount of equipment needed is determined by traffic engineering based on busy-hour traffic demand. The deployment parameters determine the number of switches and cell sites needed to support forecasted traffic. The number of switches, BSC, and cell sites along with their associated costs determine the capital expenditure for building the wireless telecommunications network.

**TABLE 11.12    Equipment Engineering Parameters**

| Parameters | Deployment values |
|---|---|
| Switch deployment | |
| Maximum ports | 20 |
| Maximum traffic capacity (BHCA) | 75,000 |
| BSC deployment | |
| Maximum cell per BSC | 16 |
| Start-up cells per BSC | 4 |
| Cell deployment | |
| Overhead channels | 10% |
| Start-up channels | 10 (cellular), 24 (PCS) |
| Channels per cell | 45 (cellular), 110 (PCS) |

**TABLE 11.13    Cost of Equipment**

|                      | Costs ($ thousands) |
|----------------------|---------------------|
| Switch               |                     |
| Base                 | 2500                |
| Per port             | 250                 |
| BSC                  |                     |
| Base                 | 80                  |
| Per port             | 7.5                 |
| Cell sites           |                     |
| Base                 | 210                 |
| Additional channel   | 3                   |

The equipment is expected to reduce in price throughout the analyzed time horizon due to cost reductions associated with the learning curve and increased volume. The expected cost reduction used in this business analysis is 5 percent per year.

Based on the cell sizes, traffic demand, grade of service of 98 percent, and coverage zones defined in the previous section, the number of cell sites needed for each network is shown in Table 11.14. The deployment parameters for each cellular and each MTA operator are the same, therefore the cell site information is provided only for one of the two cellular and PCS operators.

The 1900-MHz frequency requires more cell sites for coverage; however, as the network design moves from a coverage-based to a capacity-based, the number of cell sites needed for digital networks is smaller

**TABLE 11.14    Number of Cell Sites in Each Network**

| Operators | Phases                  | Year | Initial cells | 2010 cells |
|-----------|-------------------------|------|---------------|------------|
| CellOp A  | Initial coverage starts | 1985 | 8             | 403        |
|           | Phase 1 starts          | 1987 | 4             | 161        |
|           | Phase 2 starts          | 1989 | 7             | 273        |
|           | Phase 3 starts          | 1991 | 6             | 157        |
|           | Final buildout starts   | 1993 | 7             | 88         |
|           | Total                   |      | 32            | 1,082      |
| MTAOp A   | Initial coverage starts | 1996 | 38            | 163        |
|           | Phase 1 starts          | 1997 | 16            | 65         |
|           | Phase 2 starts          | 1998 | 30            | 110        |
|           | Phase 3 starts          | 1999 | 21            | 63         |
|           | Final buildout starts   | 2009 | 32            | 35         |
|           | Total                   |      | 137           | 436        |
| BTAOp A   | Initial coverage starts | 1997 | 38            | 175        |
|           | Phase 1 starts          | 1998 | 16            | 70         |
|           | Phase 2 starts          | 1999 | 30            | 119        |
|           | Phase 3 starts          | 2000 | 21            | 68         |
|           | Final buildout starts   | 2010 | 32            | 32         |
|           | Total                   |      | 137           | 464        |

than those for analog networks. The number of radio channels for both the networks is the same, but the number of radio channels per cell site is much larger in digital networks than those in analog networks.

There are operational expenses associated with operating and maintaining a network, in addition to the capital expenditures. These operation expenses consist of fixed and variable overhead based on the number of subscribers. These expenses are shown in Table 11.15.

## 11.7    Finance Department

The finance department is probably the most important group in the preauction phase. This department prepared the business case and raised the necessary capital for the auction and building of the network. The finance department keeps track of finances and provides information that helps plan and control the organization in the postauction phase. It provides other departments with specialized service including advice and counsel in budgeting, analysis of variances from budgets, pricing, and making special decisions (i.e., mergers and acquisitions that have tax and other financial implications).

It is assumed that BTAOp A is formed by an initial-equity investment of $200 million. For this equity, it has distributed 2 million shares at $100 per share. Additional capital needed for ongoing operations is financed by borrowing either through the commercial market or through vendor financing at an interest rate of 8.00 percent. A summary of the key financial input is shown in Table 11.16.

It is assumed that the BTAOp A qualifies as a small business. Interest is paid on the license cost over the first five years. The principal of the license cost is amortized over the remaining five years.

## 11.8    Business Analysis

The inputs described in the previous section are used to determine the business case for the new operator BTAOp A operating in the Dallas–Fort Worth BTA. BTAOp A competes with existing two-cellular operators and two new MTA operators.

The market share for the different operators is shown in Fig. 11.20. It is clear cellular operators hold an advantage because of a head start.

**TABLE 11.15    Operating Expenses for All Operators**

|  | CellOp A | CellOp B | MTAOp A | MTAOp B | BTAOp C |
|---|---|---|---|---|---|
| Operating expenses |  |  |  |  |  |
| Fixed overhead ($ thousands) | 1,500 | 1,500 | 2,000 | 2,000 | 1,500 |
| Variable overhead ($/subscriber) | 50 | 50 | 60 | 60 | 50 |

TABLE 11.16 Financial Factors for BTAOp A

| Financial factors | Value |
|---|---|
| Equity investment | $200 million |
| Number of shares outstanding | 2 million |
| License-acquisition cost | $54 million |
| Inflation rate | 3.0% |
| Interest rate on debt | 8.0% |
| Interest rate on cash | 6.0% |
| Depreciation (average) | 7 years |

However, the head start of cellular gets minimized as time progresses and the historical weight or legacy effect seems to slow its growth.

The overall market is expanded due to increased competition. The resulting competition leads to lower prices and increased awareness of wireless telecommunications. The market penetration would be approximately 25 percent by year 2010 without competition from the new PCS operators. The presence of the three new PCS operators increases the market penetration to 42 percent. Of course, cellular operators are impacted by increased competition. The market share of each cellular operator is reduced from 12.5 percent to 10.3 percent due to the presence of the PCS operators. Not only is the cellular operator's market share lower than expected without competition from PCS operators, its revenue is reduced due to lower prices. It appears from the market-share graph that the growth rate of cellular stabilizes while the new operators appear to be growing rapidly. The three new PCS operators will each have 7 percent market share by year 2010. This translates to 21 percent for all PCS operators. The cellular market

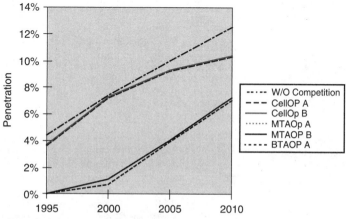

Figure 11.20  Market share graph.

share is also 21 percent. Thus, the wireless telecommunications services appear to have reached a commodity status with each provider being evenly placed competitively by the year 2010.

The proforma financial statement for BTAOp A is shown in Table 11.17.

The BTAOp A reaches positive operating income in the year 2001 or four years after start up. It attains positive after tax income and positive cash flow in year 2002 or 5 years after start-up.

A modified cash flow statement is shown in Table 11.18 to perform a business valuation of the new operator BTAOp A.

One of the most common business valuation models is based on a multiple of operating cash flow plus cash for some future year and discounted back to the present. The multiple used in this analysis is 12. The present value (1996) based on the value of BTAOp A at years 2005 and 2010 is shown in Table 11.19 at a discount rate of 12 percent.

**TABLE 11.17    Pro forma Financial Statement of BTA Op A**

|  | 1997 | 1998 | 1999 | 2000 | 2005 | 2010 |
|---|---|---|---|---|---|---|
| Subscribers |  |  |  |  |  |  |
| End of year ($ thousands) | 0 | 3 | 14 | 39 | 241 | 497 |
| Penetration (% of population) | 0 | 0 | 0.3 | 0.7 | 3.9 | 7.0 |
| Market Share (% of wireless subscribers) | 0 | 0 | 2 | 4 | 13 | 17 |
| Revenue |  |  |  |  |  |  |
| Service revenue ($M) | 0 | 1.2 | 7.1 | 21.6 | 151.0 | 304.1 |
| Other revenue ($M) | 0 | 0 | 0.5 | 1.7 | 18.2 | 47.3 |
| Total revenue | 0 | 1.2 | 7.6 | 23.3 | 169.2 | 351.4 |
| $/subscriber/month | 0 | 80 | 77 | 74 | 65 | 62 |
| Market Share (% of wireless revenue) | 0 | 0 | 1 | 3 | 12 | 15 |
| Operating expenses | 1.6 | 1.9 | 2.7 | 4.7 | 30.5 | 71.4 |
| $/subscriber | 0 | 61 | 16 | 10 | 11 | 12 |
| Gross profit | −1.6 | −0.6 | 5.0 | 18.5 | 138.8 | 280.0 |
| % of revenue | −52 | 65 | 80 | 82 | 80 |  |
| G&A | 9.4 | 7.2 | 7.8 | 9.1 | 23.3 | 46.7 |
| % of revenue | 588 | 102 | 39 | 14 | 13 |  |
| Marketing | 0 | 0.8 | 4.1 | 9.8 | 33.7 | 56.5 |
| $/new subscriber | 322 | 294 | 282 | 314 | 369 |  |
| % of revenue | 67 | 54 | 42 | 20 | 16 |  |
| Income from operations | −11.1 | −8.6 | −6.9 | −0.3 | 81.8 | 176.7 |
| % of revenue | −707 | −91 | −1 | 48 | 50 |  |
| Other income expenses |  |  |  |  |  |  |
| Depreciation | −3.8 | −4.3 | −5.2 | −6.0 | −11.4 | −22.0 |
| Interest income/expense | −0.6 | −2.5 | −4.0 | −5.6 | −3.6 | 16.2 |
| Total other income/expense | −4.5 | −6.8 | −9.1 | −11.6 | −15.0 | −5.8 |
| Income before taxes | −15.6 | −15.5 | −16.1 | −11.9 | 66.8 | 171.0 |
| % of revenue | −1266 | −211 | −51 | 39 | 49 |  |
| Taxes | 0 | 0 | 0 | 0 | −24.0 | −61.5 |
| Net income | −15.6 | −15.5 | −16.1 | −11.9 | 42.7 | 109.4 |
| % of revenue |  | −1266 | −211 | −51 | 25 | 31 |

**TABLE 11.18  Cash Flow Statement for BTAOp A**

| | 1996 | 1997 | 1998 | 1999 | 2000 | 2001 | 2002 | 2003 | 2004 | 2005 | 2006 | 2007 | 2008 | 2009 | 2010 |
|---|---|---|---|---|---|---|---|---|---|---|---|---|---|---|---|
| Equity investment ($M) | 0.0 | 0.2 | 0.0 | 0.0 | 0.0 | 0.0 | 0.0 | 0.0 | 0.0 | 0.0 | 0.0 | 0.0 | 0.0 | 0.0 | 0.0 |
| License acquisition ($M) | -8.1 | -3.7 | -3.7 | -3.7 | -3.7 | -3.7 | -12.8 | -11.9 | -11.0 | -10.0 | 0.0 | 0.0 | 0.0 | 0.0 | 0.0 |
| Cash flow from op. ($M) | 0.0 | -11.1 | -8.6 | -6.9 | -0.3 | 12.3 | 28.2 | 45.1 | 63.1 | 81.8 | 100.8 | 119.9 | 139.1 | 157.8 | 176.7 |
| Interest Inc./exp. ($M) | 0.0 | -0.6 | -2.5 | -4.0 | -5.6 | -6.8 | -7.1 | -7.0 | -5.6 | -3.6 | -1.4 | 2.0 | 5.9 | 10.6 | 16.2 |
| Capital expenditure ($M) | 0.0 | -8.0 | -3.4 | -5.9 | -5.6 | -5.0 | -7.1 | -8.8 | -12.5 | -15.8 | -17.7 | -19.2 | -20.5 | -21.1 | -28.0 |
| Cash required for taxes | 0.0 | 0.0 | 0.0 | 0.0 | 0.0 | 0.0 | 0.0 | 0.0 | -9.9 | -24.0 | -31.1 | -38.5 | -46.0 | -53.7 | -61.5 |
| Free cash/ (fund needs) | -8.1 | -23.1 | -18.2 | -20.5 | -15.1 | -3.1 | 1.3 | 17.4 | 24.1 | 28.3 | 50.6 | 64.2 | 78.4 | 93.6 | 103.4 |
| Cash balance/ (debt) | -8 | -31 | -49 | -70 | -85 | -88 | -87 | -70 | -45 | -17 | 33 | 98 | 176 | 270 | 373 |

TABLE 11.19    Valuation of BTAOp A

|  | 2005 | 2010 |
|---|---|---|
| Operating cash flow ($M) | 82 | 177 |
| Terminal value at multiple 12 ($M) | 984 | 2,124 |
| Cash balance ($M) | (17) | 373 |
| Future asset value ($M) | 967 | 2,497 |
| PV (1996) at 12% discount rate ($M) | 348 | 510 |
| PV (1996) at 12% discount rate ($/population) | 80 | 118 |

The current cellular companies have share prices that lead to a valuation in the range of $150 per population to $300 per population for most companies. There are several extremes on either side of the range. Most cellular companies fall within this range. GTE recently completed the purchase of the remaining 10 percent of Contel Cellular for $190 per population. Another example is the purchase by Lincoln Telephone of RSA properties at $186 per population. However, Lin Broadcasting was appraised in connection with the acquisition by AT&T at $127.50 per share. The per-population value is $320 based on this share price of Lin Broadcasting. However, this valuation may not be a true one. It was probably impacted by the contracts between AT&T and McCaw and the contract between McCaw and Lin Broadcasting. Most valuations have been in the $150 per population to $300 per population range.

The BTAOp A valuation is lower than the current one of cellular operators. It should be noted that these numbers will depend upon the input values. The input values will differ for each new PCS operator because of the marketing, administrative, and technical skills of the company, and the strategy and emphasis it places in the different market segments. Comparison with cellular may not be appropriate because cellular operated under a duopoly market structure while the new operators are in an oligopoly market structure with all the cellular operators having a 10- to 15-year head start. In addition, the cellular carriers did not have to pay for licenses nor for microwave relocation. With the change in industry structure from a duopoly to an oligopoly it is expected that the valuation of all wireless telecommunications operators, including cellular, will be lower than the current valuation of cellular operators in the transaction market.

## 11.9   Summary

The purpose of this chapter was to discuss the different elements that impact the profitability of a new operator. Organizational requirements from a human-resource, marketing, administration, and operation perspective were presented. The WIST model was used to simulate the wireless telecommunications market environment of Dallas–Fort Worth BTA (where the competitor consists of five operators, two of which are the existing cellular operators and the remaining three are the PCS operators). The income statement of a BTA operator was presented using generic deployment parameters. The income statement was hypothetical and did not reflect the income statement of any particular company. All information presented was for discussion purposes only.

The valuation of the new BTA operator is determined and compared with the market valuation of cellular operators from recent acquisition and merger activities. The valuation, based on the assumptions presented in this chapter, appears to be different from the high valuation of cellular operators. The introduction of PCS has changed the competitive rules and the industry structure. It is likely that the valuation of all wireless operators, including cellular operators, will be similar, and the valuation of wireless operators might be lower than what has been seen in recent years.

# 12

# The Wireless Network and Beyond

## 12.1 Introduction

These are unique times where all technology is converging. Voice and data no longer have to be transmitted over copper wire. Video no longer has to be broadcast over RF and coaxial cable. In the digital world, neither switching, transmission, the network, nor devices differentiate whether a signal is voice, data, or video. Everything is handled as a "bit stream." Only the coder and decoder are concerned how the data is to be treated. These signals, whether voice, data, or video, may traverse traditional copper wire, fiber, or the air as radio waves. And regardless of the application, everything must go over the network which is a hybrid of all the media types.

Wireless communication has been used for many decades by many of the network segments in the form of microwave hops, satellite carrier, and mobile device to base station. Frequency, available bandwidth, coordination, the high cost of deployment, and regulation kept the wide use of wireless from being deployed. With the telephone companies having a captive market, there were few justifiable reasons to go through the pain of an alternative network.

Americans, long known for their mobility, became even more mobile in the mid-1970s. Also, America was in transition from a manufacturing-based economy to a service-based economy. Commute time eroded productive time, especially for sales and service. For example, a salesperson had to spend several hours in crosstown traffic to make a sales call. Returned business calls had to wait sometimes until the next day, or several days, until the salesperson returned to the office or called in for messages.

The wireless industry has grown in North America and Western Europe because of the need for mobility. While in the rest of the world cellular grew to meet pent-up demand for basic telephone service due to the fact that the buildout of infrastructure just couldn't keep up with demand for telecommunications.

Problems arise when someone wishes to go beyond their home area: frequencies, air protocols, and numbering plans may all be different; there may be no service; there may be no roaming agreements; and so on. These are just a few of the problems with voice messaging. Users also need data, fax, telemetry, paging, and short message service.

Many futurists talk about the proverbial crystal ball. Some decline to predict the future because it is an imprecise science which has more often led to fallibility. This chapter will not predict the future but will present the regulatory and subscriber trends as well as some of the enabling technology that could determine the wireless network of the future.

The reluctance to predict does not prevent a tongue-in-cheek prediction about the evolution of the human being. Humans could evolve with antennae for brain wave to brain wave thought transmission, or they could become like a packhorse carrying multiple wireless devices. Figure 12.1 is a science-fiction depiction of humans that have evolved with antennae and communicate via brain wave with implanted smart cards.

Figure 12.2 is a caricature of a person carrying several wireless devices in order to communicate with multiple wireless networks; an AMPS/CDMA cellular phone, AMPS/TDMA cellular phone, a CDMA PCS phone, a TDMA PCS phone, a GSM phone for international travels, a DCT phone for international travel, a J-TAC phone for Japan, a two-way pager, a PCSS satellite terminal, and a wireless computer with a suitcase full of batteries, chargers, and adapters.

## 12.2   The Evolution of Lifestyle

The global economy means that people will need to be available 24 hours a day, 7 days a week to accommodate the business habits on the

**Figure 12.1**  Characterization of evolved people who communicate via brain wave.

**Figure 12.2**  A traveler prepared for multiple wireless access technologies.

other side of the globe. As the world's economy becomes global, people will travel to all parts on a moment's notice. Air travel time will become shorter. A business person could be making an overnight trip from Dallas to Tokyo. It may be necessary to contact the individual for an important decision, an urgent message, information or other business dealings whether that person is in transit or at his or her destination.

In order to contact the called party, a worldwide page must be sent for the called party's location. Then the call must be delivered to the individual's portable, the air phone, hotel-room phone, car phone, office phone, or the phone nearest to where the individual is located.

Habits and lifestyle will change with the evolution of network and vice versa. There is a constant feedback loop that stretches the envelope of lifestyle and network evolution. There will no longer be the concept of 8:00 a.m. to 5:00 p.m. Sales and services will be conducted face-to-face, while administrative and management tasks will be conducted from wherever one is currently located via wireless voice, data, and video. Connection to an individual's computer, a company's mainframe, and the Internet will be accessed from anywhere on the globe.

The information and communication age is here. Those who are part of it may be more fortunate than those who are not. Of course it depends on one's definition of fortunate. Manufacturing jobs have gone offshore, robotics is performing more and more processes, retail is done on the Internet, services and professionals are becoming information intense and highly specialized. The need for communications is an essential part of the job. The information-based organization is less bureaucratic (having eliminated the need for traditional, middle management). The future organization could be a collection of professionals, as in a law firm, a hospital, or an orchestra. In an orchestra, each person plays his or her own musical instrument coordinated by the conductor without intervening layers of management. Education is absolutely necessary to participate in the information and communications age.

## 12.3    The Evolution of Regulation

There is a worldwide trend to deregulate and privatize telecommunications. Deregulation will spur competition, drive prices down, and create demand for new technology. In the United States, deregulation will lead to the creation of the pre-1980s AT&T but unlike the pre-1980s, there will be multiple companies offering those services. Wireless, with promises of new businesses, new revenue, a better economy, and a part in the new global economy, will eventually move other countries to open their doors to competition and open communications.

Once unrestricted wireless communication is implemented in a country, the widespread use of the service will make it very difficult for governments to control the thoughts of its population. Voice, data, and video information will flow in and out of countries.

The development of wireless networks will change the nature of regulation. In the 1980s, effective competition in the long-distance market allowed the deregulation of AT&T. The development of wireless networks will provide effective competition in the local telecom market. The monopoly rights granted to the LEC will be irrelevant, hence it will lead to the deregulation of the LECs.

Regulation will evolve with the change in industry structure. The current roles of the regulatory bodies of approving rate changes and forbidding certain types of businesses may not be relevant in a competitive environment. Competition in a free market will assure fair prices. In the future, it will be possible that cellular revenues might exceed the LEC's, thereby making the issue of LEC subsidy of other business a moot point.

Regulators will have to deal with privacy and human rights issues. It will be easier for individuals and the government to track users. And, as previously discussed, users can be pinpointed to their exact locations. A totalitarian government can use this capability as a means to control its people. There has been some resistance by the industry and civil rights group in the United States to the justice department's wish to have the "clipper" chip installed in all electronic devices. The current intent of the government is for control of organized crime, espionage, and drug trafficking. This intent could evolve unless regulators and lawmakers deal with these issues up front.

For public safety reasons, future government may require all its citizens to carry a wireless communications device or smart card. An electronic device or smart card could end up replacing paper identification. There is a proposal to provide pagers to all the homeless. The intent is good but could be subject to future abuses if the issues of privacy and human rights are not addressed up front. Similarly what appears to be a good idea now with optional vehicle-tracking devices for the sake of deterring auto theft, could be misused. Regulators and lawmakers will take careful steps to preserve privacy and human rights while at the same time make use of the available technology to make human beings more at ease.

## 12.4   Wireless Network Evolution

Devices will become smaller, lower-powered, have more features and multiple functions and all for a much lower price. To the user, technology and network complexity will be transparent. Devices will be voice activated and be the size of a wristwatch or decorative piece of jewelry. Batteries will have to be changed monthly. Each device will be able to provide telemetry, two-way voice communications, and access to the information superhighway. One number will follow a user anywhere, anytime; or, be treated as the user defines. The communications device will be seamless (able to roam regardless of the infrastructure technology serving the roamer anywhere in the world).

However, technology will continue to be a mix-and-match set of technologies, standards, and protocols. The PSTN and PLMN will merge with the information superhighway, obtaining access to every database and piece of information in the world. Information access and delivery will be instantaneous. For the sake of communication and information exchange, borders will become invisible. Current wireless technologies, such as TDMA, CDMA, and GSM will eventually evolve to newer and better technologies.

### 12.4.1   Third-generation wireless systems

Wireless systems have seen tremendous advancements since their first introduction in the eighties. The first-generation wireless systems were the analog cellular and cordless telephones, AMPS and analog cordless telephony constituting that group. The second generation of wireless systems was digital. D-AMPS, GSM/DCS 1800, CDMA, DECT, CT2, and PACS technologies make up that generation. These technologies while being successful, will reach the limits of their capacity by the end of this century. As the second-generation digital technologies make the transition from urban to rural areas and gain wide acceptance, they will very quickly run out of capacity to serve the huge subscriber base. Additionally, the second-generation technologies are severely limited by the number of services they can offer. To support worldwide mass acceptance of wireless along with broadband services, there has to be a third generation of systems which can meet these requirements.

**12.4.1.1   Universal Mobile Telecommunication System (UMTS).**   Universal Mobile Telecommunication System (UMTS) is a third-generation mobile telecommunication system under development in Europe. It is scheduled to start service in the early part of the next century. It is being developed to overcome the shortcomings of the second-generation wireless systems. Its goal is to provide telecommunication services (paging, cordless telephony, cellular telephony, location services, navigation services, traffic systems, etc.) to a large number of users. Some of the services being proposed to be supported by UMTS are telephony, teleconferencing, voice mail, video, videoconferencing, fax (group 4), remote terminal, message broadcast, navigation, and location services. Figure 12.3 lists some of the proposed services in UMTS.

The goal for UMTS systems is to meet the following requirements:

- Flexible and open-network architectures
- Digital system using the 1.8–2.2 GHz band
- Support of different radio environments such as paging, cordless, cellular, and so on
- Multimode terminals to provide terminal mobility
- Variety of telecommunication services
- High quality and integrity comparable to fixed networks
- Use of intelligent network (IN) concepts for mobility management and service control
- Enhanced security and privacy

In order to meet some of these goals and proposed services the UMTS air interface has to support higher-data rates than can be supported with today's wireless technologies. It is proposed that UMTS support up to 2 Mbps data rates in order to be compatible with ISDN and B-ISDN (broadband ISDN services). The frequency band designated for UMTS is the 2-GHz band. There are several network configurations currently being considered for UMTS deployment. The UMTS network could be a stand-alone network interfacing to other networks via interworking gateways. It could be integrated with the mobile network or integrated with the wireline network. Each of these network configurations has its own advantages and most of them leverage

| TELESERVICE | THROUGHPUT (kbps) |
|---|---|
| Telephony | 8–32 |
| Teleconferencing | 32 |
| Voice mail | 32 |
| Program sound | 128 |
| Video | 64 |
| Videoconferencing | 384–768 |
| Remote terminal | 1.2–9.6 |
| User profile editing | 1.2–9.6 |
| Telefax (group 4) | 64 |
| Voiceband data | 64 |
| Database access | 2.4–768 |
| Message broadcast | 2.4 |
| Unrestricted digital information | 64–1920 |
| Navigation | 2.4–64 |
| Location | 2.4–64 |

**Figure 12.3**  Proposed teleservices in UMTS.

the existing network infrastructure. Time-Division Multiple Access (TDMA) and Code-Division Multiple Access (CDMA) technologies are currently being investigated as possible UMTS solutions. To function in the wide variety of environments, UMTS will have to support different cell structures, such as macro, micro, and pico. Macrocells provide coverage for fast-moving vehicles and low-terminal-density areas, microcells will provide coverage for low-mobility subscribers and support high-terminal-density areas, picocells will be used primarily indoors in office environments.

**12.4.1.2  Distributed-switching architectures.**  The current switching products were designed for high capacity and dense areas. As telecom networks migrate from an urban to a rural focus, the current mainframe-type switching architecture is no longer economical or relevant. The current infrastructure products would require that operators deploy remote cell sites in the rural areas and then backhaul them via terrestrial or microwave links to a centralized MSC in the urban area. These represents a major cash flow due to high capital, operating, and maintenance cost.

Celcore and other innovative companies have developed next-generation switching products that are based on distributed architecture. Traditional switching functions such as signaling, roaming, and switching are logically separated to reduce capital and operating costs. This allows the switching functions to be moved out toward distant subscribers while still preserving the centralization of common functions such as signaling and roaming. The remotion of switching functions eliminates the need for the backhauling and thereby lowers the capital and operating costs. The concepts of the distributed switching and centralized signaling is shown in Fig. 12.4.

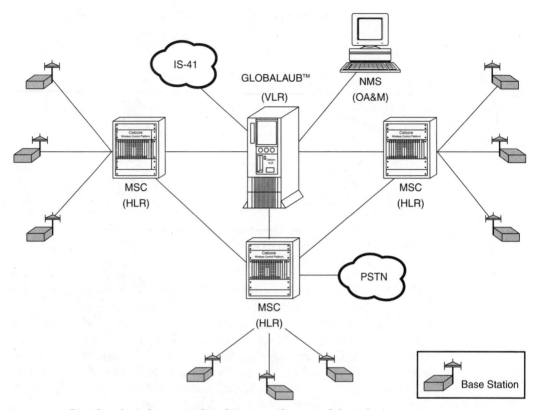

**Figure 12.4**    Distributed wireless network architecture. (*Courtesy: Celcore, Inc.*)

The next-generation architecture is a paradigm shift in the way traditional switching is done today. Similar to the transformation of the computer technology from being mainframecentric to LAN-centric, the telecommunication technology will migrate from huge centralized switches to distributed switches. These distributed switches will be networked together similar to a LAN to provide interconnectivity. Distributed switching radically reduces the infrastructure and operating costs for an operator. Since the switches are distributed, they can provide localized switching in their areas and terminate calls to the other networks only for calls outside their region. This is in contrast to current switching technology wherein all traffic is backhauled to a centralized switch.

Similarly the GSM hierarchical architecture of MSC, BSC, TRAU, and BTS are not economical for rural markets. Celcore and other innovative companies have optimized the GSM architecture by merging the functions of MSC, BSC, and TRAU into a single platform. The twin implications of this optimization are reduced capital and operating cost and ease of maintenance.

### 12.4.1.3 RF-independent switches.

Another trend in third-generation wireless systems will be the independence of switching equipment from radio equipment. This section will concentrate on the commonality among the different technologies and present a network infrastructure architecture and technology that can be used by operators to build a network independent of network and RF-access technology. In spite of the claims and counterclaims among the proponents and opponents, there is commonality among the different technologies. The differences in the technology can be solved by innovation in the area of infrastructure product design. Availability of such products can reduce the risk of a technology decision made by operators. The operator can make decisions based on availability of spectrum and infrastructure equipment without worrying about obsolescence. If at a later date some other digital technology appears attractive and makes business sense, then operators can migrate to the new technology without impacting the infrastructure equipment. Thus, additional capital costs and equipment churn can be prevented. This allows operators to make a decision today and not be locked in to an RF and network technology only to find the marketplace preference is for one of the other contending technologies. Most of the RF technologies we have discussed are very different from each other; however, these differences are typically limited to the radio interfaces. Thus, instead of having a GSM, MSC, CDMA MSC, or D-AMPS MSC, there is a trend toward having a single MSC support several radio interfaces. Most of the MSC functions are similar—call setup, database query, billing, translations, and routing. These functions are also independent of radio-access technology, which makes it easy to have a single MSC support several radio interfaces. This approach reduces the operator risk in choosing a single technology. If an operator has chosen GSM today and later on wants to migrate to UMTS or a different technology, then only the radio equipment will have to be changed and the operator's investment in switching and other network equipment is not affected. Having a single switch support multiple RF interfaces also helps an operator in deploying optimized RF solutions for each target market. For instance, an MSC which supports DECT and GSM can be used by an operator to support DECT networks in buildings, shopping malls, factories, and a GSM cellular network in the city. On the network side the vocoder is an element which is closely tied in to the RF interface. These vocoders will also have to be changed to support different radio protocols. Figure 12.5 provides a new third-generation wireless architecture. The infrastructure switching products remain the same for all RF technologies. The different RF technologies are implemented by parameterized software.

### 12.4.1.4 Service signaling integration.

Currently, several types of protocols are used by the different technologies for service signaling.

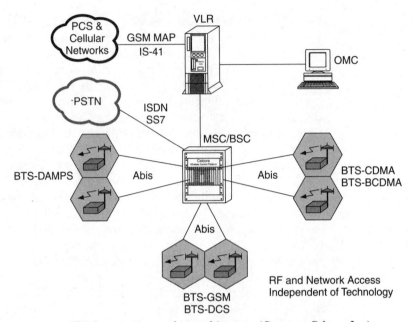

**Figure 12.5**   Third-generation product architecture. (*Courtesy: Celcore, Inc.*)

AMPS, D-AMPS, and CDMA use IS-41-based protocol, GSM/DCS 1800 use GSM MAP, and PACS favors Advance Intelligent Network (AIN)–based approach. The third-generation wireless systems will have a single integrated service signaling protocol. The networking protocols, such as IS-41 and GSM MAP, have a lot more similarities than differences. The underlying network layers of the two protocols, that is MTP1, MTP2, MTP3, SCCP, and TCAP, have minor differences. Industry is migrating toward an IS-41/GSM-MAP protocol converter. This converter will enable operators to get roaming revenue from both IS-41 and GSM subscribers. The driver for this activity will be the availability of dual-mode/dual-band phones. For instance, a dual-mode GSM/AMPS phone will enable a subscriber to get service in the GSM service area and wherever it is unavailable the phone will automatically access the AMPS cellular network. Such kinds of roaming agreements will enable the operator to not only gain roaming revenue from both networks but also provide potential nationwide coverage since AMPS is widespread in the United States. The approach taken by PACS is to try to integrate the wireless service access protocols with the wire-line service access protocols. The benefits are tremendous. A new service created for the wire-line subscriber can be offered simultaneously to the wireless subscriber also. A true one-number kind of service becomes a reality. A user will have only a single number for both wire-line and wireless phones and the intelligent network will terminate the call to the right device.

## 12.5   Impact to the Network

The effects of wireless on the network can already be seen. Major metropolitan areas have run out of telephone numbers because of the blocks of telephone numbers required for cellular, PCS and paging, and Internet access lines. Network planners are looking only at the number drain from cellular and PCS. Paging will consume even more numbers. Also, both wire-line and wireless telephone numbers for Internet access are increasing; for example, many homes are installing multiple lines for home faxes, computer lines for Internet access, lines for interactive multimedia, lines for security systems, and business lines for home offices. Businesses, too, are using more numbers as their PBXs grow (more fax lines, lines for Internet access, and for in-building wireless systems).

A quick fix is to subdivide geographic area codes. Of course, this is only a temporary solution until the subdivided area code will have to be divided again. Each change has a tremendous economic and inconvenience factor. Businesses will have to print new stationery, business cards, and business forms, and update their advertising. People will have to dial ten digits to call a wire-line or a wireless user. The long-term solution is that a new area-code plan (NPA) with ten-digit directory numbers is the solution for at least the next half century.

The intracity networks are receiving more traffic due to all these new and enhanced services. This causes a tremendous demand for local infrastructure investment. With deregulation, multiple LECs will be permitted, taking the pressure from the incumbent, the traditional telco. However, the cost to install multiple LEC infrastructures, and the time to build multiple LEC infrastructures which allow access to every home and business is quite a task.

The impact to the IXCs for both domestic and international traffic will be increased due to a mobile and a global economy. Businesses are centralizing their operations to cut operating and capital investment. This means that sales, service, and support personnel will be mobile over a greater geographic area. The global economy will require employees to traverse the globe trying to capture its share of the that market. As people travel, they will need to communicate with the office, with home, with customers, return calls, send and receive correspondences, and access databases. Whether the call is wireless or wire-line, the call traffic on the IXC's network will increase.

With more people traveling and accessing their company's LAN, or accessing the Internet, a great deal of infrastructure will be required to handle remote-access traffic demand. Many systems will have to be replaced because they are insufficient to grow and handle increased remote access. Companies will have to invest millions of dollars to upgrade their company's access equipment.

The 911 Emergency Network has already been impacted. With one out of every ten people having a cellular phone, crime is reported as it happens, and lives are being saved as accidents and fires are reported immediately. As more and more people obtain a cellular or PCS phone, more crime will be reported immediately and will be stopped while in progress; fire trucks will be dispatched sooner, aborting more serious loss; and ambulances will be dispatched sooner, saving many lives. All this good has its cost. 911 Emergency Call Centers are being inundated with calls. These call centers, designed in the 1960s, are not designed, nor equipped for the increasing load on the system. Emergency calls are held in queue until a person becomes available to handle the call. Many calls are blocked and never make it in to the call center. And, even if a call makes it into the call center, and there is an available person to handle the call, there isn't an available line going out to dispatch the appropriate resource. Cellular service is being used to augment the current emergency networks.

## 12.6   The Wireless Business

The wireless industry will be the shining star in both the United States and world economics. Virtually every corporation, small business, household, and individual will be touched. Business, family, and individual behavior will be radically changed as wireless allows an individual to become untethered from the traditional ways of each culture (i.e., education, work, shopping, communicating). The wireless business sector will grow tremendously over the next two decades, with new and innovative technology. New businesses will begin, old businesses will fade away. Employment patterns will change. Intense competition will drive vendors to change, drive down prices, and develop new technology in an unprecedented way.

### 12.6.1   The wireless operator

The wireless operators today are feeling their vibrance, money, power, and growth. The cellular operators have learned to become telephone companies with whole infrastructures necessary to support the business. The PCS operators will learn how to become a formidable competitor to the cellular companies.

Now, with deregulation of telecommunications, the cellular operators are competing not only with the PCS operators, they are competing with telephone companies, interexchange carriers, paging companies, cable companies, utility companies, CAPs, ALCs, and anyone else who wants to provide some type of communications service. The public will become even more confused as to whom they should select to provide service, even more so than with equal access of the interexchange carriers.

Like the IXCs, paging, telephone, and cellular companies are consolidating in order to position themselves for the oncoming competitive battles. There simply aren't enough subscribers for everyone to have a viable business. With the high cost of infrastructure, start-up, and operation, and lowering prices of service, operators will be forced to consolidate in order to cut these costs and to demand new technology at lower prices from the vendors.

Control of operational costs and capital expenditures will beg the following question to the wireless operators: "Are we in the business of building a network and marketing service, or are we in the business of becoming a real estate, facilities management, and support organization?" The telephone companies, including AT&T, realized that they are no longer competitive because of their large cost of operating, of managing large real estate holdings, and of maintaining their facilities. They have began to sell-off real estate holdings, outsource facilities management and infrastructure maintenance, rely on their vendors to provide installation and maintenance, and to outsource other support functions.

Cellular companies have developed large noncore business support organizations to mirror a telco operation, and PCS operators are planning to do the same. If they continue on this path, the cost structure will not permit lower-subscriber price, payment of debt, and a return on investment to the investors. Again, they will have to ask themselves, "Are we building and managing a network and marketing service as a core business or burdening bureaucracy with noncore businesses impacting the profit margins and revenue stream?"

Some futurists predict that the telecommunications industry is "going back to the future" (i.e., combining and consolidating into a few, very large companies that provide either wire-line or wireless local service, interexchange service, Internet access, cable service, mobile and paging service). This may indeed happen, but the structures of the monoliths will be such that they will have a large bureaucracy. Much of the noncore business will be outsourced. The new giants will be lean and concentrate on managing their network, fighting competition, marketing service, and customer service. Anything not directly involved in these aspects will be outsourced to companies who are professionals in areas of managing real estate, maintenance, reprographics, mail room, accounting and financial services, information systems management, billing services, human resources, purchasing, and so on. Other companies can perform these tasks and services much more cost efficiently and with a better grade of satisfaction. Also, pay scales will be different. Today, companies pay a similar range of pay to a mail room supervisor as to an engineering supervisor because they are both the same labor-grade level. With outsourcing, the outsourcing company will pay their supervisor on the market level of a mail room supervisor, while the client company will pay their engineering supervisor on the market rate for an engineering supervisor. The cost savings for both salaries and benefits, including a management fee, can be substantial. In addition, the client company would save the hassles of human resource issues, benefits administration, management overhead, union issues, training, support, and the like.

### 12.6.2   The wireless vendors

The large, wireless manufacturers will find themselves outflanked by the small, entrepreneurial companies that can develop new products and features faster and more economically than the large companies. The large companies such as Lucent, Motorola, Ericsson, Nortel, and Hughes/Alcatel are competing with one another, and not paying much attention to the "in the garage" developers. The big manufacturing companies have large, well-funded R&D groups with a huge bureaucracy. Making these R&D organizations responsive to the market is like maneuvering a battleship in a bathtub. This has been seen with IBM, DEC, and the other computer giants with the emergence of the PC market. They were too big to maneuver. The newcomers and small companies were able to give the market what it wanted, and new PC companies sprang up into multibillion-dollar companies overnight.

There will always be a place for the large companies in the wireless market, primarily for the infrastructure equipment. The fastest-growing segment of the wireless market will be for features and new technology. The operators demand a transparent, seamless, and ubiquitous network. It appears that entrepreneurial companies will develop a host of networking equipment that will make the different systems transparent to the users. Eventually, cellular, PCS, wireless local loop, PCSS, and others, will appear as a wireless telephone transparent to the subscriber. Service will be seamless and ubiquitous, not only in North America, but throughout the globe.

The entrepreneurs will drive technology, features, and even the market. They will have a steady pulse on the market and will develop new wireless devices that will drive the operators to keep up with what the market wants. The emerging wireless-device industry will mirror the personal-computer industry of the 1980s. Competition for new technology, new devices, new applications, and new features will drive the prices down. Technology will be the only way manufacturers will be able to make profitable margins on their products, and technology will spawn even more technological advances. Wireless products will become smaller with more features and a lower selling price. The survivors will be those who can respond quickly to the market, those who can make a quick change, and those who keep on the driving edge of technology.

### 12.6.3  Wireless support organizations

The wireless support organizations consist of those companies that are not directly involved in the manufacture or the delivery of wireless. These companies support those that are directly involved. For example, wireless operators should staff only those who are directly involved in their mission—to operate a wireless network, and to sell and market wireless service. Any task that is not within this mission statement should be outsourced to those who specialize in the support tasks.

The support tasks that an operator should outsource are those that are necessary to run the operation but are clearly not within the core competency of the business. Real estate, cell site installation, facilities maintenance, billing, accounts receivable, and information systems are a few of the operating necessities that are not part of their mission. It makes good business sense to outsource these types of functions to a reputable firm that can handle them efficiently and cost effectively.

The whole area of outsourcing has been overlooked by the telephone operating companies. Each telephone company self-performed all aspects of operations. This led to gigantic bureaucracies, excess head count, out-of-control budgets, union issues, personnel issues, benefits and pension issues, employment litigation, restructuring, layoffs, bad morale, inefficiency, and so on. These companies were more tangled in their own people problems and inefficiencies than they were in delivering service. And, it was all protected by the monopolies granted by the PUC and the FCC. The cost to deliver service included all the inefficiencies and were artificially high. With deregulation, telephone companies and wireless operators can no longer afford inefficiencies. Price will determine who will or will not survive. The successful operators will be those who shed cost of operation and concentrate on their mission. Those who fail will be those who make large bureaucracies and self-perform noncore tasks.

This situation will lead to a whole business sector within the wireless industry: those who support operations of those who directly serve the wireless industry. These companies have as their core competency support functions, such as real estate, property management, facilities management, administrative support, information systems management, billing services, and so on. They can provide the manpower to perform the support aspects of operations more efficiently and at a lower cost by paying their specialized staff at the market rate rather than the client company rate due to equivalent job grade classifications where individuals of different skill sets will get the same rate of pay because each job is classified the same. There will be a host of companies that have positioned themselves for outsourcing for the expanding wireless industry—CBIS (Cincinnati Bell Information Systems) and EDS for billing, Axiom Real Estate Management for real estate and property management and facilities management, IBM for system integration, and more. There are over five hundred companies offering products and services to the wireless industry.

### 12.6.4   The wireless user

In the initial years, the wireless user will be more confused than ever. There are many decisions to be made, such as analog or cellular, cellular or PCS, TDMA or CDMA, hybrid or standard, which carrier, type of handset. Ultimately, the wireless user will go with whomever provides the most features and best roaming at the lowest price. Wireless users will expect the same transparent and ubiquitous service that they have been accustomed to for decades. Anything less will be unacceptable. The wireless users will come to expect more features and services, and the price to continue to decrease. There will be a lot of changeover from carrier to carrier, because if one carrier doesn't have a feature, or if the price is lower elsewhere, or if customer service isn't responsive, the subscriber will simply go to someone else. The cost of switching from one carrier to another is low.

The wireless user's life will change dramatically. Information and communications will be available anywhere, anytime. The confines of a workplace will disappear, and it will become wherever a user is. The economy will go global. An individual will need to be available 24 hours a day, since conducting business knows no time zone. However, the freedom to communicate anywhere, anytime has its trade-off; that is, the right to privacy, the ability to be tracked and monitored by others, and acceptance of the health risks associated with RF emitted everywhere, anytime.

Virtually every person will use wireless technology in the workplace or in the home. People will report crime faster, survive accidents because of quick notification of rescue resources, and will be able to roam untethered to home or work. Individual lifestyle at work and at home will be redefined, and will be radically different than it is today.

Information and communication will no longer be limited to urban areas. Without this limit, the population will be free to leave urban congestion and problems and resettle into rural areas. Wireless will mean that a user is untethered and free with access to information or anyone, anytime.

## 12.7   Summary

There is no doubt that wireless will touch everyone. New technology and features will be unleashed faster than ever. Personal freedom will become greater as individuals will be able to communicate with any human or machine. The associated trade-off is the potential loss of privacy.

The wireless revolution will make the economy robust with new companies forming, companies merging, and with the revenue that the wireless industry will put back into the economy. Operators who don't keep up with technology, features, and a high quality of service will wither away. Individuals who don't get on the wireless bandwagon will be left behind and be lost from information, communication, and technology. What the world sees of wireless today is just the tip of the iceberg. There will be applications that aren't even dreamed of yet.

The environment will change. There will be cell site towers and microcells everywhere. At first, sight pollution, but, eventually, towers and microcells will become environmentally friendly and aesthetically pleasing. The sky will no longer appear static. Many satellites will be traversing the sky, moving from one horizon to the other every ten minutes providing thousands of voice and data communications throughout the globe. Wireless will touch and affect every person on earth.

# Index

## ABOUT THE AUTHORS

RAJAN KURUPPILLAI, CEO, Celcore ITI Limited, is a technical and business management professional with extensive experience in business and market development, organization development, product management, network planning, and software development in all areas of telecommunications, including cellular, PCS, long distance, and data networks. Previously, Mr. Kuruppillai was Senior Manager in the Wireless Division of Northern Telecom/BNR Inc. in Richardson, Texas.

MAHI DONTAMSETTI is Chief Technologist at Lockheed Martin, where he is responsible for strategic product development and management. His current projects range from satellite-based broadband networks, Internet, and wireless systems to multimedia. Previously, he was with Celcore and Nortel, where he was involved in various areas of wireless business. Mr. Dontamsetti has several patents pending.

FIL J. COSENTINO is a telecommunications professional who specializes in technical marketing, business development, program management, strategic planning, and engineering and technical management in the wireless, switching, and networking industry. He recently joined AT&T Wireless Services after working at Axiom Telecom and Wireless Group, where he was Vice President of Business Development.